W·W·NORTON & COMPANY

NEW YORK·LONDON

NORMAN MYERS

Updated for the 1990s

The Primary Source

TROPICAL FORESTS AND OUR FUTURE

THE TEXT OF THIS BOOK is composed in the typeface Avanta. The display typeface is Typositor Delphin. Book design by Marjorie J. Flock.

First published as a Norton paperback 1985; reissued in expanded form 1992

Library of Congress Cataloging in Publication Data
Myers, Norman
    The primary source: Tropical forests and our future.
    Includes index.
    1. Forests and forestry—Tropics.   2. Rain forests.   3. Forest ecology—
Tropics.   4. Rain forest ecology.   5. Forest products—Tropics.   6. Forest
conservation—Tropics.   7. Deforestation—Tropics.   I. Title.
SD247.M94   1984   333.75'0913   83-13494

ISBN 0-393-30828-6

W. W. Norton & Company, Inc., 500 Fifth Avenue, New York, N.Y. 10110
W. W. Norton & Company Ltd., 10 Coptic Street, London WC1A 1PU

    3 4 5 6 7 8 9 0

To all those who, at the end of a public lecture on Tropical Forests, ask, *"What can we do to help the situation?"*

Nobody makes a greater mistake than he who did nothing because he could do only a little.

— EDMUND BURKE

Contents

III.   CONTRIBUTIONS TO OUR WELFARE

IV.   WHAT WE CAN DO

Foreword

BY H. R. H., THE PRINCE PHILIP, DUKE OF EDINBURGH

It is altogether too easy to think of the tropical forests as vast impregnable and immutable features of our globe. The facts, as this book makes abundantly clear, are very different. They are only too vulnerable to modern technology and world demand for timber and agricultural land.

The natural and renewable resources of the world have always been exploited, as many man-made deserts bear testimony, but this is the first time in the world's history that all the remaining forests are in serious danger of over-exploitation at the same time. This is sufficiently worrying for future human generations which will be denied these natural resources, but it is altogether catastrophic for the—literally—millions of species of animals and plants which are wholly dependent on the forests for their existence.

The great value and importance of this book is that, apart

from showing their significance for wildlife, it demonstrates the vital influence of tropical forests on the daily lives of all people on earth. It makes it abundantly clear that they form an essential part of the complex web of life on and around our planet.

While Norman Myers explains with great clarity why the tropical forests are in such a critical situation, the book ends on a note of optimism as it shows the way to turn disaster into opportunity.

Buckingham Palace
1983

Acknowledgments

THIS WAS the most difficult of my books to write. While the other four contained masses of facts plus some linkages between facts, this one required many more such linkages. Facts are difficult enough to handle, but linkages—relationships, interactions—are even more so. You have first to discern them and then to describe them correctly. Thus a book can convey not only errors of fact, but also, more insidiously and more damagingly, errors of interpretation. So in writing this book I depended more heavily on support from friends and associates. I called on many professional colleagues for their time and advice in checking through portions of the manuscript. Without the willing help of these friends in many lands, the book would have been far different.

All the more, then, I regret that I cannot list everyone who gave me the benefit of his or her professional expertise. I must limit myself to those who offered exceptionally large portions of their busy work schedules in order to keep my text on the right track. I am specially indebted to Dr. Peter S. Ashton, Harvard

University; Dr. Mark N. Collins, Conservation Monitoring Unit, Cambridge, U.K.; Professor William M. Denevan, University of Wisconsin; Dr. Terry L. Erwin, Smithsonian Institution, Washington D.C.; Professor John J. Ewel, University of Florida; Dr. Alwyn H. Gentry, Missouri Botanical Garden; Dr. Kenneth M. Green, Systems and Applied Sciences Corporation, Hyattsville, Maryland; Mr. Jeremy Harrison, Royal Botanic Garden, Kew, U.K.; Professor Susanna B. Hecht, University of California at Los Angeles; Dr. Richard A. Houghton, The Ecosystems Center, Woods Hole, Massachusetts; Professor Daniel H. Janzen, Department of Zoology, University of Pennsylvania; Dr. Carl F. Jordan, University of Georgia; Dr. Clive Marsh, The Sabah Foundation, Malaysia; Professors Robert May and John W. Terborgh, Princeton University; Mr. Jeffrey A. McNeely, International Union for Conservation of Nature and Natural Resources, Gland, Switzerland; Professor Gordon H. Orians, University of Washington; Dr. Ghillean T. Prance, New York Botanical Garden; Drs. Ira Rubinoff, Donald M. Windsor and Henk Wolda, Smithsonian Tropical Research Institute, Panama; Drs. Stephen H. Schneider and Robert Dickinson, National Center for Atmospheric Research, Boulder, Colorado; Mr. John S. Spears, The World Bank, Washington, D.C.; Dr. Matthew Suffness, Natural Products Division of National Cancer Institute, Bethesda, Maryland; Dr. Peter E. Waterman, University of Strathclyde, Scotland, U.K.; and Professor Edward O. Wilson, Harvard University. All these persons checked at least one chapter. A few others went so far as to check several chapters, and of course my gratitude to them is that much greater: Professor Herbert G. Baker, University of California at Berkeley; Dr. James A. Duke, Economic Botany Laboratory, Agricultural Research Service, Beltsville, Maryland; Dr. James D. Nations, University of Texas; Dr. Peter H. Raven, Director of Missouri Botanical Garden; and Dr. Michael Ross, Oxford University, U.K. None of these persons is responsible for any errors, whether of fact or interpretation, that remain. The responsibility for these lies with the person who perpetrated them in the first place.

The one who supplied the most protracted help throughout the exercise is Jennie Kent, my principal secretary. This book, around 110,000 words, went through several drafts; and before I started work on the manuscript, I dictated hundreds of sheets of notes. All the typing was undertaken by Jennie, who reduced a typewriter to a heap of exhausted metal before she moved on to learn the secrets of a word processor in just a couple of days. She typed much of the material during hours closer to midnight than to the normal end of a working day. In addition, she found ways to work with an author who does not bear frustration readily: when I became disconsolate that a chapter was not working out, she cheerily assured me, "Never mind, it will all fall into place tomorrow." For doing much more than beat the typewriter keys, thank you, Jennie.

Far away from my office in Oxford, across the Atlantic in New York, another person has sustained me with her belief that I would survive my battles with the book—a book that I eventually came to view as That (expletive deleted) Book. Many moons ago, I enjoyed an exploratory lunch with an editor at W. W. Norton, Mary Cunnane, and over dessert she asked me when I thought I could deliver the manuscript. I airily responded that she could look for it in just a few months: I had already assembled the background material, no problem. Mary flickered not an eyelash: tact of extreme degree. She had to wait more than a little before the manuscript was finally delivered; throughout the delays she displayed extraordinary patience, while carefully—and ever cordially—escalating her warnings that Norton would not wait until I was suitably fired by inspiration. When I eventually dumped a box of typescript on her desk, she was firmly constructive in her suggestions about transferring this section of text to that other place, and so forth. I could hardly ask for a more congenial colleague to work with on the other side of the Big Ditch.

Finally I encountered patience of a different sort here in my own household. While laboring along the finishing straight with the manuscript, I was allowed to contract out of my usual contributions to the daily round. Now, as I make up lost ground by chauffering the children to school, dance class, swimming club, and parties, I realize what a burden fell on the shoulders of my wife, Dorothy. I also think back to the occasion when I excused myself from the supper table, saying I had to go on and get on with my book, and our youngest offspring remarked that it must be a very long book and I must be a very slow reader.

Oxford
December 1983

Prologue to the 1992 Edition

I WROTE THIS book almost ten years ago. I well remember my mood. I was gripped by the challenge of giving the reader a sense of tropical forests' grandeur. Not just the physical grandeur, startling though this still is every time I set foot among those formidable trees. Rather it was the grandeur of nature at its most exuberant, with more species than the rest of the Earth combined, living together in an ecological complexity that baffled the intellect and the imagination alike.

I was also seized by the challenge of describing what was at stake in the forests—and of how fast they were disappearing and who was doing the damage. These were easy questions to answer. Not so simple was to track down the root causes of the destruction, why the forests were being put to the chainsaw with such zest. It seemed to me that the problem of deforestation epitomized many other environmental problems that afflict our planet:

we, all of us, demand too many things of our Earth—and there are too many of us given our appetites for ever-more natural resources. So there was little profit in pointing an accusing finger at the commercial logger, the cattle rancher, and the slash-and-burn cultivator. I had to hunt around for the true source of deforestation, the myriad forces that impel the hands of destruction. These forces include a host of pressures that ultimately reflect our economic lifestyles and political systems right around the world. The search became a kind of environmental "whodunnit." Conclusion: however little we may be aware of it, we all have our hands on the chainsaw, and we are wielding it with ever-greater energy.

The sleuthing of my ten-years-ago work made me so engrossed that when I closed down my word processor in late evening I could hardly wait for next morning to get on with the task.

How different today as I write this Prologue, and an Epilogue, for the book's new edition. I have not one mood but two. I feel dismayed that the forests are being destroyed more rapidly than ever. I also feel exhilarated at a recent sunburst of interest in their fate. This new enthusiasm is demonstrated by all sorts and conditions of people, from grassroots activists to top-flight politicians, both in tropical-forest countries and the rest of the world. So while there is some dreadful news, there is some better news too. Not that it ranks as good news yet: it is plainly far from enough to turn the situation around. But it surely marks a start on a start toward a brighter prospect.

What has fired this interest? What is the magic chemistry that has brought tropical forests to the front pages of our newspapers and to top slots on certain political agendas? There are many factors. One is the public awareness that tropical forests are something super-special, the apogee of evolution. Another is the accelerating rate of destruction, and the realization that tropical forests are indeed doomed unless we mobilize our conservation muscle with far greater clout than in the past. Another is the dawning fact that we shall all lose in many day-to-day senses if they disappear, if only through new medicines, foods, and industrial raw materials that could support our economic welfare into perpetuity and through the forests' linkages to global warming that promises to dislocate everybody's climate. Yet another is the recognition that we are all involved, however unwittingly, in the impending demise of the forests. All of this argues that we should bestir ourselves into a global campaign to help the forests at their time of terminal threat.

So in this Prologue we shall consider a number of "new news" items. We shall look at areas where large concentrations of species face imminent extinction—areas that qualify for priority treatment from conservationists. We shall look at new scientific findings about the workings of the forests and how they help to stabilize climates around the world. We shall consider some economic analyses that show we can often derive greater revenues from the forests by harvesting their non-timber products, a mode of exploitation that scarcely disrupts forest ecosystems. And we shall look at an unlikely-sounding "output" of the forests, their contribution to national security.

DEFORESTATION WORSE THAN EVER

First, the appalling news. Let's get it over with.

In 1980 the National Academy of Sciences published my report on deforestation rates. This was followed a year later by an assessment on the part of the UN Food and Agriculture Organization. We both found that deforestation in the sense of outright destruction of tropical forests—i.e., excluding gross disruption that leaves some trees standing—amounted in 1979 to 75,000 square kilometers. In 1989 I undertook a further survey, this time for Friends of the Earth in London. More recently still I have updated my findings again. It turns out that by 1989 the annual destruction rate had soared to 142,200 square kilometers, for an 89 percent increase in just a decade. (At least as much again was being grossly degraded without being destroyed outright.) The total represented 1.8 percent of the 8 million square kilometers of remaining forests, which in turn amounted to roughly half of the original expanse.

Moreover, the annual destruction rate seems set to accelerate yet further, and could well double within another decade. This is partly due to sheer growth in human numbers, and especially the pressures they generate to clear the forests for agriculture. During the 1990s the global population is projected to expand by another one billion people, and of this total 60 percent will come on board in tropical-forest countries. Within a further three decades the global total is projected to reach 8 billion people, of whom a full 80 percent will be living in tropical-forest countries—or many more than the global total today. Consider some individual cases: Ecuador's population is projected to increase from 11 million today to 26 million (136 percent greater) before it attains zero growth in about a century's time; Brazil's from 153 million to 304 million (99 percent); Ivory Coast's from 13 million to 85 million (554 percent); Nigeria's from 123 million to 580 million (372 percent); Madagascar's from 12 million to 54 million (350 percent); India's from 859 million to 1,876 million (118 percent); Burma's from 42 million to 96 million (129 percent); Indonesia's from 181 million to 371 million (105 percent); and Vietnam's from 68 million to 160 million (135 percent).

But we must be careful not to oversimplify the situation. Let's take a closer look at Brazil, a country with one of the largest populations of all. It turns out there is a host of related factors at work in addition to population. They include the share-out of farmlands in areas far outside Amazonia: 5 percent of farmland owners possess 70 percent of all farmlands, while 70 percent cultivate 5 percent. Some of the large land holders cultivate only half their thousands of hectares, while over the fence the hard-scrabble peasant tries to make out with only a few hectares. This skewed situation is growing worse as another 1.7 million Brazilians enter the job market each year, over half of them failing to find enough work to support themselves and hence inclined to seek their livelihood through the only option they see left, a slash-and-burn lifestyle in Amazonia. In Rondonia, a Brazilian state in southern Amazonia, the population grew at an annual rate of 15 percent between 1975 and the late 1980s, by contrast with the growth rate for all Brazil, 2.1 percent.[1]

There are similar mass migrations into tropical forests, albeit at lower rates, in Colombia, Ecuador, Peru, Bolivia, Ivory Coast, Nigeria, India, Thailand, Vietnam, Indonesia, and Philippines. In many of these countries a key factor is the maldistribution of existing farmlands, compounding the impact of population growth. Often enough, there are still further pressures to squeeze smallscale cultivators out of traditional farming areas, pressures such as inequitable land-tenure systems, inefficient agricultural technologies for subsistence peasants, and insufficient attention by national leaders to subsistence agriculture generally. These might sound to some observers like overly technical matters that surely have little to do with tropical forests many horizons away. But they lie at the heart of tropical deforestation, and they matter much more than traditional measures to protect the forests, for example the setting up of extra protected areas.

There are also factors that derive from countries outside the tropical forest zone, notably foreign debt. Tropical-forest countries owe well over half of the $1.3 trillion of debt oustanding on the part of the developing world. There is scant prospect that the burden will be relieved any time soon. Brazil alone owes $120 billion. As long as this foreign millstone hangs around the neck of the Brazilian economy, there is all the less scope for increased funds to be channeled into subsistence agriculture and thus reduce the motivation for the impoverished peasant to pick up his machete and matchbox before heading off to Amazonia. Equally to the point, and with regard to developing nations as a whole, note that when we add up all overseas aid and public loans from the North to the South, and factor in the debt issue, we find there has long been a net annual flow of funds from the developing world to the developed world totaling almost $50 billion.

The new forest farmers are altogether different from the shifting cultivators of tradition. They practiced a form of forest exploitation that was clearly sustainable; otherwise they would not have been able to continue with it for centuries. As long as there were not many of them and they had plenty of forest to shift around in, their system worked fine, as the forest ecosystems had enough time to recover before the next cultivation cycle began. But today they are joined by many times as many of the displaced peasants, there is less forest left to move around in, and the newcomers have little ecological understanding of how to make a sustainable livelihood from the forests.

"Shifted cultivators" as they are sometimes known, they are now accounting for well over half of all forest destruction, a proportion that is increasing fast. This means that if we put an end to all commercial logging and cattle ranching forthwith, we shall have solved only a small and indeed diminishing part of the problem overall. Nor are they to be reproached for what they do. Driven by forces they hardly understand and cannot resist, they are no more to be blamed for their activities than a soldier is to be held responsible for starting a war.

The phenomenon of the shifted cultivator represents the foremost change in the tropical forests' situation since the time I wrote this book. The challenge now is to tackle the problems giving rise to the shifted cultivator. Yet the nature of the challenge is scarcely recognized. We shall return to the whys and wherefores in the Epilogue.

MASS EXTINCTION OF SPECIES

As the forests go, so do their species. When we hear about a mass extinction overtaking the Earth, it is largely occurring in tropical forests. According to several sets of recent calculations,[2] and if we suppose there are at least 30 million species in the forests, it turns out we could be losing between 5000 and 10,000 species each year right now, and the final fallout could amount to 10 million species, possibly many more. This alone would constitute a greater biological debacle than any since the demise of the dinosaurs and associated species 65 million years ago.

True, the bulk of these species are insects. Some readers may feel inclined to ask, "What have creepy-crawlies ever done for us?" We know that tropical forest plants supply materials to counteract cancer and several other leading diseases. A Queensland plant looks as if it may supply a response to AIDS. But insects—what can they offer?

Consider the situation as concerns oil-palm plantations in Malaysia. Until the early 1980s the pollination of millions of oil-palm trees was done by human hand, an inefficient and expensive way of performing the task. Finally the plantation owners asked themselves how the oil palm got itself pollinated in its native habitats of West Africa's forests. Researchers went off to Cameroon, where they found the job was undertaken by a tiny weevil. Start-up stocks of the weevil were taken back to Malaysia, where they were released into the plantations. (There was no problem of ecological complications with other species, since the weevil confined its attentions to the oil palm alone.) The pollination is now entirely accomplished by the weevil, with savings of $140 million per year.[3] We can reflect on that the next time we utilize cosmetics or other products that may contain palm oil, and the next time we hear of insects becoming extinct in large numbers.

"HOT SPOT" AREAS

During the course of recent research I have found that certain sectors of tropical forests feature exceptional concentrations of endemic species, i.e., species that are found nowhere else—and these areas also face exceptional threat of destruction. I have come up with a list of fourteen such "hot spot" areas: western Ecuador, western Amazonia, the Choco strip of Pacific-coast forest in Colombia, the Atlantic-coast forest of Brazil, southwestern Ivory Coast, montane forests of Tanzania, Madagascar, the Western Ghats of India, southwestern Sri Lanka, eastern Himalayas, Peninsular Malaysia, northeastern Borneo, Philippines, and New Caledonia. Altogether they feature 311,000 square kilometers of untouched forest (about the same as New Mexico), or less than 4 percent of remaining tropical forests.[4]

Unless massively increased conservation efforts can be mounted, these forests will be largely eliminated by the end of the century or shortly thereafter. They contain well over 37,000 endemic plant species, or twice as many as all plant species in North America, and 15 percent of all plant species on Earth, in just 0.2 percent of Earth's land surface. In addition they contain at least 750,000 endemic animal species, probably several times more.

By concentrating on these areas where needs are greatest and where the payoff from safeguard measures would also be greatest, conservationists can engage in a "silver bullet" strategy for saving large numbers of species.

THE CLIMATE CONNECTION

Ten years ago I outlined a number of ways in which tropical forests appear to regulate climate. Today we have a clearer idea of the mechanisms at work—and they demonstrate that as the forests disappear, we can expect climatic hiccups aplenty.[5]

First, local effects. Deforestation sometimes results in reduced rainfall, which can be unusually significant for agriculture and the prospects of feeding two billion people in the humid tropics. A number of tropical crops are more susceptible to climatic changes than temperate-zone crops usually are, and a marginal decline in rainfall can sometimes cause a substantial decline in staple crops.

In Peninsular Malaysia, the Penang and Kedah states have experienced disruption of rainfall regimes to the extent that 20,-000 hectares of paddy ricefields have been abandoned and another 72,000 hectares have registered a marked production dropoff in this "rice bowl" of the Peninsula, leading to an overall shortfall of well over one quarter of the Peninsula's rice harvest.

Similarly in the Western Ghats of southwestern India—a country that depends upon the monsoon season for 70 percent of its agricultural water—deforestation seems to have led to such disruption of the monsoon that tea plantations have had to be abandoned. In southern Ivory Coast, the period during which there has been extensive deforestation has been marked by a drying-out of agricultural lands, and areas that were formerly excellent for cocoa growing are being abandoned because of less rainfall, less humidity, and longer and harsher dry seasons, as well as lowering of water tables. Similar deforestation-related changes in rainfall have been documented in Panama, Costa Rica, southern China, and Philippines.

But a far bigger climate dislocation will surely ensue from deforestation's contribution to the greenhouse effect. Most deforestation is caused by burning rather than by activities such as logging that leave wood intact. A tree is half carbon, and when it burns it releases its carbon into the atmosphere where it combines to form carbon dioxide. This gas accounts for almost half of global warming, the phenomenon that threatens to cause massive disruption of climates far and wide, notably in North America and Europe. True, most of the carbon dioxide in the global atmosphere comes from combustion of fossil fuels on the part of developed nations. But forest burning in the tropical developing world contributes roughly 30 percent of the problem. Moreover, forest burning is expanding so rapidly that its proportion may well rise to a level where it almost catches up with fossil-fuel combustion by early next century. It will not stay at that level for long, however, since there will soon be hardly any more forests left to burn.[6]

Through the climate connection, then, the future of the American wheat grower is tied in with the future of Amazonia, and the welfare of corn consumers everywhere is intimately related to the welfare of forests in Borneo. Let us bear this in mind next time we sit down to a breakfast bowl of cornflakes or enjoy a packet of popcorn with the evening television show.

THE SECURITY DIMENSION

Strange as it may sound, tropical forests contribute to national security.[7] Consider Ethiopia, for example. The Ethiopian highlands used to feature much forest cover, but in recent decades it has declined by nine-tenths. As a result, there has been widespread soil erosion in the fertile farmlands that have traditionally supported the bulk of the country's populace. As a further result, the highlands have become less and less able to sustain their human communities. It was primarily environmental impoverishment, with too little food for too many people, that caused the fall of former emperor Haile Selassie's regime in 1974.

More recently, there has been a largescale spillover of peasantry from the denuded highlands to Ethiopia's lowlands. They have headed notably toward the Ogaden region spanning the border with Somalia, where the land-hungry throngs have come into conflict with their neighbors, leading to the war of the late 1970s. If this conflict had been confined to Ethiopia and Somalia, it would have been regrettable enough, these being among the most impoverished countries on Earth. But because the Horn of Africa lies adjacent to the strategic oil-tanker lanes from the Persian Gulf, the superpowers intervened, pouring in $2 billion worth of weaponry and other forms of military support. If a small part of the sums eventually expended—$500 million during the 1960s, according to the United Nations—had been earlier assigned to reforestation in the highlands, among other kinds of environmental rehabilitation, the migration of peasantry could have been stemmed, and the conflict avoided.[8]

We can discern a similar connection in El Salvador. This benighted country suffers more economic difficulties, social disorder, political turmoil, confrontation, and violence than any other country in Central America. It is also the country where once-abundant forests are now a matter of history, their demise leading to disruption of watershed systems, declining water supplies for irrigation agriculture, siltation of hydropower dams, and frequent flooding. Surely no coincidence? Of course many other factors contribute to El Salvador's parlous state, including repressive governments, population growth, and an oligarchical society. But deforestation appears to have played a sizeable part too.

In short, there is an emergent connection between tropical deforestation and security concerns. For sure, the links cannot be readily demonstrated in strict cause-and-effect terms, and all kinds of other factors are usually mixed in. Nonetheless the links will become increasingly significant as deforestation becomes more widespread. They are well summed up in the words of a former leader of the United Nations Environment Programme, Peter S. Thacher: "Trees now or tanks later."

This latter assertion is further demonstrated by India and Pakistan. The first urgently needs to restore forest cover in 350,000 square kilometers, the second in 96,500 square kilometers; their respective governments consider the tasks national priorities. In both cases the reforestation would help to rehabilitate watersheds that are critical to agriculture and hence to food security, political stability, and international equanimity in a volatile region. In both cases the job could be done for about one-fiftieth of what is now spent on military activities.[9]

THE ECONOMICS OF FOREST EXPLOITATION

As was made plain when I wrote this book ten years ago, tropical forests can confer many benefits without having to endure the destruction that accompanies commercial logging, cattle ranching, smallscale farming, and other conventional modes of exploitation. These benefits include a host of non-wood products (fruits, fibers, gums, oils, and the like), raw materials for drugs and medicines, genetic resources for new and improved crops, and a number of environmental services such as watershed functions. This list has now been joined by still more important benefits, notably climate regulation and security safeguards. They can all be "exploited" with only trifling, if any, disruption of forest eco-systems, by contrast with the predominant types of exploitation that result in elimination of forests.

So why don't governments involved do more to enjoy the non-destructive benefits? This key question has been illuminated by some pioneering research during the late 1980s that has produced one of the most profound changes in our understanding of tropical forests and their fate. Much of the research is economic analysis and policy appraisal—hardly the stuff of traditional conservation with its emphasis on beating back the bulldozers. But resisting false economics and perverse policies can be more productive, while less spectacular, than resisting bulldozers. Dr. Robert Repetto of the World Resources Institute and Dr. Malcolm Gillis of Duke University have demonstrated that it is in such core factors, abstruse and dull as they may appear, that much of the deforestation problem lies.[10]

It turns out that certain tropical-forest governments have been engaging in covert subsidies and other financial inducements that induce gross over-exploitation of tropical forests. When exploitation of forests began several decades ago, it was fair enough for governments to offer all manner of commercial supports in order to get their logging industries and ranching programs underway. But after the exploitation patterns became firmly established, the supports were not removed. Long after they have served their purpose, they have continued to work their insidious impact, with the result that they have fostered runaway deforestation. In Philippines the government has persisted with reduced royalties and fees on timber harvests, tax-free holidays for logging concessionaires, and restricted export taxes on processed timber, to the extent of forgoing revenues five times greater than it could have collected. In Indonesia the government has been losing more than $1 billion a year in stumpage fees, royalties, rents, and other harvest revenues that it should have derived from the recent timber boom.

The subsidies problem is particularly acute in Brazil, where much deforestation has occurred through cattle ranching in Amazonia. The government has spent $2.5 billion in subsidizing ranchers' investments through long-term loans, tax credits and other fiscal incentives, monetary inducements, and duty-free imports of capital equipment. This support has meant that many ranches, no matter how inefficient, have made a quick killing by courtesy of the government's give-away inducements—to the extent that hardly any ranchers have bothered to sell the timber felled to make way for their pasturelands, collectively torching $5-billion worth of timber. Virtually every ranch has been a financial success for the individual entrepreneur, while an economic disaster for the national economy. Fortunately the new government of President Fernando Collor de Mello has decided there will be no subsidies for additional ranches, though such supports continue for established ranches covering 120,000 square kilometers—ranches that have already cost the Brazilian treasury more than $2.5 billion in revenues forgone.

Before people in the outside world raise a critical eyebrow about this government support for deforestation (unwitting as it mostly is today), they might recall that some of the most striking forests outside the tropics, those of the Tongass National Forest in Alaska, are being felled at a rate even more rapid than most tropical forests, with inducements supplied by U.S. government subsidies even greater than in tropical-forest countries.

This economic assessment by Bob Repetto and Malcolm Gillis has been matched by some analysis of the value of non-wood products available from western Amazonia. Dr. Charles Peters and his colleagues at the New York Botanical Garden, Yale University, and the Missouri Botanical Garden have documented the use made by local people of wild fruits and rubber from the forest around Iquitos in Peru.[11] Of 193 fruits, no fewer than 139 are harvested for human consumption, and 57 are marketed in the urban area. At least one-tenth of an average peasant's diet comes from these fruits. Together with the rubber, this enables a patch of forest to generate revenues several times greater than could be derived through the main alternative form of exploitation, commercial logging; and whereas logging tends to be practiced on a once-and-for-all basis, depleting many of the forests' other products, harvesting of the non-wood products is undertaken in sustainable fashion and without harm to forest ecosystems. This should not be taken to mean that an exploitation strategy of this sort could be replicated right across Amazonia. There would soon be a limit to the amount of fruit and rubber that markets could accommodate. But it highlights the scope for people to make use of the forests without knocking them over.

The same spirit lies behind the extractive reserves pioneered by the late Chico Mendes in Brazilian Amazonia. Rubber tappers in Acre State regularly earn between $1000 and $1500 a year by selling wild rubber, fruits, and nuts, among other non-wood products. There are now 70,000 households occupying 100,000 square kilometers of extractive reserves.[12] Again, one should not go over the top with the idea. There is no prospect of converting Amazonia into one vast network of extractive reserves, since the strategy is tightly tied in to local circumstances of products available, marketplace demand, and the like. But the local people have tapped not only rubber, they have tapped the innovative idea of making a self-renewing profit from forests anywhere through exploiting their rich array of products apart from conventional materials such as timber.

DEBT-FOR-NATURE SWAPS

Another imaginative measure of recent years lies with debt-for-nature swaps. Many tropical-forest countries are so heavily indebted that they cannot hope to pay off all they owe. So a "secondary market" springs up among the banking community, offering debt notes at a small part of their face value (the banks in question are not averse to this maneuver since they would rather get back a portion of their loans than none at all). A conservation organization—often the World Wildlife Fund–U.S.—can then work with a private philanthropist to purchase part of a tropical-forest country's foreign debt at a sizeable discount, while gaining the country's agreement to set aside a sum equivalent to the full debt amount but in local currency, for use on conservation activities. This device of "creative financing" has achieved some success in several countries, notably Ecuador, Costa Rica, Madagascar, and Philippines.

While debt-for-nature swaps certainly help, and we need

many more such imaginative ideas, they have hitherto been on too small a scale to make much difference to the overall problems of foreign debt and forest destruction. An expansion of the concept has been proposed by the British entrepreneur Sir James Goldsmith, envisaging an international company that would finance deals as large as $50 billion, writing off tropical-forest countries' debts at less than face value against undertakings for forest protection. Both sides would come out ahead. Creditor-country governments already disburse as much as 34-percent tax repayments on losses incurred in the secondary market; and they may eventually have to abandon large amounts of debt anyway. Note moreover that the West German government has written off DM850 million (equivalent to half a billion dollars) of debt in support of reforestation measures in Kenya.

The debt-for-nature concept has been vigorously opposed by Brazil, among a few other countries. They protest that it is ill conceived in principle since it infringes sovereignty (though the new government of President Collor de Mello appears to look on it more favorably). This obstacle could be surmounted, however, if the initiative were undertaken by tropical-forest countries themselves.

A SUNBURST OF NEW INTEREST

Finally, the best news of all. There has been an extraordinary groundswell of interest and support on the part of people right around the world. At long glorious last, it looks as if lots of people are ready to get on top of the problem before it gets on top of us.

First of all, there has been one series after another of citizen activities by on-the-ground groups in tropical-forest countries. In India the Chipko people or "tree huggers" have beaten back the loggers who planned to fell some local forests in disregard of local needs. The community in question valued the forests for their fruits, fodder, medicinal materials, and fuelwood, all of which could be harvested sustainably, and they saw little benefit for themselves in watching the trees being converted into board feet of timber for remote commercial interests. Their tree-hugging efforts took several years to persuade the loggers to look elsewhere; but finally the Chipko people, mostly semi-literate peasants, won the day. Their success was subsequently replicated by another grassroots initiative at Silent Valley in southern India, when local people halted a dam that would have flooded a large part of their "forest." As it turned out, the beleaguered forest contained a wild variety of rice with genetic resistance to a disease that was threatening to ravage much of Asia's rice crop.

In many other countries too there have been grassroots measures to counter deforestation. In Kenya the Greenbelt Movement, run entirely by women, planted more trees in its first year of operation than the government achieved during the previous ten years. This exercise relieves excessive exploitation pressure on remaining forests. There is a similar Greenbelt movement, achieving similar success, in Colombia. In Indonesia there are four hundred local conservation groups, with sufficient collective clout at the national level that they regularly gain the ear of government ministers. In Costa Rica, Ecuador, and a lengthy list of other countries there is a similar burgeoning of local-level activism. Even in Brazil there is a flourishing community of nongovernmental bodies that the government feels it must heed.

Similarly, citizen supporters of tropical forests are on the march in the United States. In 1985 the Rainforest Action Network was almost alone in its campaign. Today there are nearly two hundred Rainforest Action Groups around the country, mainly on college campuses, and another springs up every ten days. A parallel story can be told in Britain, Germany, Holland, Sweden, and a number of other countries way outside the tropics, all recognizing their common responsibility for the common heritage in those far-off forests.

The American activists have recorded a remarkable breakthrough as concerns the notorious "hamburger connection." In the mid-1980s the Rainforest Action Network mobilized a consumer boycott against the largest importer of Central American beef, Burger King. It took two years of persistent effort on the part of millions of individual Americans before success came their way. Many of them must have had occasion to ask themselves whether they would ever win. But they stuck at their task, determined to register their conservationist vote in the marketplace. Eventually Burger King caved in, vowing to import no more of the artificially cheap beef. This exceptional success story should hearten us whenever we, as individuals, suspect the deforestation problem is so big that little can be achieved through the one-by-one actions of citizens.

As for political leaders—the people who are often so ecologically illiterate that they think a food chain is a line of supermarkets—they now sense they have no option but to follow the spirit of the times. The heads of governments in Philippines, Thailand, India, and Kenya have declared that deforestation constitutes a "national emergency." President Collor de Mello in Brazil has appointed a long-time environmentalist, José Lutzenberger, as Secretary for the Environment. The Colombian government has handed over half of its Amazonia forest to the care of tribal peoples in light of their demonstrated capacity to make a living out of the forests without knocking them over. As for developed nations, the British government has assigned an extra £100 million to tropical forests. The German government is mounting a largescale program in support of the cause. In Holland, Sweden, Norway, Italy, and Canada, there are signs that governments are ready to play a much more solid part in the global campaign. The Intergovernmental Panel on Climate Change has urged that for purposes of stemming the greenhouse effect there should be a vastly increased effort to halt the torching of tropical forests.

So even as more forests than ever are dispatched in smoke, there is a worldwide movement to confront the challenge. For sure, we must not overplay the new initiatives. They are no more than a start on a start to tackle a superscale problem. But it all looks far more promising than in the dismal days of ten years ago, when tropical forests were on the agenda of hardly anybody but loggers, ranchers, and their like.

In conclusion, we can take hefty heart that there are many, many more people today who are aware of many more reasons to save the forests. But that raises a big question: Why aren't we doing a better job? Why are the forests falling almost twice as fast as ten years ago? What more do we need to do? Why aren't we getting on with it? It turns out there are solid reasons for our failures: not acceptable reasons, but powerful reasons to explain why we are allowing them to be destroyed before our eyes. We shall look at these reasons in the Epilogue. The clues lie in obscure areas that easily escape our attention, and it is among the shadows that we must search. But then, tropical forests are not places of blazing light for the most part. It is their shady sectors, the sectors that make up the bulk of their structures, that contain the secrets of what makes them tick.

To return to where I started, I have mixed feelings as I write this Prologue and bring us up to date on tropical forests. I remind myself I am heartened by the wave of recent support from people right around the world. But we have lost a lot of forest since I wrote this book, an area one and a half times the size of Texas. If some of the deforestation had served to promote the human cause in the long term, and that could not have been achieved any other way, I would feel better about it. But virtually all the destruction has contributed nothing of lasting benefit to human-kind's lot. It is unnecessary and avoidable. It is a record that, to quote Hamlet, is "flat and unprofitable."

But we still have half of all tropical forests that ever existed. So I should not see the glass as half empty, but rather as half full. There is much left to battle for. Immediately that triggers the thought that these forests are indeed the apogee of four billion years of life on Earth, the finest celebration of nature ever to grace the face of the planet. My mind and spirit flood with astonish-ment that we should be so lucky as to share the planet with such a spectacle. I feel stirred at memories of those forests where I have spent some of my finest days, and I recall the sense of exhilaration when I first sat down to write this book. I felt so driven to tell my story that the book almost wrote itself, chapter after headlong chapter, setting out the story of the Great Forests. Read on.

The Primary Source

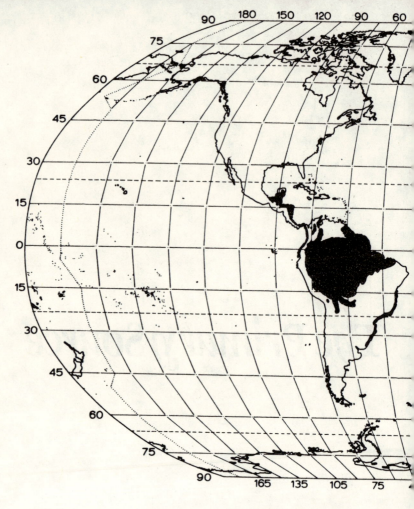

THE DISTRIBUTION OF TROPICAL FORES

CENTRAL AND SOUTH AMERICA

Belize	Ecuador	Guiana	Panama
Bolivia	El Salvador	Honduras	Peru
Brazil	French Guiana	Mexico	Surinam
Colombia	Guatemala	Nicaragua	Venezuela
Costa Rica			

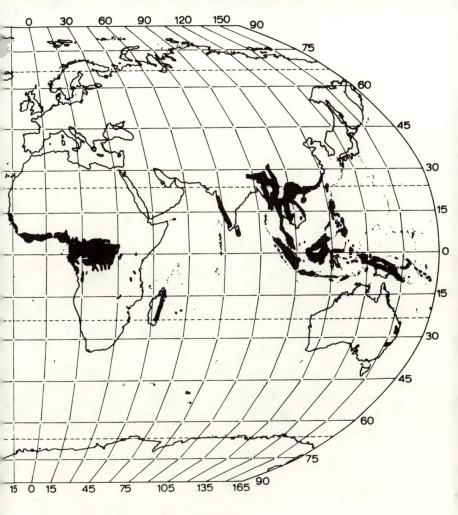

MONG THE COUNTRIES OF THE WORLD

	AFRICA		ASIA
Berunda	Kenya	Bangladesh	Pakistan
Cameroon	Liberia	Brunei	Papua New Guinea
Central African Republic	Madagascar	Burma	Philippines
Congo	Nigeria	India	Sri Lanka
Ethiopia	Rwanda	Kampuchia	Thailand
Gabon	Sierra Leone	Laos	Vietnam
Ghana	Tanzania	Malaysia	
Guinea	Uganda	Melanesia	
Ivory Coast	Zaire	Nepal	

Introduction: The Primary Source

TROPICAL FORESTS rank foremost among my professional pre-occupations. They fascinate me, they awe me. I probe their workings, investigate their makeup, undertake research of a dozen sorts, and am enthralled at the insights that must lie ahead. I likewise feel I have no more important professional endeavor than my inquiries into the status of tropical forests—how many we have left, how fast they are disappearing, what their outlook is for the year 2000, for the year 2100. I further feel that there is no more challenging task for me than to contribute to such conservation activities as assist the cause of tropical forests. In short, tropical forests excite me like no other subject in my work as a scientist and a conservationist.

Even more to the point, I feel, as an individual—not as a scientist or a conservationist, but as a human being—that this is a splendid time to be alive, when, at long, glorious last, we can penetrate to the furthest reaches of tropical forests, not only their remote heartlands but also the zone that, paradoxically, has remained more concealed from us than have the depths of Amazonia or Borneo and that forms the forest canopy scores of meters above the ground: the last, great unexplored frontier of life on Earth. The thought that we shall learn so many new secrets of life within the next few years appeals to me more than most things I can think of. There is one notable exception: I feel lucky to be alive at a time when I can take part in a great enterprise by people in many lands to safeguard this, the finest celebration of nature ever known on the planet.

Thus for me tropical forests serve as a primary source of inspiration, whether as a professional scientist and conservationist or a private citizen. I know the same applies to many of my friends and colleagues. Probably rather few people feel so excited by tropical forests, even though they might agree that tropical forests represent the outstanding sector of our planetary patrimony and deserve an exceptional effort to be saved. But tropical forests are a primary source of welfare for people everywhere, in ways they never suspect—especially through their material contributions to our everyday lives. Forming a green band around the equator, tropical forests cover only little more than nine million square kilometers—an expanse less than that of the United States, a mere one-sixteenth of Earth's land surface. Their limited expanse notwithstanding, they supply far more to our well-being than does any other ecological zone of similar size. We benefit from them dozens of times each day and have done so for many years, many millennia: in fact, since man first climbed down from his tropical forest trees.

Moreover, the best times may yet lie ahead of us. If we manage to make full, sustainable use of the tropical forests' myriad products, we shall enjoy the fact of their existence in more ways than most people now dream. However little we may realize it, our future is likely to become ever-more closely linked to that of tropical forests. As a wellspring of products to sustain our sojourn on Earth, they are a primary source.

MATERIAL GOODS

When we drink coffee, we enjoy a product that derives in part from wild plants of tropical forests via their genetic resources (this link and others will be explained further along in the book). The same when we help ourselves to a banana split: we enjoy a fruit, ice-cream ingredient, and chocolate additive that reach us through the courtesy of other plants. Similarly, when we enjoy a cookie, or make cinnamon toast, or pour mayonnaise on a salad. Each time we take a glass of orange juice, we might reflect that the orange grower, and his associates in other sectors of the citrus-fruit industry, may well owe the continued productivity of his crop to a wasp or other predator from tropical forests that serves to keep down populations of insects destructive to the orange grove. Those of us who like to chew gum can thank the sapodilla tree of tropical Latin Americas for an essential raw material: the gumlike substance called chicle.

Compared with the number of our foods and beverages that in one way or another derive from tropical forests, these represent a mere first course—and nothing like the cornucopia that awaits us. Despite their limited area, tropical forests harbor well over two-fifths of the five to ten million species of plants and animals on this planet. As agriculturalists investigate more of the abundant stock of foods in tropical forests, we can look forward to an even greater selection for our tables. In New Guinea alone, over 250 kinds of trees bear edible fruits; only a few of them are consumed by local communities; less than one dozen reach the marketplaces of Southeast Asia; and only two or three of them reach the supermarkets of North America. That leaves plenty of scope for the future.

When we pick up a medication or a pharmaceutical at our neighborhood drugstore, there is roughly one chance in four that the product owes its existence, either directly or indirectly, to raw materials from tropical forests. The product may be an analgesic, an antibiotic, a diuretic, a laxative, a tranquilizer, or a cough pastille, among hundreds of other items. Over-the-counter sales of these products amount, worldwide, to $20 billion per year.

In 1960 a child suffering from leukemia faced only one chance in five of remission. Now that child enjoys four chances in five, thanks to two powerful drugs prepared from biocompounds found in a tropical forest plant, the rosy periwinkle. At least another fourteen hundred kinds of plants from tropical forests contain principles active against cancer. Still more plants reveal the potential to help with another pressing health need of the modern world: effective and safe birth control.

As for industrial goods, we need only look around our homes to note a dozen ways in which we benefit from the hardwood timbers of tropical forests. If we were to figure out exactly how many times we use natural rubber during a day, we might need that entire day to track down its diverse applications. A host of other latexes are available, together with gums, resins, waxes, dyes, tannins, and essential oils, to name just a few leading categories of industrial materials. We use one or another of these ingredients each time we apply a deodorant, an after-shave lotion, or a lipstick, each time we use cellophane or dynamite, each time we read a glossy magazine or mail a letter, each time we apply varnish to our furniture or our fingernails, each time we listen to a record or wield a squash racquet, and each time we get out our Gucci accessories or pull on our jogging shoes. The global chemicals industry, worth more than half the global armaments industry and thus truly big business, is concerned that petrochemicals with their soaring costs will cause plastics, among other petrochemical products, to be priced beyond reach of the consumer. So they are seeking alternative supplies of organic raw materials. What better place to look than among the vast stocks of botano-chemicals in tropical forests?

We can even turn to tropical forests for future sources of energy. Technology is now available to prepare bio-crude from plants at competitive prices, provided that mega-amounts of vegetable matter can be assembled at the processing facility, month in and year out. Again, no other sector of the plant world presents so many options as do tropical forests, with their communities of plants that specialize in turning solar energy into stored energy. Fossil fuels are themselves no more than plant biomass that has been slowly converted into petroleum, coal, and natural gas. The day is coming when we shall be able to replace petroleum with "phytoleum": a plantation for growing gasoline need never run dry like an oil well.

Since food experts, pharmacologists, and industrial chemists have taken a cursory look at only one in ten of all plant species in tropical forests and a close look at a mere one in a hundred, the forests may well contain exceptional stocks of natural resources to help us confront the challenges of the future.

ENVIRONMENTAL SERVICES

Tropical forests supply a variety of services, including environmental functions. Through their watershed effects forests act as a "sponge," soaking up rainfall before releasing it slowly and steadily into rivers. As long as the forest cover remains intact, rivers run clear and clean—and regularly throughout the year. When the forest is cleared away, rivers start to turn muddy with eroded soil; they become swollen and then shrunken. As a result of deforestation in the Himalayan foothills, the Ganges plain, with its five hundred million people, grows more and more vulnerable to flooding. During the 1978 monsoon season, India suffered losses of buildings, crops, and livestock worth $2 billion, plus hundreds of people swept away in floods. Because farmers in Southeast Asia can no longer depend on regular amounts of irrigation water from their deforested watersheds, the Green Revolution is proving less revolutionary than many had hoped it would be. As the hinterlands of Manila and Panama City lose their tree cover, urban communities find their supplies of domestic water declining in quantity and quality, bringing on a risk of contaminated-water pandemics. In Ecuador, Kenya, and Thailand, washed-off sediment from deforested territories leads to silting up of hydroelectric dams: loss of greenery in the hills causes brown-outs in the cities. Through their many environmental services, then, tropical forests serve agriculture, public health, energy, and several other economic sectors. By virtue of their ecologic link-ages, they are a primary source of support for virtually every major development activity in emergent regions.

But tropical forests offer a still greater environmental service through their regulatory effect on climate, whether regional or global. In Amazonia, more than half of all moisture circulating through the basin's ecosystem remains within the forest zone, thanks to the forest's evapotranspiration function: rainwater is absorbed by plants, before being "breathed out" by them again into the atmosphere. Were a large part of the forest to disappear, the remainder could well become less able to retain so much moisture, meaning that it would become a drier forest. As more of the forest was cleared, the desiccating process would become more pronounced, reducing the moisture stock of the entire basin and undermining the unparalleled vigor of Amazonia's plant life. The repercussions could extend further afield, even drying out the climate for croplands in the southern half of Brazil.

Tropical forests also help to stabilize climate at the global level. Their mass of vegetation absorbs much solar radiation, soaking up the sunshine. When forests are cleared, there is an increase in the "shininess" of the planet's surface, leading to a so-called albedo effect. More of the sun's energy is radiated back into space. To visualize the mechanisms involved, consider the reflectivity of an open desert or a snowfield; a deforested land-scape has moved a stage across the albedo spectrum. The defore-station-caused increase in albedo would disrupt convection pat-terns, wind currents, and rainfall in lands far outside the tropics —one consequence could be less moisture for North America's grain-growing territories.

While we now realize that tropical forests do not affect

Earth's oxygen balance in a way that need concern us, we are finding they play an important part in the carbon dioxide budget of the global atmosphere. Tropical forests represent one of Earth's main stocks of stored carbon. When the forests are burned, they release considerable quantities of carbon into the skies, just as our autos do when they burn fossil fuel. The buildup of carbon dioxide in the global atmosphere looks as if it is triggering a greenhouse effect, bringing warmer climates for everyone but drier climates for some—again, especially for North Americans. What if the great grain belt starts to come unbuckled? We can envisage no greater environmental upheaval, short of nuclear war, than climatic dislocations of this type and on this scale. More than we may appreciate, tropical forests contribute to the basic regulation of global climates.

Tropical forests have been vigorous contributors to the Earth's vitality for a long time and are one of the planet's oldest ecosystems. Some forests of Southeast Asia have enjoyed continuous existence for at least seventy million years, ever since the heyday of the dinosaurs. Other forests in parts of equatorial Africa and Amazonia probably appear much the same now as they did aeons ago. It is due mainly to their relatively stable history that tropical forests have developed into the most diverse and complex ecosystems on Earth. A patch of forest in Amazonia or Borneo reveals ten times more life forms than a patch of forest in New England or Montana—and hundreds if not thousands more ecological interactions. Because of their rich makeup, tropical forests are a powerhouse of evolution. Not surprisingly, they inspire some of the most basic research attempted by scientists. In light of their unique concentrations of organisms of myriad sorts, they make other portions of the planet seem impoverished aberrations. For biologists, tropical forests represent the ultimate—biology's moon shot. Insofar as tropical forests constitute a kind of benchmark for life processes, we shall not understand life properly until we understand tropical forests. Darwin learned as much about evolution from the plethora of creatures in Amazonia as from the famous finches on the Galapagos Islands. Notwithstanding the small sector of the planet occupied by tropical forests, they are throwing up more basic findings on what makes evolution tick than are all other natural environments put together.

During the course of their emergence tropical forests have produced abundant manifestations of life's diversity. We have noted that they harbor millions of species—and we are discovering many more new kinds of organisms here than in all the other areas of the planet combined. If the reader were to go into a tract of forest with a net, he would need only a few hours to catch an insect not yet known to science—he could even name it after himself. Thus the biological role of tropical forests as a primary source of "new" animals and plants. Following the glaciations of the Ice Ages, when much of the temperate zones became barren, tropical forests supplied a reservoir of life forms by which the sterilized areas recovered much of their biological health.

Tropical forests have even been postulated as a source of intelligence. As primordial monkeys became ever-more adept at swinging through the trees of their convoluted habitats, the greater coordination of limbs, greater keeness of senses, and greater all-around competence developed in conjunction with the enhanced quality of their gray matter. By the time one precocious primate descended from the trees and adopted a ground-level life, becoming a tool user and then a tool maker and finally a full-fledged human, his brain had become more advanced than that of any other creature to have appeared since the first flickerings of life three and a half billion years ago. More remarkable still, his complex brain was giving rise to what we call "mind": he was becoming the first creature to gain consciousness of himself and his environs.

A few million years later, tropical forests were becoming primary sources of civilization. In several sectors of Asia, and in at least one locale of Central America, sizable communities developed a way of life built on agriculture, drawing on the water supplies and foods provided by the forests. In what is now Sri Lanka, Thailand, Kampuchea, and Vietnam, there emerged dynasties and empires that persisted for centuries as societies of sophisticated endeavor, exemplified by, for example, the temples of Angkor Wat. In Central America, the Mayas established a culture founded on forestland farming and bonded by intricate irrigation systems. They too pursued their agricultural way of life for extended periods without environmental deterioration, thanks to the support of their forestland ecosystems.

Virtually all these civilizations confined themselves to limited portions of lowland forests. They apparently recognized the need to safeguard, among other crucial resources that underpinned their lifestyles, their forested environs as sources of moisture. In those few areas that were hilly, the symbiotic relationship between people and forests was generally fostered yet more assiduously. On lower slopes, farmers built terraces to husband their precious supplies of water, enabling them to grow two crops of rice a year, while on upper slopes and mountain ridges, they preserved the forest cover as the ultimate source of their livelihood.

During lengthy periods of its formative stages, then, humankind's prosperity has been closely connected with tropical forests. To these forests—which act as "green machines" that perform an alchemy of sunlight, water, and nutrients to produce an array of benefits—communities of all sorts and conditions have looked for their food, shelter, fuel, forage, and medicines. So pervasive has been the forests' influence on certain of these societies that we still find, among the longer-established cultures such as those in Java, Bali, and Thailand, a sense of numinous awe associated with their forest homelands. In contrast to the folklore of temperate zones, which often regards forests as dark places of danger, traditional perceptions of forests in the humid tropics convey a sense of intimate harmony, with people and forests equal occupants of a communal habitat: A primary source of congruity between man and nature.

DECLINE OF TROPICAL FORESTS

Only in recent times has the sense of concord been widely compromised. After the end of World War II, man started to exploit tropical forests in a manner that they cannot sustain. In many places, excessive harvesting of commercial timber is taking its toll. Undue demands for fuelwood are degrading one tract after another. Small-scale cultivators engage in slash-and-burn farming with an intensity that allows the forest little chance to renew itself. Large-scale ranchers cut down entire sections of forest, put a match to the felled timber, and then plant pastures on which to raise beef for the fast-food trade in the United States and elsewhere.

At present rates of depletion—and the rates are accelerating—our grandchildren may well witness the end of tropical forests, except for a few isolated patches of degraded remnants. Instead of having a band of greenery around the equator, the Earth may eventually feature a bald ring. The integrative relationships between humankind and tropical forests are being ruptured in a far more profound manner than has ever occurred in the past—and the effects may persist for thousands, if not millions, of years. These unique forests, which have served humankind as a fountainhead of welfare for many millennia, look likely to be all but destroyed in a single century. The primary source will be gone for good—a loss in more ways than many of us apparently care to think about right now.

Citizens of tropical forest countries will not be the only ones to suffer from the demise of the forests. We shall all be losers. Just as we are about to embark on a Gene Revolution to surpass the finest achievements of the Green Revolution, in Iowa as well as India, and many other parts of the world besides, the richest reservoirs of gene resources on the face of the planet are being eliminated. Similarly, a uniquely diversified stock of raw materials for innovative medicine and industry is being rapidly extinguished. Our lives and the lives of our descendants for untold generations will thereby be impoverished. Climates in all parts of the global ecosystem will undergo changes for thousands, perhaps millions, of years. The economic and political upheavals ensuing may well exceed those of any military conflicts we have known. If, as is not unlikely, a greenhouse effect caused by carbon dioxide leads to the ultimate melting of the polar ice caps, sea levels will rise and shorelines will retreat at a far faster rate than any we can discern in the prehistoric past, with adverse repercussions for the one-fifth of humankind that will probably be living in coastal zones. Species will disappear in a mass extinction greater in its concentrated spasm than any since the emergence of cellular life, and the course of evolution will be fundamentally modified if not debilitated. So far as we can tell from the paleontological past, the planetary ecosystem will need between ten and twenty million years to restore the damage done to the fabric of life.

Thus the losers will include all inhabitants of Earth—no more than appropriate, some observers may reflect, since all are involved in the depletion of tropical forests. Misuse and overuse of the forests through timber harvesting reflects international marketplace demand for tropical hardwoods. The consumerist appetite of developed-world citizens for specialized timbers of tropical forests, with their many luxury applications in virtually every home of an affluent-world citizen, has expanded almost twenty times during the past three decades, until developed-world consumption has recently surpassed that of the developing world. Hence a key question: Whose hand is on the chainsaw at work in tropical forests? Furthermore, much of the excessive logging is conducted by giant timber corporations based in North America, Western Europe, and Japan. All the more we may ask, Whose hand is on the chainsaw?

The recent surge of cattle raising in forestlands of Latin America has been stimulated by the Americans' search for "cheap," that is, noninflationary, beef for hamburgers, frankfurters, television dinners, and other convenience foods. This trend is being replicated in both Western and Eastern Europe, Japan, and some of the newly rich nations of the Middle East. At the same time, certain of the ranching interests are based in those same nations that foster excessive demand for cheap beef. People around the world are involved in the decline of tropical forests.

COLLECTIVE CAMPAIGN IN SUPPORT OF TROPICAL FORESTS

Since everybody will lose by the decline of tropical forests, and since everybody contributes to the process, it makes sense that everybody should become involved in safeguarding the forests. Of course we should not seek to keep them locked away, declaring them off limits to development. Sustainable use is feasible in many ways, as this book will demonstrate. We can derive far greater benefits from tropical forests than we have in the past, and derive them in perpetuity. We can even turn to tropical forests as a more significant primary source, of multiple sorts, than was thought possible by our forebears. Yet the task of safeguarding tropical forests represents a challenge as great as that of any other resource issue confronting humankind. The very prospect of trying to save tropical forests raises hosts of problems in one's mind. But problems also imply opportunities. A time of crisis, as the Chinese say, is a time for breakthrough, a time when we can move beyond the patterns and practices of the past and attempt bold, new strategies for the future. We shall need a whole series of initiatives by the tropical forest countries themselves, matched by a similar set of initiatives by the developed nations. Even more, we shall require a set of measures taken by the community of nations acting as a whole: a concerted endeavor of a nature and on a scale such as is rarely envisaged, let alone implemented, by the collectivity of citizens around the world.

Taxing as such measures will be, they do not compare with the steps we would have to take to compensate for the harm done to society if tropical forests continue to decline. In this prospect should lie our motivation. What better impulse for collective endeavor by the global citizenry than an effort to safeguard the extraordinary natural phenomena and the exceptional stock of natural resources inherent in tropical forests? When we consider what is at stake, how can we question whether we are up to the task?

If we seek further incentive, we need only contemplate these same forests. Surely they offer a primary source of inspiration for us to act together in defense of a salient segment of our One-Earth home.

Such thoughts often occupy my mind as I stand amid the grandeur of a tropical forest. Something of the scene outside me serves to stretch something inside me. It makes my faculties operate with a sharpness I do not generally sense; a host of additional nerve endings comes alive. I seem to sprout antennae to probe this new and fascinating world. At my finest moments in wild nature, such as I experience hour by hour in a tropical forest, I feel as if my whole being is standing on tiptoe. I shall not be the same again—thank goodness. I shall look at things differently, and not just the natural world, but cities, newspapers, friends, and unknown faces. I have grown a little—and what better place to do it than in that luxuriant community where growing, ever-more growing, is the essence of it all.

Perhaps what most stimulates me about a tropical forest is the notion that its diversity, its interactions, its sense of wholeness, all are so advanced that it is far beyond the scope of my imagination—let alone my intellect—to grasp how advanced they are. I feel cut down to size, even as I feel I stand taller than before.

I. BIO-ECOLOGICAL

BACKGROUND

I

Nature's Powerhouse

I WELL REMEMBER the day I took a hike through a forest in Borneo. It sticks in my mind because that proved to be the most impressive patch of tropical forest I have encountered during twenty years of roaming around these exuberant expressions of nature.

I had joined a field trip of three Indonesian colleagues from the University of Samarinda, the largest port city on Borneo's east coast. These local scientists had kindly offered to show me a sector of their famous dipterocarp forests, named after the predominant tree type. In Southeast Asia and centered on Borneo there are more than five hundred species of the plant family Dipterocarpaceae—the Latin name referring to the fact that the tree's fruits have two "wings." Esoteric as this scientific designation might seem to some readers, we can reflect on the relevance of dipterocarps to our daily lives when we note the number of items around our homes that derive from those same forests of Southeast Asia: specialist hardwoods for our parquet floors, fine furniture, decorative screens, fittings and cabinets, and the like. A large share of international trade in hardwoods comes from the dipterocarp forests of Southeast Asia. Gazing around me, I could see why dipterocarp forests are so renowned in the world of biologists, conservationists, and foresters. If tropical forests are the richest and most complex ecosystems on Earth, the dipterocarp forests are surely the grandest and the most prolific, as well as the oldest, of them all.

On looking back, I feel that that forest represented a more striking spectacle than any other I have come across during my travels in almost one hundred countries. At other times and in other places, I have watched a wildebeest migration in the Serengeti Plains of northern Tanzania—half a million animals in a single stretch of savannah. Off Baja California, I have seen large groups of whales taking their ease in their calving grounds. In the Caribbean, I have enjoyed some of the most splendid skin-diving I could imagine, fishes of every shape, size, and subtle hue, all within a few meters of my face mask. In southern Greenland, I have looked out over glaciers of such size that a single blue-green crevasse could swallow up an apartment block. At Lake Nakuru in Kenya, I have looked out on two million flamingos; and I understood why ornithologists call this the greatest bird sight on Earth. But a dipterocarp forest is in a class of its own. It is akin to other super-size spectacles on Earth, such as the Grand Canyon and the Victoria Falls. However much you read about the scene, however many photographs or films you see of it, nothing prepares you for the phenomenon itself with its sheer scale and impact. You gaze on it and you feel your life has started on a new phase. Things will not be the same for you again after setting eyes on something that exceeds all your previous experience.

STATURE OF A DIPTEROCARP FOREST

So it was with this dipterocarp forest. In the main, I suppose, I was affected most by the trees, giants on every side. An above-average dipterocarp, by no means an outsize specimen, can measure 5 meters around, making it one of the bulkiest boles anywhere. To encircle a large trunk, I and my three friends had to join hands and stretch as far as we could. Most dipterocarps are not that big, many measuring only half as much and many more even less than that. But a large dipterocarp is at least as big as a good Douglas Fir, and a lot more stately. Dipterocarps are tall trees, and their trunks, if slender by some standards, are impressive growths, straight and unencumbered for dozens of meters, soaring sheer from the ground to the canopy that is so far overhead that it seems to belong to a different world.

Still more important, it was the entire community of dipterocarps that impressed me, so many of these giants towering in one area. Out of a total of 400 to 700 trees per hectare,* at least 50 were giant-sized, while one dozen out-towered the rest above the forest canopy. My imagination failed when I tried to picture these huge plants extending across the island of Borneo for 1,000 kilometers, a greater quantity of impressive trees than anywhere else. Moreover, as a tree is more than just wood, so a forest is more than just trees. On every trunk and branch I saw a variety of herbaceous (non-woody) plants, and on some of these plants, still more plants. As for animals, the forest interior supported monkeys, birds, frogs, insects, and snakes, among many other kinds. Although I saw only a single snake that day, and generally I sight none at all, I am sure that dozens turned a beady eye toward me. The insects, by contrast, are not hard to find: I can well believe those scientists who tell me there are at least as many insect species in tropical forests, covering a mere one-sixteenth of the planet's land surface, as there are on the entire rest of the Earth.

When my friends and I tried to encircle a tree with our arms,

---

*For those readers who are unfamiliar with the international system of metric measurement, one hectare is equivalent to almost 2.5 acres, or roughly one-third as big again as a football field; 1 kilometer is 0.6 of a mile, or roughly the extent of twelve city blocks; 1 square kilometer is equivalent to almost 0.4 of a square mile, and Central Park in New York City is about 3 square kilometers. Furthermore, one cubic meter is equivalent to 1.3 cubic yards; and one metric ton is fractionally less than one "regular" ton.

we could not choose just any tree. Many support buttresses at their base—"plates" of ultra-hard wood that occupy the triangle between the bole and the main roots. A typical tree can have three or four such buttresses, and I found several with twice as many. Although a buttress is thin, it usually reaches at least a man's height up the trunk and sticks out as much along the ground. I have seen similar flanges in tropical forests of Africa and Amazonia, and never outside the moist tropics. My Indonesian friends explained that so many dipterocarps have these buttresses because their shallow roots must remain near the surface to scavenge such nutrients as come their way from above ground, there being next to no sustenance in the infertile soils of tropical forests. The shallow root system offers little anchorage for the vast trunk, especially when 100-kilometer-per-hour winds assault the tree's crown and send a whiplash down the bole to the base. Without the buttresses, the tree might be uprooted during one of the storms that sweeps over the forest.

Looking up, I could see what my friends meant. A typical dipterocarp soars 50 meters, making it as tall as a twenty-story building, taller than the Statue of Liberty. Occasionally one stands twice as high, or almost as tall as the highest redwoods. This makes the dipterocarp a sizable plant. Indeed, a 50-ton dipterocarp is not unduly large, and a 100-ton tree is not unknown. Again, we can note for comparison that an average dipterocarp weighs about as much as an average redwood—though a giant sequoia, the bulkiest tree in the world, can readily top 1,000 tons. But whereas the redwoods and sequoias of California are limited to an area of just a few thousand square kilometers, the dipterocarp forests of Southeast Asia cover at least 1 million square kilometers.

Through the columns of dipterocarp giants, I could just make out the topmost layer of the forest, the canopy. The trees generally grow for two-thirds of their height with no branches, presenting an uncluttered bole of striking simplicity before starting to sprout branches that themselves are as big as many trees I have

seen in my native Britain. One particularly splendid dipterocarp
seemed to reach well above its neighbors. This exceptional specimen, explained my Indonesian colleagues, could have a canopy
extending dozens of meters across. Again, I thought, dipterocarps
are no ordinary trees; they are an expression of plant life in a
league of its own. A moderate-sized dipterocarp, with a girth of
two meters or so, will be at least 100 years old, possibly 200, while
an "emergent," such as the one I had noticed soaring above the
canopy of the main forest, can be anywhere from 400 to 600 years
old.

As I looked around the forest, I could scarcely grasp the fact
that plant communities like this one, with similar architectural
grandeur, still extend not only across Borneo but also across much
of Southeast Asia. What a capacity, I thought, for growing wood,
more wood, ever-more wood. Whereas a temperate forest generally represents a few sprouts, so to speak, of timber sticking above
the ground, a tropical forest amounts to a veritable dynamo for
generating wood. My colleagues told me they had measured the
amount of "woody biomass" in typical patches of dipterocarp
forest and often found 400 cubic meters per hectare, or between
three and five times as much as we would expect to find in a
similar-sized patch of average forest in New England or Great
Britain. Occasionally a dipterocarp forest measures as much as
600 cubic meters of woody biomass per hectare, while record
totals surpass 750 cubic meters. This is only the wood of the
forest; when we include herbaceous plants as well, the total occasionally approaches 1,000 cubic meters. To put the measurement
into perspective, 1 cubic meter of wood weighs, roughly speaking,
rather less than 1 ton.

PROFUSION OF PLANTS

But these statistics, as is frequently the case, tell less than the
entire tale. What matters more to the workings of a tropical forest
is the variety of the parts that make up the community. The trees

around me appeared, at first glance, to be more or less alike. They all had smooth bark, rather light in color. Freshly fallen leaves on the forest floor were all dark green, leathery to touch. Yet the sector of forest that we could take in with our eye, roughly half a hectare, contained dozens of different sorts of trees. My Indonesian guides pointed out at least twenty species within just a few strides, these all being plants that "officially" qualify as trees by virtue of having a diameter of at least 10 centimeters (about four inches).

I thought back to the forests of Britain that I knew as a youth; they have a mere three or four species in a similar-sized patch. In the Appalachian forests of the United States, among the richest in tree species of any temperate forest, one would find no more than 25 species. But, my Indonesian colleagues told me, a tropical forest has, including its few bushes and shrubs, at least 100 woody species per hectare—often as many as 150, sometimes even 180, and occasionally, in the wettest lowland areas, more than 200. In an area of 10 sq. km. we can certainly expect to find 300 species, often twice as many or even more. British ecologist Professor Peter Ashton, who once worked in a sector of Brunei on the northwestern coast of Borneo, checked some 30,000 trees in an area of only 45 hectares and found 760 species, while neighboring Sarawak, a composite area totaling less than 10 hectares, revealed more than 780 species.[1] In Brunei and Sarawak together, a territory totaling 126,000 sq. km., there are at least 2,500 native tree species, whereas in Great Britain, with its 244,000 sq. km., there are only about 35 such species. Generally speaking, we find a far greater variety of woody species in a tropical forest than elsewhere. Almost every order of plants includes trees or treelike growths among its representatives. Bamboos, for example, which are strictly speaking grasses, often reach 20 meters in height, with so much woody tissue that a thicket of bamboos will turn aside a herd of elephants. Milkworts appear in the form of tough twiners as tall as any true tree, and with strands so tough that the strongest man could not hope to break one off. Even violets, dainty growths in temperate lands, can look like trees.

As for non-woody plants, we may expect to find 1,000 species within just a few hectares of dipterocarp forest. Especially numerous are climbers, as I learned by looking around me. Almost every tree, it seemed, was festooned with vines. Some of them were thin, clinging to tree trunks like ivy and having tiny aerial roots that grow along their stems. Others, more striking in appearance, were great woody vines called lianas, some of the larger ones being as thick as a football player's thigh and occasionally twisted together like ropes. Although I could not see the forest-top lianas, my Indonesian friends told me that a network of interlacing loops links the treetops to one another. Sometimes these canopy lianas form such tangled networks that they can even hold a tree upright after its base has been cut by a chainsaw.

This phenomenon of plants growing on plants is a common characteristic of tropical forests. It is not difficult to find at least 30 species growing on a single tree, including ferns, vines, epiphytes, bromeliads, nettles, orchids, and passion flowers. Among the epiphytes ("air plants", from the Greek *epi* for "upon" and *phyton* for "plant"), particularly prolific are the orchids, of which almost 50 species have been found blooming on a single tree. All told, there are a full 4,000 orchids in Southeast Asia, or one-quarter of the Earth's total. While other plants in the region are not so numerous and diverse, Southeast Asia supports one in ten of all plant species in just 3 percent of the planet's land surface. I noticed that plants occupy almost every crack and crevice of the forest. They flourish not only on trees but on the forest floor, on rotting logs, on rocks of every shape, wherever there is a cranny or rough surface to adhere to. In that Borneo forest, I found at my feet a plethora of ferns, mosses, liverworts, and lichens. There were few of the flowers that make a carpet on the floor of a temperate forest. But wherever there was enough light from shafts of sun penetrating the canopy, there was undergrowth aplenty.

Yet despite the variety of plants, the undergrowth was far from thick. There was hardly any of the tangled vegetation popularly associated with tropical forests. I and my companions could

walk about almost as we pleased, and in many places we could have marched in a straight line for dozens of meters. We carried machetes in order to mark our trail but did not need them to hack away a single strand of vegetation. All this represented a far cry from the scene so often portrayed by writers and movie-makers —a tumult of vegetation where Tarzan could easily lose his mate. This type of undergrowth is caused by the amount of light at the forest floor, only one-tenth of the sun's brilliance above the canopy. What explains the jungle-like accounts of tropical forests? Naturalist explorers from North America and Europe have traditionally roamed forest regions by boat along waterways, where the forest is exposed to lots of sunshine—which stimulates the dense curtains of foliage that the explorers saw along river banks. Hence the accounts of impenetrable masses of vegetation that presumably persisted through the forest interior—hence too the impressions portrayed by oldtime movie-makers, who used the same waterways.

Far overhead, the canopy offers splendid scope for sun-loving plants of every kind. Especially plentiful are epiphytes, which, needing no soil, depend on their tree hosts for nothing more than support: they derive all their sustenance from the air. Many of them are small ferns and other nonflowering plants, but bromeliads, members of the pineapple family, can be more showy even than orchids, blossoming in every size and hue. Skimming above the canopy in an airplane, I am astounded at the color of the spectacle. This sector of the forest is far more than a mass of greenery. So diverse is the world of the forest canopy that it can be considered the last great frontier of biology. No other habitat sustains a greater abundance and concentration of species, both plants and animals. Nor does any other habitat remain so rela-

---

A strangler fig among other climbing plants, plus epiphytes, illustrate the abundance of plants that grow on plants in tropical forests. The example shown here is from northern Sulawesi, Indonesia. Compare this with termperate-zone forest, where virtually all plants grow separately from each other. (*Michèle Dépraz, World Wildlife Fund*)

tively unknown. Here an entire realm of life awaits our attention, with a richness that I could never have guessed from the forest floor. Around me I sighted about one dozen different kinds of birds during the course of the day, but my Indonesian friends told me that in the middle reaches of the forest there could be between 50 and 70 species, with well over 100, perhaps 130, flitting around at topmost level.

To date we have obtained only a few specimens, and our knowledge of the forest canopy remains scanty. Yet this, the greatest array of life known on Earth, is not so much on Earth as 30 to 50 meters above it—much too far overhead for the forest-floor observer to see in any detail, even with binoculars. Yet because forest trees generally rise sheer from the ground with no branches until their upper reaches, they offer no handholds for the would-be explorer. Moreover, the bole timber is often so hard that you cannot drive a climber's spike into it: the metal simply bends. And the higher the explorer climbs, the more he has to cope with insect attacks, not to mention vertigo. Fortunately, the last few years have seen a breakthrough in research techniques. There is now an assortment of rope slings, pulley systems, and other fancy gear to enable scientists to explore the canopy more freely than they dreamed of only a decade ago. We can confidently expect that, within the next few years, the pioneering research of the upper forest will surpass, in its rich findings, the probings of deep-ocean trenches and other recent advances in frontier biology.

One of the main conclusions we can already anticipate is the discovery of a great many species, almost all of them very sparse in number. Of the Borneo plants that have already been identified by science, my Indonesian colleagues told me that I would generally find only a few representatives in a whole hectare. In some instances, I would come across only one in 20 hectares. In a typical hectare of Amazonia, with its 100 or more tree species, as many as half may be different from those in another hectare only 1 kilometer away. These distribution patterns mean that many tropical forest plants are comparatively rare, which makes some of them liable to local extinction when a forest is subject to intensive exploitation. So extremely rare are the individuals of certain plant species in these forests, notably orchids and legumes, that we know a few of them from only a single encounter several decades ago. In the Malay Peninsula, a begonia was discovered during the 1940s, and, because of its ornamental appearance, it has become a popular cultivated plant in Europe, with millions of specimens in gardens, nurseries, and so forth. But it has not been found in the wild again. In Amazonia, numerous tree species were identified between 50 and 100 years ago that have not been located since.

Much the same applies to butterflies and birds, among other animals. In the Corcovado Park in Costa Rica, there are eight species of *Heliconius* butterflies, which researchers hardly ever encounter at a rate of more than three adults per hectare, often only a single one. Many butterfly species, together with large numbers of other insect species, appear to occur still more rarely. As for birds, in Amazonia the black-chested, tyrant flycatcher recently appeared 800 kilometers from the nearest place it had hitherto been seen, while the red-shouldered parrot had been recorded only through sightings of seven isolated individuals in all the 6 million square kilometers of the Amazonia region before scientists recently encountered an entire flock.

This extreme sparseness contributes to the unexpected diversity amongst tropical forests. Contrary to popular impression, they are far from being one homogeneous mass from horizon to horizon with each patch of greenery much like any other. When I travel from Borneo to New Guinea, a mere 1,700 kilometers eastward, I know that I shall find far fewer dipterocarp species, only 15 instead of 262, with entirely different families of trees in their place. The same applies, of course, to associated plants, together with animal life. Something similar, moreover, happens in a single land mass. The director of research at the New York Botanical Garden, Dr. Ghillean T. Prance, has found that Amazonia has eight phytogeographic zones, or "plant areas," each with a distinctive assembly of plants and animals.[2] This diversity of formations in tropical forests is in contrast to the pattern of forests elsewhere. In Alaska, for example, we find a type of forest that is virtually identical to one in northeastern Canada—4,500 kilometers away.

Such divergence between tropical and nontropical forests alerts us to the fact that what makes one tick is very different from what makes the other tick. If we are to keep tropical forests ticking at all, we shall have to recognize that the strategies we have developed in temperate zones, whether for exploitation or for management or for preservation, do not work nearly so well, if at all, in tropical forests. Our scientific surveys must be different, our logging practices must be different, our planning of parks must be different. In fact, we must operate differently from start to finish. The more I become acquainted with tropical forests, the more I believe that using the term "forest" for a bunch of trees in the tropics and a bunch of trees elsewhere is misleading. The two categories may reveal a few immediate similarities, but from there on they go their separate ways. How much better we might understand tropical forests if we gave them a new name, indicating that they are a fundamentally different state of affairs from the forests with which we are more familiar.

THE FOREST ENVIRONMENT

The basic dissimilarity between tropical and nontropical forests was further apparent to me later that day in Borneo. By mid-afternoon, the atmosphere became heavy, while the rare glimpses of blue sky overhead gave way to dark clouds. I knew we could expect a thunderstorm—routine toward day's end in a tropical forest. I also knew that Borneo receives 5 meters of rainfall, spread more or less evenly throughout the year, or five times more than that in New York City. I knew, moreover, that a half-hour downpour could bring 50 millimeters, or twenty times as much water as an average shower in New York. But when the heavens opened over backwoods Borneo, I was surprised that the rain did not deluge us. Instead it came down in a warm film, leaving me feeling for all the world as if I were taking a fine-spray shower. Way above, in the canopy, I could see tree crowns being tossed by the wind, and when we briefly detoured along the bank of a stream, I noticed the surface roiled by sheets of rain. Yet within the interior of the forest, the foliage served to break the impact of the downpour. My colleagues told me that, according to their measurements, one-quarter of the rain was reaching the ground by trickling slowly down branches and tree trunks, the rest reaching us in a drizzle. Another revelation for me.

I was also surprised by the tranquility of the forest while the storm raged above. On the forest floor, my colleagues told me, the wind speed would be only one-hundredth what it would be above the canopy—while in the canopy it would be only one-tenth what it would be a few meters higher. This insulation of the forest interior, I surmised, must help to maintain the equable climate, with its stable warmth and moisture levels throughout the day and night.

Following the thunderstorm, the forest released a smell of earthy fertility. A musty odor, like that in a greenhouse, it was strangely satisfying even though it spoke of decomposition. I had been vaguely aware of it earlier in the day, and the humid atmosphere seemed to make it stronger. I knew where it came from but scarcely thought it could be so pervasive, or so pleasant. Then I remembered that a forest is home to hosts of decomposers, notably organisms of the topsoil, such as mites, nematodes, ants, and termites. In 1 square meter of leaf litter, Professor Daniel Janzen, of the University of Pennsylvania, found 800 ants belonging to 50 species, while a similar square meter may contain as many as 2,000 termites.[3] In the dipterocarp forests of the Gunung Mulu National Park in Sarawak, Dr. Mark Collins found between 4 and 7 grams of soil fauna per square meter, an amount twice the likely weight of all mammals and birds in the region put together.[4] Of these creatures, at least one-third and sometimes one-half are termites; the weight of termites in some dipterocarp forests of Peninsular Malaysia can be even higher, up to 10 grams per square meter, or over half as much as in the greatest wildlife spectacle on Earth: the herds of Serengeti National Park in northern Tanzania with their four million wildebeest, zebras, and gazelles.

Perhaps more important still, the topsoil contains multitudes of fungi, especially the mycorrhizal fungi, which recover and recycle nutrients from fallen vegetation and dead creatures. When a tree, for instance, dies, and its decaying parts are made available to support the lifestyles of other organisms, the mycorrhizae flourish. In other words, the smell of the fungi and other decomposers is the smell of life. My Indonesian friends added that when a patch of forest is cleared away, the smell disappears with it—as do the nutrients that the fungi and rootlets once garnered to maintain the healthy working of the ecosystem.

By the same token, I was struck to find only a meager amount of litter underfoot—and by "litter," I refer to fallen leaves and the like. A popular image of the tropical forest suggests that the ground is covered with age-old accumulations of decaying vegetation. Yet there was barely enough litter to conceal the soil—a result of the forest ecosystem. With its warm, moist climate, it provides a fertile environment for the multitudes of mini-organisms that, together with the fungi, break down the fallen leaves, logs, fruits, and other detritus far faster than would be the case in temperate zones. In fact, leaf litter can decompose within six weeks, as compared with one year in a temperate, deciduous forest, and seven years in a boreal, conifer forest.

As I tramped around the almost litter-less forest floor, I reflected on other sterotypes of tropical forests: the green vastness of the "jungle," the lurking snakes, the choking vines, the clouds of mosquitoes, the crushing heat. In fact, I found the dipterocarp forest, like any other tropical forest I have visited, to be not only an extraordinarily interesting place to explore, but a very beautiful one to admire. I also found it a fairly pleasant place, and safer than the streets of the capital of Indonesia, Jakarta—or of Rio de Janeiro or Tokyo or New York.

While a visitor can readily see through the popular stereotypes of tropical forests, we still know next to nothing about what makes the forest continue on its quiet, complex way. Scientific researchers have barely made a start on their task of understanding this secluded world. We have probably not discovered more than 20 percent or so of the several million species that likely exist there. In the largest forest region of all, Amazonia, scientists have made a detailed study of no area larger than a few dozen hectares. Just recently we discovered several large tributaries of the Amazon River, and we now realize that we have drawn entire mountain ranges on our maps of Amazonia in places hundreds of kilometers away from their actual location. Moreover, as recently as the early 1970s, scientists noted the Tasaday tribe in the Philippines pursuing their Neolithic way of life, after having been cut off from the outside world by a strip of forest that is a mere 25 kilometers wide. Perhaps most telling of all, we do not have a precise idea of just how much tropical forest still exists—and how fast it is being cut down.

I ended my day feeling as I generally do after sojourns into tropical forests, sensing that I had picked up more basic biology in this patch of forest than possible during a day in any other ecological zone. I had achieved it in one of the few ways that really matter, through first-hand experience. At the same time, I was absorbing my biology "lesson" both through the scientific information that my Indonesian friends fed to me, and through a process of "imaginative osmosis" that I find stirs within me whenever I am confronted with a major phenomenon of nature. A day of field research in a tropical forest is not work for me; rather, it is recreation in the sense of re-creation.

2

Forests and "Forests"

JUST WHAT KINDS of forests are we talking about when we use that catch-all label, "tropical forests"? Do they vary much from one region to another? What causes one to differ from another? What, in short, makes them "work"?

There are all sorts and conditions of tropical forests. We cannot even say that they are all confined to the tropics, because in several sectors of Asia they extend for quite a few hundred kilometers north of the Tropic of Cancer. So let us start with a definition of the tropics that takes account of the way nature works, rather than the way people draw arbitrary lines around the globe. In a literal sense, the tropical zone encompasses a band around the Earth beyond which the sun does not wander during its annual "migrations" (*tropos* is Greek for "turning point"). Within this zone, the sun's rays beat down from a high angle, more or less directly overhead, throughout the year. In temperate zones, by contrast, the sun shines on the Earth for much of the year while lying closer to the horizon, meaning that the energy from each sunray is spread over a larger area of the Earth's surface and is filtered through a thicker layer of atmosphere; hence, it generates far less light and warmth per square meter of ground. In the tropics, the sun generates more warmth during the year by virtue of its invariably lofty angle. To be precise, 1 square centimeter of ground in the tropics receives more than twice as much solar energy as does 1 square centimeter at a latitude of 50 degrees north or south of the Equator—locations such as Vancouver, Paris, the southern tip of New Zealand, and the Falkland Islands. Because of the enhanced solar energy in the tropics, a few places just outside the Tropic of Cancer and the Tropic of Capricorn appear disproportionately warm; when they are allied with sufficient moisture, they produce the "tropical" forests of northern India, Bangladesh, and Burma, plus southern China.

Approaching the subject from a different direction, we can characterize the tropics as those areas that never become really cold. Scientists believe that the cut-off line is a mean temperature level in sea-level equivalents of 75 degrees Fahrenheit* around the year. A related factor deals with the differences between very warm and not-so-warm temperatures. Looking at those territories where the difference is less than 10 degrees Fahrenheit, and marking them with boundary lines, we find that appreciable portions of the geographic tropics fall outside these "temperature tropics": for example, southern Brazil, which even receives occasional frosts. We find, in short, that the mean difference between warmer and cooler periods of the year is less than the temperature difference between daytime and nightime. All in all, we can consider the tropics a zone where there are no notable seasonable fluctuations in temperature: we need never get out our winter clothes. These various considerations mean that the tropical boundary should be about 4 degrees latitude closer to the equator than are the 23.5-degree lines of latitude that indicate the Tropic of Cancer and the Tropic of Capricorn.

*Throughout this book, in reference to its initial publication in the United States, I have used the Fahrenheit rather than Celsius temperature.

Several other factors complicate the situation. Oceans and their currents exert a moderating effect on temperatures, as do atmospheric conditions, convection currents and wind patterns, and altitude and other features of physiography. Mountains especially have a marked impact on temperature: with every 1,000-meter increase in elevation, the mean annual temperature is likely to fall by at least 10 degrees Fahrenheit. In short, it is more realistic to define the tropics by temperature characteristics than by straight lines on a map.

Tropical forests, however, require not only warmth, but lots of moisture as well. Generally speaking, the wetter an area, the more vegetation it produces—increasingly complex vegetation (Fig. 1). In tropical areas with less than 500 mm. of rainfall per year, we find only scrub brush and thorn woodland. In areas receiving between 500 and 2,000 mm., we find broad-leaved deciduous forest, sometimes qualifying as "moist" forest. When rainfall exceeds 2,000 mm., we generally find evergreen forest. In super-wet areas that receive more than 4,000 mm. (four times as much as New York City), we see true rainforest. Thus we arrive at the famous forests of the humid tropics. Strictly speaking, and for the sake of a technical definition, we find tropical lowland, broad-leaved evergreen forests—the most widespread and most diverse type—in those areas of the tropics where the amount of year-round rainfall (plus other forms of precipitation such as mist and dew) equals or exceeds the amount of moisture that evaporates from the land surface, plus the amount that passes through plants before evaporating via their leaves.

Surprisingly, however, the overriding influence of rainfall does not stem from the total amount, important as this is, rather from its distribution throughout the year. In the wettest parts of Asia —Mount Cherrapunji in Assam, and Baguio in the Philippines (both of which receive almost 12 meters per year)—there is a dry season lasting several months. These areas have only about one-tenth as many tree species as does an area with only one-eighth as much rainfall but with plenty of moisture in every month throughout the year. In parts of Latin America, a patch of dry forest is likely to contain only one-third as many species as a patch of moist forest, which in turn is likely to contain only half as many species as does a patch of rainforest.[1] At the Kra Isthmus, in southern Thailand, located at about 11 degrees N., the moist forest zone gives way to the rainforest of Peninsular Malaysia and Indonesia; the area north of the Isthmus has about 200 genera of plants, while to the south there are 375 genera.[2]to So marked are these variations between tropical dry and tropical wet forests that we find they reflect a greater difference in species abundance than we find between tropical forests as a whole and temperate forests.

FIGURE 1. This diagram illustrates the relationship between moisture and plant life in a tropical lowland forest. The diagonal line from bottom left to top right represents a gradient of mean annual precipitation. We see that as the amount of moisture increases, the forest becomes more complex; there is greater biological diversity and ecological stratification.

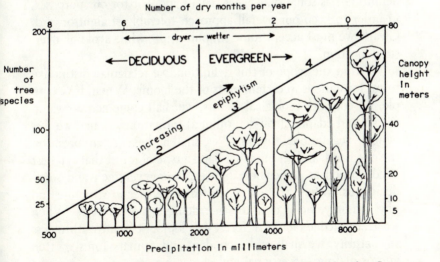

Source: *Ecological Aspects of Development in the Humid Tropics*, 1982, National Academy Press, Washington D.C.

Having defined the humid tropics in terms of temperature and rainfall, let us look at the definition of tropical forests that is used in this book: forests that occur in areas that have a mean annual temperature of at least 75 degrees Fahrenheit and are essentially frost-free—in areas receiving 2,000 mm. or more of rainfall per year and not less than 100 mm. of rainfall in any month for two out of three years. They are mainly, if not entirely, evergreen.[3] We generally find such forests at altitudes below 1,300 meters, though sometimes in Amazonia up to 1,800 meters, and usually in Southeast Asia up to only 750 meters. In mature tracts of forest, there are several more-or-less distinct strata, and the canopy is made up of almost continuous interlocking tree crowns. A generalized definition, it does not take into account many significant sub-classifications that have their own intrinsic merits. But this book deals with many other levels of complexity, especially with respect to the many ways that humans disrupt forests, again with varying degrees of impact. So, although our definition has some drawbacks, let us adopt it for our purposes, in order to keep our overall approach tolerably straightforward. Clearly we must accept some simplification, while avoiding over-simplification.

Within the scope of this definition, we recognize distinctive forest formations in certain parts of the biome. When, for example, we consider ecological factors, we shall come across swamp forests and edaphic (soil-determined) formations. When we recognize "vegetational rhythms," we shall find evergreen forests on the one hand, and on the other hand a type of forest that can best be described as somewhere between semi-evergreen and semi-deciduous. When we take note of divergencies in structure and physiognomy, we shall find either dense or open forests. When we acknowledge factors of forest development from youthfulness to maturity, we shall find secondary and primary forests, often approximating successional and climax forests.

So much for the basics. Let us move on to that prime differentiating influence, seasonality.

SEASONALITY

To reiterate a central point, forest ecosystems are hypersensitive not only to the amount of moisture but to its seasonal distribution. So important is seasonality that it is a main "break-point" factor between rainforest and seasonal forest, and between evergreen and only partly evergreen (semi-deciduous) forest.[4]

The term "rainforest" is loosely applied by some writers to all forms of tropical forest. But as noted above, we shall use it here with regard to those forests that receive at least 4,000 mm. of rainfall a year. To incorporate the seasonality factor, let us add that a rainforest must receive at least 200 mm. in ten months out of the twelve. The rainiest rainforests receive at least twice as much rainfall in total, and at least 200 mm. in every single month. In appearance, moreover, a rainforest is a closed community of trees and other plants of abundant forms, the vegetation characterized by exceptional luxuriance and complexity. Many of the trees sport large buttresses and smooth-barked columnar boles. On trunks and larger branches, we often find flowers and fruits borne on short stalks—a phenomenon known as "cauliflory." The forest features several layers of trees and shrubs, each layer with plenty of epiphytes and vines, so that the whole looks like forest piled on top of forest. The tallest and richest forms of rainforest, notably the dipterocarp forests of Borneo, can feature at least 650 tons, and occasionally 1,000 tons, of plant biomass per hectare. All in all, a rainforest presents a specialized aspect of its own to the observer. A rainforest in Southeast Asia is so similar in structure and function to a rainforest in equatorial Africa or tropical Latin America that it has now come to be classified as a category of forest on its own: we no longer speak of "rain forest," but of "rainforest."

As we move away from the equator, we approach the climatic limits of rainforests. The formations start to diverge, until they need to be sub-classified as semi-evergreen rainforest, a variation characterized by a lack of water for at least a few weeks each year.

We find a belt of this "compromised" rainforest in a transitional zone on the way to semi-deciduous forest in parts of Thailand, Burma, and India.

After this intermediary belt, we come to seasonal forest, sometimes known as monsoon forest. "Seasonal forest" is a convenient term for those forest types that develop in areas with a marked dry season, that is, where plant growth seriously slackens for want of water during several months of the year. In appearance, seasonal forest is less lofty and luxuriant than rainforest. Toward the end of the dry season, it may cease to be green altogether, turning brown as it loses most of its leaves. Since seasonal forests do not have continuous canopies as do rainforests, they allow much sunlight to penetrate to the forest floor, where it promotes dense undergrowth. But apart from these junglelike tangles at ground level, seasonal forests contain far less vegetation. The biomass rarely exceeds 80 tons per hectare, or only one-fifth as much as in a regular rainforest.

Seasonal forests of various sorts are widespread in Southern and Southeast Asia, notably in Indo-China, Thailand, Burma, India, and Sri Lanka. They are also extensive in Africa, where rainforest as we have defined it is actually limited to just a few patches. In Latin America, where climates tend to be wetter, we find that rainforests predominate, and seasonal forests are comparatively rare.

PRIMARY AND SECONDARY FORESTS

Let us consider two other important categories of tropical forests. When a patch of forest has been left undisturbed for a long time, at least decades and sometimes centuries, it is regarded as primary forest. Of all forest types in the biome, none is richer or more complex than primary forest. Sometimes known as virgin forest in order to distinguish it from disturbed (secondary) forest, primary forest generally constitutes forest that, having worked its way through successional stages of development as a young forest, has reached a state of maturity—often considered a state of ultimate diversity and complexity. Such a formation has been described by some scientists as a climax forest, one representing a vegetation type that exists in dynamic equilibrium with its environment, particularly with its climate.

When a patch of primary forest is cleared, and the plant life left to regrow, new biomass accumulates rapidly until it peaks at around fifteen or twenty years. After this building phase, there is little change in the volume of plant life, only in its characteristics. Through a process that occurs slowly at best, pioneer species and low-ground vegetation fade away, and their detritus provides nutrients for larger and longer-lived trees. Eventually the secondary forest develops toward a formation akin to that of primary forest. In time, and provided all the necessary complex of factors is available, primary forest reestablishes itself. But the process can be protracted. Certain tree species of primary forests do not flower for the first time until they reach fifty years of age. Equally important, a good one-third and even one-half of a primary forest's plant species may be woody plants (in contrast to a temperate-zone forest, where only one species in ten is likely to be woody). This means that a tropical forest community shows only slow turnover among many of its component organisms: changes in the ecosystem are slow.

Thus a secondary forest features a community of plant species with a makeup fundamentally different from that of a primary forest. Equally important, the structure and functioning of the community are different, and the ecosystem is less diverse. Trees grow quickly, with life spans that may not extend beyond a few decades. They start to produce numerous seeds early in their lives, seeds that are small and easily dispersed, often by wind rather than by animals, and capable of germinating in full sunlight on bare soil. In all these characteristics, a secondary forest is a world apart from a primary forest. Even the animal community, with teeming species, may not overlap much with that which emerges when the primary forest becomes established.

A typical tree of secondary forest, in the Tai region of the Ivory Coast, West Africa. Successional trees tend to be tall and spindly, traits that, together with leaf design, equip the trees to capture all the sunlight they can reach. Note that the forest community reveals far less diversity at the early pioneer stage than we find in a mature or primary forest. *(Professor U. Rahm, World Wildlife Fund)*

Secondary forests tend to persist for long periods. Around Angkor Wat, in Kampuchea, the forests of large areas that were reputedly cleared six hundred years ago do not yet fully resemble surrounding patches of undisturbed primary forest. Some scientists believe that lowland rain forest ecosystems, after being grossly disrupted or destroyed, may need a successional process of several centuries, even one thousand years, before a final-form, primary-forest ecosystem is restored.[5] Generally speaking, human activities are the cause of primary forests giving way to secondary forests. But nonhuman factors can contribute. In the western Pacific, notably in Melanesia, the Philippines, and parts of New Guinea, typhoons and other exceptional weather conditions cause so-called catastrophic damage to primary forests. Thus forests in the region that have been little touched by humans may not resemble what we would consider primary forests elsewhere. The storm forest of Kelantan, in Malaysia, which originated in November 1880, has remained recognizable as such until the present time. During the course of this century, islands of Melanesia have experienced cyclones on average once every 2.4 years; a cyclone that hit Santa Isabel Island in the Solomon Islands in 1972 destroyed more than 330 square kilometers of forest, a phenomenon from which the forest is unlikely to make a full recovery for some fifty to one hundred years.

It is important to differentiate wherever possible between primary forest and secondary forest, on the grounds that the former is much the more valuable in bio-ecological and economic terms. In a strictly scientific sense, a primary forest can be disturbed by even the lightest logging, perhaps by a harvest of just half a dozen trees per hectare, as a result of which it loses some of its former richness of life forms. As we shall see in chapter 4, most tree and plant species of primary forests rely on birds, monkeys, bats, insects, and other creatures to transport their pollen and to disperse their seeds. In some localities of Southeast Asia, teak tree pollen is carried by only two bee species. Since certain of these pollinators operate across wide stretches of forest, many tree species could easily be threatened by disruptions on the scale that man inflicts. The chance extinction of a small population of pollinators, through, for example, logging that selects mature specimens of a particular tree species, could lead to the local elimination of a tree species, which in turn could cause the elimination of certain insects that depend on it for food—and so on. Thus this book will repeatedly emphasize that we must be alert to the biological integrity of forest ecosystems. When we unwittingly degrade this integrity, the repercussions can eventually prove profound and virtually irreversible short of a period reckoned in centuries.

EXTENT OF TROPICAL FORESTS

Of the 12 percent or so of Earth's land surface that enjoys a warm, damp climate all year, tropical forests still occupy well over half. Probably once covering about 16 million sq. km., however, primary forests now amount to little more than 9 million sq. km. at most. A maximum of 5.1 million sq. km. are in Latin America, 2.1 million in Asia (largely in Southeast Asia), and 1.8 million in Africa, with patches in a few Indian Ocean and Pacific Ocean islands. So far as we can determine, Latin America and Asia have each lost almost two-fifths of their original forests, and Africa a little over half.

Although tropical forests comprise only one-third at most of the world's forests, they contain four-fifths of the Earth's land vegetation. One hectare of primary forest may support plant material weighing anywhere from 300 to 500 tons, sometimes from 600 to 800 tons (the maximum recorded was in a Panama forest: 1,189 tons).. With year-round growth, a forest can produce as much as 70 tons of plant material per hectare per year, or twice as much additional biomass as is generated in Earth's forests as a whole, and a higher level of productivity than for any other vegetation type (except a few forestry plantations and high-yielding agricultural crops such as sugarcane and irrigated rice). Because organic matter is speedily decomposed, however, annual net increment in an undisturbed forest is usually nil.

Such, then, are the forests of the tropics. Hitherto we have talked about plants and other living things. Let us conclude this chapter with a short account of a nonliving component of tropical forests, a component that gives life to the millions of plants and insects: water. We shall examine it through its most remarkable manifestation, the Amazon River.

THE AMAZON RIVER

The Amazon first became known to white explorers in the year 1500 when a Spanish sea captain, sailing in the western Atlantic Ocean at the equator, found, to his surprise, that he was sailing in a patch of fresh water. Heading westward to investigate, he found he was actually in the outflow of a giant river. The Amazon drains an area of 6.2 million sq. km. It flows 6,720 km. before reaching the Atlantic, where it disgorges one-fifth of all riverwater on Earth, or four times more than the second-largest river, the Zaire, and about eleven times more than the Mississippi —or as much in one day as the Thames delivers in one year. For a fair distance, the Amazon is wider than the English Channel, that is, a good 50 km.; from a boat in the middle of the river a passenger cannot always see the shore on either side. The mouth of the river is 250 km. across, embracing in its delta an island called Marajo, which is the size of Switzerland. The outflow stains the ocean a muddy color for hundreds of kilometers out from land.

The Amazon system includes more than 1,000 sizable tributaries of one sort or another, 17 of them more than 1,500 km. long. The total number of tributaries large enough to deserve the name is some 10,000. Their collective length is more than 80,000 km., or twice as much as would stretch around the equator. The two largest tributaries of the Amazon, the Negro and the Madeira, are each as large in water volume as the Zaire River.

This vast river system is almost completely flat. After its source in the Peruvian Andes, less than 200 km. from the Pacific shoreline, the river falls almost 4,900 meters in its first 1,000 km., yet in the remaining 5,700 km. it drops to sea level by only slightly more than 300 meters, a scarcely discernible slope. The Brazilian city of Manaus, 1,600 km. upstream, is a mere 32 meters above sea level, while remote Yurimaguas, in Peru, is only 179 meters above sea level. This flatness has a favorable consequence in that it allows ocean-going ships to penetrate far inland—a distance equivalent to a crossing of the North Atlantic. In all, at least 24,000 km. of waterways serve as navigable "trunk rivers," representing a magnificent network of communications—and undermining Brazil's declared rationale for the TransAmazon Highway system as a device to connect the various parts of the basin.

Although the main catchment zones receive rain throughout the year, the melting snows in the Andes increase the volume of water so much that they cause downstream flooding. At Manaus, the annual fluctuation averages at least 12 meters, and high levels of almost 20 meters are not unknown. While this phenomenon makes things difficult for the burghers of Manaus, it makes life exceedingly fecund in the extensive floodplains. These "varzeas," as the floodplains are known, extend at least 50 and sometimes 100 km. from the main channel. They cover 100,000 sq. km., or almost 2 percent, of the basin and support the largest, most complex, and most variegated "flood forests" known anywhere.

An indication of the ecological complexities in this strip sector of Amazonia is the large number of fish that feed primarily off seeds that fall from trees.[6] Many of the fish apparently feed on nothing else during the low-water months, but draw instead on the fat reserves they have built up in their large stomachs. In order to adapt to their specialized diets, certain of the fish have also developed large flat-topped molars (similar to those found in sheep, cows, and other animals that graze), together with jaw musculatures that enable them to crack even tough nuts. Through their feeding activities, the fish help to prepare nuts for germination; in addition, they distribute fruit seeds, which further assists the trees in return. As many as 200 fish and tree species appear to depend on each other, and of the fish, at least 50 are commercial species, making up three-quarters of the fish catch in Amazonia. These ecologic linkages deserve recognition when development technocrats assert that, were the forest cover to be cleared, the floodplains could grow vast quantities of rice and support huge herds of water buffalo. They could—but in some localities, better-quality food could well be obtained through maintaining the flooded forests in order to harvest the sustainable bounty of the river fish stocks. The Amazonia fishery can be roughly estimated at a minimum of 1 million tons, with a renewable offtake of one-quarter of a million tons.

Similar symbiotic interactions can be seen in the Tonle Sap (Grand Lac) region of Kampuchea. When rivers are in flood, fish disperse to surrounding swamp forests, where they put on weight very rapidly, producing ten tons of first-rate animal protein per square kilometer during the course of the flood season. In a tributary of the Mekong River in Kampuchea, the fisheries of the Grand Lac have declined following deforestation in the environs.[7]

3

Biological Richness

As DESCRIBED in the previous chapter, tropical forests exhibit a great variety of life forms, somewhere between 2 and 4 million of the Earth's 5 to 10 million species, all in a mere 7 percent of Earth's land surface.

How do we know that so many plants and animals exist in these forests when scientific researchers have barely begun their job of unraveling the forests' secrets? Well, such statistical evidence as we possess is compelling. Through information from taxonomic surveys, we know that the forests support at least 90,000 of Earth's 250,000 identified plant species (scientists suspect another 30,000 await discovery, mostly in tropical forests). This total of 90,000 compares with only 50,000 in the entire northern temperate zone, with its vast territories in North America and Eurasia. The single-richest region of the tropical forest biome, Amazonia, is believed to support at least 30,000 plants, as compared with only 10,000 in all of temperate South America. Taking both plants and animals together, we can

Treetop "walkways" such as this enable biologists to explore the last great biological frontier, the tropical forest canopy. Many thousands of new species are being discovered each year—while very many more remain unknown. *(Andrew Mitchell, London)*

reckon that the tropical forests of Latin America support one million species, those of Southern and Southeast Asia three-quarters of a million (the bulk of them in Southeast Asia), and those of Africa one-third of a million. Since these figures are based on a global total of 5 million species, the regional numbers will double if we assume a global total of 10 million species.

PLANT LIFE

Since the comparative figures for plants are so revealing, let us look at them a little further.[1] Brazil possesses 20,000 plant species, the majority of them in its tropical forests covering 3 million sq. km., whereas the United States possesses only 20,000 or so in its 9 million sq. km. Colombia possesses somewhere around 25,000 plant species in its 1.1 million sq. km. (a little larger than Texas and New Mexico combined)—again, the bulk of them occurring in its moist forests, which cover only one-third of its territory. Tiny Costa Rica, with 52,000 square kilometers, possesses 8,000 plant species, in contrast to Great Britain, with 244,000 sq. km. and only 1,443 species.

Probably most remarkable of all the smaller Latin American countries is Ecuador, with at least 15,000 and possibly as many as 20,000 plant species. This surprising abundance of flora appears primarily in the country's tropical forests, both in Amazonia and along the Pacific Coast; the total is to be compared with Europe's 13,000 plants and 31 times greater area. Moreover, Ecuador's flora contains many endemics, notably epiphytes. The Rio Palenque Research Station in coastal Ecuador amounts to only 1.7 square kilometers, yet it supports 1,025 plant species—probably the highest-recorded concentration of plant diversity on Earth; an exceptional one-quarter of these are found only in coastal Ecuador.[2] Regrettably the Research Station, the last remaining patch of wet forest in coastal Ecuador, is being undermined by local people who enter the forest to cut wood for fuel and construction.

If certain of the endemic species become extinct, the loss will be felt not only by "pure science" researchers who, in the eyes of some observers, enjoy the luxury of pushing back the frontiers of knowledge at a time when humankind is beset with more urgent concerns. One of Rio Palenque's plants is a relative of the cultivated avocado and, by supplying rot-resistant root stock, could offer commercial benefit to avocado growers in many parts of the tropics and sub-tropics—and to avocado consumers around the world. A second species is a relative of cacao; it yields a reputedly tastier form of chocolate and potentially could contribute to the daily lives of people everywhere. Among the larger trees known only from the Research Station is a fast-growing species of the Lauraceae family, which used to be the tree most favored by timber harvesters in the region; who knows but that there might be, among the other large tree species of the Research Station, a species to match the record of the Monterrey pine, which, once reduced to four areas of California, was saved from extinction to become the most widespread industrial timber tree in the Southern Hemisphere.

Extending from Ecuador northward along the Pacific Coast of South America, is the Choco region of Colombia. It is superwet, with over 10 meters of rainfall per year and hence is ultra-rich in plant life. According to Dr. Alwyn H. Gentry, of the Missouri Botanical Garden, sample plots of forest in the Choco that add up to a mere one-tenth of a hectare reveal 208 tree species.[3] In an entire forest tract of, say, the Missouri Ozarks, we would find perhaps 25 different trees, and in New Hampshire about one dozen.

The next-richest region is Southeast Asia. Covering 4.3 million sq. km., the region contains 25,000 plant species. Peninsular Malaysia, accounting for only 0.09 percent of Earth's land surface, supports 8,500 plant species, or 3.4 percent of the planetary total (2,500 of these plants are trees, a total three times greater than the United States'). The second-richest sector of Southeast Asia is Borneo, an island of 725,000 sq. km. (the size of California and Nevada) and the largest tract of true rainforest outside Amazonia (the Zaire Basin is not wet enough to qualify as rainforest proper): Borneo contains 10,000 plant species. On Mount Kinabalu in northeastern Borneo, there are five times as many oaks as in the whole of Europe, together with 400 species of ferns and 800 species of orchids.

As a further measure of Southeast Asia's richness, we can consider the factor of endemism. Peninsular Malaysia, while possessing one of the most abundant floras in the world, has surprisingly few plants that do not grow elsewhere as well. By contrast, endemism is moderately high in the Philippines, and unusually high in Sri Lanka (strictly speaking, in Southern rather than Southeast Asia)—even though the total number of species in Sri Lanka is not relatively great. The moist forests of Sri Lanka reveal a level of endemism unsurpassed for a similar-sized area elsewhere in Asia, with over 800 flowering plant species discovered there and nowhere else. These plants persist, however, only in fragmentary remnants of habitat, and many species must comprise only a few individuals. Since the country's wet zone is limited in area, 15,000 square kilometers or so (smaller than Massachusetts), and since the unlogged lowland forests now total only 1,800 square kilometers and are being cut back year by year, we can suppose that entire communities of plants are at risk; a good number may well disappear by the end of the century.

In Sri Lanka's forests a concentrated effort by conservationists could accomplish much more, by saving dozens if not hundreds of plant species (plus their associated animal species), than could a similar effort in a locale where not so many threatened species are crowded together. Sri Lanka's moist forests deserve to come near the top of any conservationist's list of priorities, even though the area receives far less attention, and far less funding, than certain individual species receive elsewhere in the world. Imagine what even one-tenth of the money assigned to the California condor, $25 million, could achieve in Sri Lanka—with far better than the fifty-fifty survival prospect accorded to the condor.

As for Africa, the most notable endemism is in Madagascar, an island a little smaller than Texas, which contains an astonishing total of 12,000 plant species, over 60 percent of which are endemic. Madagascar possesses nine species of baobab, whereas the whole of continental Africa has just one. The present flora represents only a portion, and probably a small portion, of that which existed on the island before nearly four-fifths of the primary vegetation—meaning, in the main, the moist forest cover—was destroyed. While the forest along the eastern ridges of the island may once have covered at least 62,000 square kilometers, and possibly half as much again, it has now been reduced, as a consequence of farming pressures, to only about 24,000 square kilometers at most. Of this relic, more than half has been grossly degraded through migratory farming, which, having steadily grown more extensive and intensive during the past several decades, is currently reckoned to account for about 2,500 square kilometers of primary forest each year. Hardly any of Madagascar's primary forest may remain intact by the end of this decade, and we can expect that dozens of endemic plants are becoming extinct, with hundreds if not thousands of others reduced to a few sparse populations.

We have been looking at the so-called higher plants, that is, the flowering (vascular) plants. We find similar richness when we consider lower plants, for example, ferns: Of 12,000 species worldwide, 11,000 exist in the tropics, the great majority of them in forests. Much the same applies to mosses. Of 12,000 species worldwide, 9,000 are tropical, most of them occurring in forests.

Not all types of tropical forests contain similar concentrations of species. For instance, there are some marked differences according to altitude. This is, as a moment's thought tells us, no more than we might expect. The higher we go, the less warm the temperature, and the less likelihood that evolution has enjoyed favorable conditions to generate hosts of species. Whereas the mean temperature in lowland forests varies between about 75 and 85 degrees Fahrenheit year round, it rarely reaches above 67 degrees F. or so at altitudes greater than 1,500 meters, where we find that the broad-leaved trees of the lowlands start to give way to conifers and oaks, with less abundance of species. Regrettably, it is the lowland forests that are being most heavily exploited, primarily because they are the most accessible, and because they do not present hilly terrain to balk logging trucks. If the forests below 400 meters in Brunei, for example, were to disappear, it would eliminate 65 of the nation's 147 dipterocarp species.[4]

In a few localities, notably where Amazonia meets the foot-hills of the Andes, we find some of the world's greatest biological richness in a zone roughly between 200 and 500 meters altitude. These areas receive some of the heaviest rainfall on some of the most fertile soils—a combination that almost always fosters abun-dance of species. But the fertile soils not only stimulate luxuriant growth of plants, they also attract settlement farmers, who quickly reduce the biological diversity. This is all the more regrettable because this forest zone is the probable origin of many economi-cally important plants, such as the peach palm and the papaya—the latter often found on American breakfast tables.

At the same time, we can note that just a few hills and mountains are able to offer sufficient "creative disruption" in heterogeneous terrain to promote speciation. This happens espe-cially where uplands occur in isolated fragments, and hence offer an ecological "island effect." When communities of plants and animals become cut off, by virtue of up-rising hills and mountains, from other communities of kindred sort, they tend to evolve in their own particular ways, eventually producing new species. For the most part, then, because Borneo is far from flat, it features more species per unit area than does much of the Amazon basin, four-fifths of which lies below 200 meters.

ANIMAL LIFE

Greater diversity of plants increases the number of habitats for animals. Not surprisingly, tropical forest animals are more numerous, more varied, and often more flamboyant than those anywhere else on Earth. In the forests of Southeast Asia, we find not only the elephant, tiger, and rhinoceros, but strange creatures that awake at dusk for their quiet, nocturnal activities, notably tarsiers and lorises, small-scale relatives of the larger monkeys that racket around the forest by day. Still more bizarre are the gliding animals, more numerous in Southeast Asia than elsewhere, including squirrels and geckos with flaps of skin that form "parachutes," and gliding frogs that use their filmy webbed feet to float through the air. The airborne lemur (not a true lemur) easily manages a flight of 50 meters. There is a flying lizard and even a snake that takes to the air for short distances.

In Amazonia we find armadillos, sloths, and anteaters, among other curiosities that evolved during the millions of years when South America was an island separated from other large land masses. Also in Amazonia lives the Hercules beetle, the world's largest at 19 centimeters long, with a pair of barbed horns. Among other exotic insects are butterflies, some with a brilliantly metallic sheen, while among colorful birds are the largest of the world's parrots, the macaws, and tiny jewel-like hummingbirds. As for

Tropical forests feature many categories of animal life not found elsewhere, for example, this flying lemur in Peninsular Malaysia. No other ecological zone possesses mammals akin to this curious creature. *(Dennis Young, World Wildlife Fund)*

Africa, the forests contain not only the gorilla and the chimpanzee, but the okapi, and a rabbit-sized antelope.

In terms of overall numbers, the proportion of the animal kingdom that we find in tropical forests at least matches that of plants, even though the animals are not a fraction so well known as are the plants. In Peninsular Malaysia, for instance, there are 675 birds and 200 mammals, and in Southeast Asia as a whole, there are 656 mammals—plus 850 amphibians, or almost one-third of the world's stock. As for invertebrates, Southeast Asia is endowed with over 700 butterflies—and afflicted with more than 400 mosquitoes. The Sunda Shelf sector of the region, roughly the western half, supports 297 mammals and 732 birds, whereas Europe west of the Soviet Union, an area almost four times as large, totals only 134 and 398, respectively.

We find similar proportionate richness in Amazonia. The region features one in five of all bird species on Earth. It also contains 2,000 known species of fish, or eight times as many as in the Mississippi River system, and ten times as many as in the whole of Europe. The eventual total of fish species to be identified in Amazonia may reach 3,000. In a single locality of Costa Rica, the La Selva Forest Reserve, which amounts to only 730 hectares, there are 320 tree species, 394 birds, 104 mammals, 76 reptiles, 46 amphibians, 42 fish, and 143 butterflies—a tally that is, broadly speaking, half as large again as that of California with its 410,000 sq. km. On Barro Colorado Island, a 14.8-sq.-km. island in the middle of Lake Gatun on the Panama Canal—featuring seasonal and mainly secondary forest rather than primary rainforest—Dr. Henk Wolda, of the Smithsonian Tropical Research Institute, came across 1,100 species of Homoptera (leafhoppers, cicadas, etc.) and discovered that hundreds of them are new to science.

Birds abound in tropical forests. Looking only at those birds whose existence is so severely restricted to forest habitats that elimination of the forests would cause elimination of the birds, we find that they total some 2,600 species, or three out of ten of all birds on the planet.[5] Of these, about 1,300 occur in the forests of Latin America, 900 in those of Asia, and 400 in those of Africa. Some tiny sectors reveal exceptional concentrations: for example, the new park called La Amistad (friendship) in Costa Rica, with more birds in its 2,170 square kilometers than in the entire North American continent with its 20 million square kilometers. A single locality in Amazonian Peru, known as the Tambotata Reserve, supports about 530 bird species, compared with about 850 in the United States and Canada combined. Regrettably, the Reserve is not actively guarded, and illegal hunting is greatly diminishing the populations of most larger birds and mammals.

Of course certain birds inhabit both Costa Rica and North America during different stages of their annual life cycles. Here we can note that the ecologic linkage is matched by certain economic linkages. According to Dr. Eugene S. Morton and his colleagues at the Smithsonian Institution in Washington D.C., there has been a recent decline in numbers of U.S. songbirds, especially kingbirds, warblers, vireos, tanagers, and peewees, among 55 species in all, losing numbers at rates between 1 and 4 percent a year. The scientists believe that when several billion of these birds leave the United States each fall, they find, upon arriving in Costa Rica and other wintering grounds of tropical Latin America, that their forest habitats have succumbed to the machete and the match. As a consequence, there are fewer birds to head back each spring to the United States.[6]

Ironically, it is precisely at the time of the songbirds' return that a number of insect species are likewise reappearing in North America at a key phase of their life cycles (as larvae, etc.) that makes them unusually vulnerable to birds. It seems possible (though not yet proven) that hitherto the insect populations have been kept below levels at which they would be harmful to agricultural crops through the depredations of millions of songbirds returning at just about the right moment. But if the songbirds continue to decline, the insects could, within a few years' time, enjoy a population explosion every spring—which could mean bad news for America's agriculture. If we leave the situation to unfold until the outcome is clear beyond doubt, it may be too late to do much about it.

We might ponder this connection next time we munch a hamburger. The beef for the hamburger may well have come from forestlands of Central America, where trees are being burned down to make way for man-established pasturelands on which to raise extra beef cattle needed to meet demands of the insatiable fast-food trade in North America, which protests that U.S.-grown beef is now so expensive that it is a principal culprit behind inflation. Thus the "songbird connection" points up linkages of resource relationships between the rich temperate zones and the developing tropical zones—a theme to which we shall repeatedly return in this book.

Fortunately, we can note that at a time when many species are disappearing, a few, just a very few, are being discovered or rediscovered. On the island of Maui in the Hawaii group, a new bird—a honeycreeper rejoicing in the local name of po'o-uli—has come to light, occupying a range of a few square kilometers of montane forest in a canyon that is inaccessible to man except by helicopter. This bird is so exceptional that it has been assigned a whole new genus. In the forests of Paraguay, a peccary once thought extinct has been found again—though its habitat is being bisected by an all-weather road, which will leave the peccary, a piglike creature, accessible to hordes of human hunters. In upland forests of Peru, the largest primate of Latin America, the golden-tailed woolly monkey, was thought to have become extinct around the start of this century, but it has recently been discovered again with just a few remnant numbers.[7] Similarly, one of Madagascar's lemurs, the ayeaye, such a distinctive creature that it has been assigned its own zoological family (comparable to the cats with their 37 species, or the legumes with their 20,000 species), was though to have held out no longer than 1930 or so—until it was rediscovered just a few years back. Quite a few other species have survived their obituaries, especially small, obscure species that live in remote sectors of tropical forests.

INSECTS

As we might guess, the animal biomass, or the total weight of living animals, in a tropical forest is only a fraction of the plant biomass. Whereas vegetative material in one hectare of forest may well weigh 500 tons, sometimes much more, the animal biomass is likely to be only one-five hundredth as much, and sometimes only one-third of that. This is because the bulk of animal life in the forest is made up of insects and other arthropods. These small invertebrates, especially creepy-crawlies of multitudinous kinds, comprise three-quarters of all species on Earth. In many tropical environments, such as Serengeti Park in savannahland Tanzania, it is the biomass of mammals and other vertebrates, one-hundredth as numerous as invertebrate species, that constitute the greatest amounts of living protoplasm. In tropical forests, by contrast, there are relatively few mammal species (apart from bats) and relatively very few individuals of each species. It is the ants and termites, plus butterflies, spiders, and the rest that are by far the most numerous.

Butterflies, for instance. A century ago, the British explorer Henry Walter Bates collected more than 700 species of butterfly within an hour's walk of his home in eastern Amazonia in Brazil; they amounted to about 1 in 30 of all butterfly species on Earth. In a single square kilometer of Freetown, Liberia, around 300 species of butterfly have been recorded, as many as in the whole of the western United States.[8] Much the same applies, broadly speaking, to many other categories of insects and invertebrates. In the Gunung Mulu Park of Sarawak, a team of scientists has listed more than 20,000 species of invertebrates in just 529 square kilometers, including 8,000 beetles and 3,000 butterflies and moths—and the survey is far from complete.[9] In Amazonia, the fauna hidden in the soil can amount to 165 kilograms per hectare, way beyond the above-ground and better-known herbivores (30 kilograms) and carnivores (15 kilograms). Sometimes there can be a remarkable number of creatures in a very restricted habitat: for example, in the fur of a sloth. Dr. Gene Montgomery, a scientist at the Smithsonian Tropical Research Institute in Panama, found a single sloth that is literally "bugged," in that it provides living space for at least 3 species of beetles, 6 species of mites, and 3 species of moths, almost 12 species of insects altogether living in its fur; one sloth has been found to contain as many as 978 individual beetles. The sloth also supplies habitat for 3 species of algae, which add a greenish hue to the creature's fur and thus camouflages it when it roams around leafy treetops. The algae supply the sole source of food to the larvae of one of the moth species.

Of course the greatest abundance and diversity of insects is to be found in the secluded, sunlit realm of the forest canopy. Hitherto we have known next to nothing about the multitudes of tiny creatures to be found there. But we are gaining a few clues. Drs. Gene Montgomery, Yael Lubin, and Melvin Boreham, all of the Smithsonian Tropical Research Institute in Panama, followed by Dr. Terry Erwin, of the National Zoo in Washington D.C., have been taking a lengthy look at a medium-sized common tree known as *Luehea seemannii* in Panama. They have benefited from new technology that enables a researcher to roam around the uncharted territories of the forest canopy and to investigate the crowns of a number of *Luehea* trees. During the course of three seasons of sampling in Panama, they came across more than 1,200 species of beetles, while further research in Brazil reveals that these samples are by no means atypical of tropical forest canopies.

Building on this research, Erwin has engaged in some pioneering arithmetic. He estimates that beetles comprise about 40 percent of all arthropod species in tropical forests (arthropods being centipedes, millipedes, spiders, and crustaceans, among other major groups—and three-quarters of them are insects, i.e., six-legged creatures). Erwin has also found that of the 1,200 beetle species, 163 cannot live on any other sort of tree apart from the *Luehea* with the foods it supplies. Erwin goes on to calculate that there are 70 different species, or groups of species, of forest trees per hectare. He further believes that for every 3 arthropod species found in the forest canopy, there is probably another 1 species to be found on the forest floor; in addition, he believes he should include a small number of transient species in his tally. Thus there could well be more than 41,000 species of arthropods per hectare of forest in Panama. Moreover, this calculation is based on a tree species characteristic of seasonal and somewhat scrubby forest in Panama—a forest type that is far from being so rich in species as are tropical forests that receive a year-round drenching of rainfall. (At the same time, let us note that the *Luehea* is a common and widespread tree, with many specialized insects; not all tropical trees are anywhere near so numerous, and many of them probably do not feature so many specialized insects.)

All told, tropical forests contain an estimated 50,000 tree species. If we use the base-line reckonings above and apply the formula across the board for all tropical forests, we come up with a figure of some 30 million species.[10]

These calculations are preliminary and exploratory at best. Far from being definitive, they are illustrative, serving to point up the possibility that there could be many more arthropod species in tropical forests than we have so far supposed. Whereas we already realize that tropical forests possess an extraordinary richness of species, we may be a long way from grasping the ultimate scale of that richness. Our best estimates to date, based on select surveys and heroic extrapolations, suggest that the planetary total of arthropods presumably lies somewhere between 4 million and 8 million species. We may grossly err in our reckonings through being overly cautious—a reflection, perhaps, of our limited imaginations as much as of our scant knowledge.

We must regret, of course, that we shall probably never establish whether tropical forests contain just a few million arthropods, or tens of millions. Within another few decades, and very likely before we can mount a research effort to document all the species that inhabit these forests, the forests may well become grossly degraded, if not destroyed. We shall have to do a much better conservation job if we are not to lose at least 1 million arthropods by the end of the century—and possibly, according to the calculations of Dr. Terry Erwin and his colleagues, many millions.

But what, some observers ask, does it matter whether there are a few millions or dozens of millions of arthropods? If a good number of them are to be eliminated within the foreseeable future, why should we waste our time arguing about the undemonstrable point of how many exist there now? Well, there is a purpose to this abstruse theorizing—a severely practical purpose. If there are no more than, say, 2 million arthropods in tropical forests (a minimum figure), and we stand to lose a full one-half of them during the course of the next half-century, then that is a fall-out of life forms on a sizable scale, with greater concentrated impoverishment for our biosphere than any extinction spasm since the first flickering of life on Earth some 3.6 billion years ago. How much greater, then, will be the tragedy—to give it an appropriate designation—if the total loss amounts to 15 million species. This will be a biological debacle of an altogether different order.

Moreover, the repercussions will not just amount to a severe rending of the fabric of life on Earth. Let us remember that each species represents a unique manifestation of life's diversity, with its own "genetic fingerprint." Thus a mass extinction of arthropods will remove, forever, a vast stock of genetic materials, or irreplaceable natural resources of exceptional value, with which to boost our agriculture, medicine, industry, and other utilitarian activities (see chapters 10–13). In light of recent revelations from genetic engineering, with its impressive potential contribution to our material welfare, we shall eventually need all the genetic variability we can find. The demise of 1 million arthropods will leave us, and all our descendants, so much the poorer: the demise of 15 million species will be something else again.

OUR SCIENTIFIC IGNORANCE

There is so much, so much, that we do not know. As a measure of our scientific ignorance, let us consider a 1,900-meter mountain astride the Panama-Colombia border known as Cerro Tacarcuna. It had remained unexplored until a recent visit by Dr. Alwyn H. Gentry, of the Missouri Botanical Garden, in conjunction with several colleagues. These botanists concentrated only on the forest zone above 1,400 meters, far from the richest zone; during their expedition, they came across 239 plant species, of which about 1 in 5 turned out to be a species new to science.[11]

Nor is this an isolated incident. A similarly significant number of hitherto undescribed species has been turned up each time botanists have explored a new and isolated tract of forest in Panama, Colombia, Ecuador, Peru, and Venezuela. Of Venezuela's 880,000 square kilometers, an area twice the size of California, less than 2 percent has been investigated by botanists.[12] Perhaps the richest storehouse of plant life on Earth, the Choco forest along the Pacific Coast of Colombia, is considered to harbor one plant species undiscovered by science for every ten species that have come to our attention—one-quarter almost certainly found nowhere else on Earth.[13] There could be well over one thousand unidentified plant species in this region alone, plants that will become subject to final elimination as the timber cutters, hard at work, reduce this ultra-rich forest to a biological level way below its present unique status. As for Latin America altogether, there could be as many as fifteen thousand plant species still to be described by science and, hence, not yet known to modern man.

So much for on-the-ground discovery of plant species. The systematized documentation of this fieldwork falls still more lamentably behind what is needed. The flora of Central America, a region only a little larger than France, is probably three times richer than that of all Europe, yet fewer than 50 floristic inventories have been published in Central America, in contrast to more than 5,000 in Europe. Great Britain possesses a flora of around 1,400 species, of which only a handful are endemics, 15 or so. Yet Great Britain has at least 1,500 botanists of one sort or another, while Colombia, which harbors between 45,000 and 50,000 plant species, has a mere few dozen botanists who are expert enough to identify plants in the field.

Nor is the cataloguing of species a matter of mere academic importance. When new species are discovered and defined, they boost our understanding of ecosystems, what makes them tick—and what could make them tick to greater human benefit through, for example, upgraded agriculture. It is in forest ecosystems that we can best learn how to make plants grow to our advantage in the humid tropics, how to ward off insect pests, how to maintain fertility of the soil, and so forth (see chapters 8 and 10). Unless we rapidly increase our insights into the functioning of these ecosystems, we shall never learn how to modify them in the service of human welfare, with all that implies for the two billion people who, we can expect, will be trying to gain their living in the humid tropics by the end of the century.

Despite the scale of the scientific challenge that confronts us in the immediate future, we are worse than short on professional muscle to undertake the task. Worldwide there are no more than 3,000 scientists who specialize in tropical biology, not all of them possessing the expertise to discover new species. According to a recent report of the National Research Council,[14] the pool of taxonomists and systematists who study tropical organisms amounts to a paltry 1,500, a total that needs to be increased five times over forthwith.

Let us remind ourselves that, of all the problems facing humankind right now, few show such unique traits as the impending mega-extinction of species around the world—mostly in tropical forests. Almost all other environmental problems, such as the fouling of our skies and seas, the spread of deserts, the retreat of grasslands, the disruptions of climate, and so forth, can be put right if we decide to make the effort. Few environmental disorders cause irreversible damage to our biosphere. But the extinction of species is a different ballgame. When a species is gone, it is gone forever.

Before we can gain a concise understanding of the scale of the challenge to save species, we need to learn the dimensions of the problem. We must establish just how many species exist, where they are located, how far they are threatened, and so forth. An exercise of this sort could be accomplished in tropical forests for an outlay of, say, $5 million a year until the end of this decade: no great cost, the equivalent of ten minutes' spending on armaments, or the equivalent of what American communities spend each month to cover the costs of stray dogs.

Furthermore, our scientific ignorance of tropical forests extends way beyond the species that live there. Our understanding of forest ecosystems is rudimentary at best. Not only are tropical forests far richer than any other ecological zone on Earth. They are much more complex in their workings. Well might they be described as the apogee of the evolutionary process. Yet we know next to nothing about the dynamics of these ecosystems. Our ignorance is so profound that we have yet to devise a systematic classification of forest types that we can apply uniformly from region to region. In other words, we have yet to learn how to describe these forests in consistent fashion.

It is not going too far to say that we now know more about certain sectors of the Moon's surface that we do about the the heartlands of many tropical forests. Yet the Moon will stay around, virtually undisturbed, for as long as we care to look into the future. Tropical forests may not survive much more than the next half-century, except in grossly degraded form. The funds currently spent on space exploration of all kinds amount to almost $30 billion a year. Which investment will yield us the better return in terms of meeting our material needs on Earth?

4

Ecological Complexity

IF I HAVE WONDERED AT the abundance of life forms during my treks around tropical forests, I have marveled even more at the intricate ways in which these organisms go about their communal activities. This tree, that flower, those insects: what parts of the ecosystem contribute to their well-being—and whose well-being do they themselves support?

The extreme biological richness of tropical forests is related to their extreme ecological complexity. With every tick of the clock, day and night, multitudes of animals seek multitudes of plants for food—the numbers of both categories being, as we have seen, far higher than in any other ecological zone of similar size. To ward off plant-eating animals, plants deploy defenses of extraordinary variety (we shall come across many examples further in this book, notably chapters 10–12). At the same time, many organisms make use of other types of organisms: trees use animals as pollinators and seed dispersers, and microorganisms transport nutrients from one individual organism to another and on to another. All the while, creatures are competing with each other for food or light, or simply for space; simultaneously, many exploit others for protection against still others. The whole scene, it seems to me whenever I survey it, resembles an arena of incessant pushing and shoving, each organism searching for advantage of whatever sort that assists its survival—and that thus offers a chance to propagate its kind by passing on its genes, its own version of the throb of life, to a further generation.

Yet however vigorous the struggles to stay alive, there are obviously checks and balances. Obviously, in that plant-eaters consume plants but seldom eliminate them, and eaters of plant-eaters gain enough sustenance to keep going without generally causing final harm to their prey species. Not so obviously, in that it is difficult for the observer to track down the inter-relationships that maintain the system as well as its parts. As a scientist, I recognize that there is balance, or "ecological equilibrium." But just where I can discern that balance, just how I can see it actually at work, is something else again. The various strands of life clearly make up an ecological web, which in turn constitutes a fabric with its own design imprinted on it. But what is the nature of the overall picture that we may perceive in the forest's workings? That is something we have yet to discover in detail. At best we see a blurred image. Truly spoke the scientist who said that our understanding of tropical forests resembles the forests themselves, with many fast-growing components, numerous gaps, and tangles on every side.

However hard it is for me to visualize the scale of biological richness in a tropical forest, I find it far more difficult to imagine the complexity of interactions between plants and animals, and between them and their physical environs. After all, if there are 1,000 species within one particular square kilometer of forest, their relationships with each other—their comings and goings, their incessant encounters with associates and enemies—certainly number tens of thousands of interactions, probably hundreds of thousands, possibly many more. I sometimes speculate that, given our scant understanding to date, I likely fail to recognize the very scale of a forest's complex functioning. In other words, I have little idea of how massively complex it is.

Thus it is the complexity of a tropical forest that excites me when I stand amid those stately trunks and their convoluted vegetation abuzz with creatures of myriad forms. Whereas it would take me many moons to identify and document the main species (certainly not all of the species—I would need the rest of my life for that) within 100 meters of where I stand, it would take me many times longer to trace even a portion of their inter-relationships, their energy flows, their distinctive roles in the phenomenon that makes up a tropical forest. What I see is proba-bly surpassed in complexity by only one other living entity on Earth—that which is between my ears. Yet how far short is my gray matter from gaining even an intuitive sense of the full intri-cacy of the scene before me.

SOURCES OF ECOLOGICAL COMPLEXITY

What has given rise to this ecological complexity? It stems largely from the exceptional amounts of light, warmth, and mois-ture that everlastingly foster favorable conditions for the evolu-tion of species in their multitudes. It is climatic stability, with no seasonal hiatus in the form of winter each year, that seems—so far as scientists understand the situation—to make it work. In these tropical zones, we find abundant food all year long—food for plants in the first place, and thereafter for plant-eaters, eaters of plant-eaters, and so forth. In the temperate zones, by contrast, plant growth tapers off for several months of each year; as it does so, animal life endures a prolonged famine. But in a situation with food aplenty at all times, and food with lots of variety, we gener-ally find that great numbers of species evolve to take advantage of what is, in many circumstances, an unceasing feast. Thus, at any rate, runs much scientific thinking on what makes a tropical forest tick to such fecund effect.[1]

Furthermore, in a situation where there is often—though by no means invariably—an abundance of permanent food supplies, many organisms do not have to constantly roam the forest looking for their next meal. Usually they can afford to stay in one place, since there is enough food at hand: thus the remarkably large numbers of plant species, which by definition are stationary forms of life. Thus, too, the disinclination of many forest animals to wander far from home: insects, together with other arthropods that constitute the bulk of forest animals, generally confine their activities to a limited space, often just a few square meters of forest habitat. Each takes a portion of the food supplies available, generally without over-encroaching on the feeding habits of its neighbors. The result is room for all, as long as they all do their own thing. The overall result is that the "all" becomes very numerous.

It is this division of habitats into mini-habitats, followed by subdivision into micro-habitats, that offers vast scope for species and their populations to engage in adaptive radiation, that is, throwing off new forms. These new forms can lead to new subspecies and eventually to new species. Moreover, plants and animals, developing in co-evolutionary accord, devise new niches, or ecological living space, for themselves and thereby create further variation on established themes. In turn, the every-richer array of life forms finds ways for all to live with each other, on each other, off each other, through each other—building up, in short, the multitudes of lifestyles that constitute ecological complexity.

Yet another dimension to the situation emerges. Evolution works slowly; it has needed eons to produce such complexity. For further clues to the workings of tropical forests, then, we can look at their paleoecological pedigree. The generally stable background of their prehistory has allowed differentiation through processes of natural selection to develop undisturbed—and to do so for a very long time. Indeed, this last factor, the evolutionary ancestry of tropical forests, is so important to our understanding of present-day forests that we should look at what has been happening during a long line of yesterdays.

PALEOECOLOGICAL PAST

Tropical-forest ecosystems are almost certainly the oldest continuous ecosystems on the planet. Of course new species have come and gone, entire communities have emerged and disappeared in the process of becoming modified into fresh assemblies of plants and animals. But the fundamental configurations of the ecosystems have, in the main, persisted for very long periods of time. In parts of Southeast Asia, for instance, we know from fossil deposits of forest plants that the forest there has existed in more or less its present form for at least 70 million years, possibly 100 million years. We cannot assert as much for any other major ecosystems on Earth. This is not to suppose, however, that there have not been phases of change, occasionally drastic change—especially during the recent past. Yet far from impoverishing the forest ecosystems, these upheavals appear to have fostered an increase in biological richness and thus in ecological complexity. In short, tropical forests have generally experienced exceptional stability and continuity for much of their past, with opportunity for evolution to go its creative way undisturbed. At other times, they have undergone enough "constructive disruption" to foster fresh evolutionary departures.

The principal upheavals occurred during the late Pleistocene, when glaciations in the Northern Hemisphere were matched by repeated expansion and contraction of tropical forests, producing market local differentiation in climate and biogeography.[2] There were several such occasions, the last one about 10,000 years ago. During the times of greatest contraction of the forests of Amazonia, a few moister localities appear to have persisted as "forest refugia," serving as reservoirs of plant and animal species from which the forests were ultimately able to reestablish themselves when wetter conditions returned. Because the species communities in these refugia were cut off from each other, suspending their gene flow, each community could follow its distinctive track of evolution by responding to its own environmental circumstances in a manner that would allow natural selection to throw up new variations of the old stock. In short, each refuge would tend to produce an assembly of new species that would supplement or in other cases supplant the established species. In turn, the increased richness of species communities would serve to generate greater complexity of ecological functions in many parts of Amazonia.

In Southeast Asia the recent changes seem to have been less pronounced. During the drying-out phases of the late Pleistocene, the region did not desiccate nearly so much as did the tropical forest zones of Latin America and Africa, presumably due to the maritime effect (i.e., the moderating effect of the surrounding seas on climatic upheavals). Southeast Asia constitutes not a continental land mass, but an archipelago with a few large islands such as Borneo, Sumatra, and New Guinea, and thousands of smaller ones. This means that local climatic regions had to have drawn sufficient moisture from the nearby oceans to maintain a consistent expanse of forests. At the same time, the archipelago effect (i.e., the influence of so many island territories of different sizes) would have stimulated speciation by virtue of the fact that the plant and animal communities were divided into thousands of distinctive units—a situation that promotes extreme differentiation.

ISLANDS

Whereas Amazonia is split into various sectors, or ecologically differentiated islands, Southeast Asia has numerous geographic islands. The Malay Archipelago, as Southeast Asia is known to biogeographers, covers 2,894,000 square kilometers—only twice the size of Alaska. Yet the region comprises at least 20,000 islands, of which 13,000 make up Indonesia in a chain 5,000 kilometers long between Asia and Australia. The Indonesian archipelago has not always been so broken up. When the sea level has dropped, as it has several times during the relatively recent past, many of the islands have merged to form a single territory, allowing their wildlife communities to mingle. When the seas have receded again, splitting up the prehistoric forests into fragments once more, each separate sector would follow its evolutionary path in response to its own set of local environmental conditions and selection pressures. Greater diversity and complexity result. When the seas advance again, there will have been a repeated mingling of biotas and more creative tension between disparate communities of animals and plants, which often leads to some extinctions but eventually fosters more speciation. And so it continues.

The overall result is an exceptional amount of biotic diversity within Indonesia's 2 million square kilometers, almost all of it to be found in the primary forests that reputedly still cover two-thirds of national territory (roughly one-eighth of the entire biome of tropical forests). Of Indonesia's 500 mammal species, 100 exist only within the confines of that country; and Indonesia features 1 in 6 of the world's bird species, again with a good number of endemics. We know Indonesia as home to tigers, elephants, rhinos, and orangutans—but perhaps not everybody realizes that it provides living space for the Komodo dragon and a host of other bizarre-looking creatures that are native to Indonesia alone.

The insulating effect found within the Indonesia archipelago is well demonstrated by Siberut, a 4,500-square-kilometer island off the west coast of Sumatra.[3] This island, having been separated from the mainland for at least one million years, has enjoyed evolutionary scope to produce four endemic primates, more per unit area than any other island in the world, even Madagascar. Siberut is the smallest island in the world to have any endemic primates of its own—and of its 31 mammal species altogether (apart from bats), 25 are native to the island alone. More remarkable still is the endemism of Maluku, the group of islands formerly called the Moluccas and sometimes known as the Spice Islands.[4] The several hundred islands of the group support 450 bird species, including 89 endemics, and 80 mammals, 14 of them endemics. In addition, at least 25 butterflies and several reptiles and land snails are limited to these islands. Among the endemic birds, 94 percent are confined to a single island or a small group of islands. We might also note that nine-tenths of the forest in the wettest and hence the richest two-thirds of Maluku have been assigned to timber concessionaires.

An additional 20,000 islands are scattered across 29 million square kilometers of the southwestern Pacific, known collectively as Micronesia and Polynesia. Again, the mind can hardly entertain the uniquely varied bioecological character of these islands. On New Caledonia, for instance, an island of only 19,000 square kilometers, or smaller than Vermont, there are three thousand plant species (as many as in Canada), four out of five of them endemic. Moreover, New Caledonia's flora contains high concentrations of primitive species that illuminate our understanding of basic evolutionary processes.

So much, then, for some formative factors that contributed to the ecological complexity of tropical forests. It is important to understand the dynamic origins of today's ecosystems so that we can do a better job of ensuring them a future. At the same time, we need to balance our grasp of the bioecological components of these ecosystems with some insights into their geophysical underpinnings, particularly the mineral nutrients that feed the ecosystems.

MINERAL NUTRIENTS

Central to the workings of tropical forests is the cycling of mineral nutrients through their ecosystems. As we have noted, tropical forest soils, with the exception of recent alluvial and rich organic soils, are very old and, hence, impoverished.[5] In Amazonia, for example, where there has been no recent geologic activity (such as mountain building) to throw up new materials, the soils have long been weathered, leaving them deficient in phosphorus, potassium, sulphur, and other key minerals. Plenty of Amazonian topsoils possess only minor quantities of phosphorus, less than is generally needed for a region to qualify as fertile; or the phosphorus is chemically fixed in forms that leave it unavailable to plants. Moreover, such few minerals as have survived the lengthy periods of topographic stability in Amazonia's past tend to have lost their nutrients through leaching deep into the ground as a consequence of heavy rainfall.

All in all, three-quarters of Amazonia contains soils with high acidity and aluminum toxicity (not unusual in the humid tropics, where the overall proportion is two-thirds). A further 15 percent of Amazonia comprises younger soils, often alluvial materials, many of them fertile; while 4 percent consists of mature soils that are unusually fertile and have great agricultural potential. But many of these better soils tend to become water-logged. Only about 300,000 square kilometers of Amazonia, notably in the foothills of the Andes with their high-nutrient soils and in the alluvial territories of the floodplains, can be considered to offer both the fertility and the drainage suitable for agriculture. To grow crops in the rest of the region, farmers need to spend large sums on lime and other fertilizers.

True, some regions of the moist tropics have nutrients in underlying mountain rock, which become available as the exhausted topsoils erode. We find examples not only in the lower Andes, but in Central America and some of the Caribbean islands, where hills (defined as areas with greater than 8-degree slopes) and highlands often account for one-half of the terrain.[6] We also find nutrient-rich soils in volcanic areas of Central Africa and Southeast Asia. Soils in these territories, with their fine stocks of nutrients, are akin to those of temperate-zone forests, where nutrients are common—in contrast to the tropics as a whole, where they are rare.

In these circumstances of infertility, it is "inefficient" for forest organisms to allow the nutrients to stay in the soil. Evolutionary processes have responded by developing a virtually leak-proof system for cycling nutrients through the ecosystem.[7] When the main source of nutrients, rainfall, hits the forest canopy, it can bring with it (as measured at Manaus in Amazonia) 3 kilograms of phosphorus, 2 kilograms of iron, and 10 kilograms of nitrogen per hectare per year.[8] The multilayered structure of the forest then serves to filter out nutrients from through-falling water—a process in which epiphytic organisms, such as lichens, bryophytes, and algae, plus bacteria, play an important role. These findings are based on only a few experiments scattered through southern Venezuela, but they illustrate the many devices adopted by tropical forests to safeguard their precious stocks of nutrients.

At ground level, we encounter still more nutrient-conserving mechanisms. Tree roots, some of which extend 100 meters along the surface of the ground from the tree trunk, form a network that can be three times as dense as it would be in a temperate forest —occasionally as much as one-quarter of the tree biomass. (Some roots even emerge from the soil to climb up tree trunks, thus enabling the trees to capture nutrients before they enter the soil.) The root mat, as much as 30 centimeters thick, is extraordinarily efficient at absorbing nutrients washed into the soil, whether from rainfall or from rotting vegetation. Field experiments in southern Venezuela show that when calcium and phosphorus are sprinkled on these root mats, 99 percent is absorbed quickly, within six months at most.[9]

Closely associated with the root mats are fungus-root combinations known as mycorrhizae.[10] These micro-organisms of the soil, together with their associated bacteria, are essential to the well-being of many tropical trees in that they are the main way tree roots recover phosphorus, zinc, copper, molybdenum, and other minerals from leaf litter. The mycorrhizae grow on or near the surface layers of feeder roots of plants, where they work in close symbiotic relationship with their plant hosts. In a particle of soil the size of a sugar lump, there may be several meters of delicate fungal filaments (and as many bacteria as there are humans on Earth). By colonizing the roots, the fungi enhance the workings of the rhizosphere, or root world. They not only enable plants to absorb more minerals from the soil, they also help them to resist root pathogens, to withstand drought, and to tolerate other adverse conditions. In return for this support, the fungi obtain energy from their plant hosts in the form of fixed carbon. I have often gazed at a tall tree in all its stately splendor and reflected that much of the giant plant's welfare depends on tiny fungi that operate out of sight below ground level—and each dipterocarp species may have its own particular fungus.

So widespread are these mycorrhizal fungi, and so critical are they to tree growth, that some scientists wonder whether a forest, once cleared, can ever reestablish itself without them. When an extensive patch of forest is eliminated, the soil becomes drier and warmer, causing the mycorrhizae to die out. In other words, the tree species and their symbiotic fungi are entirely dependent on one another. Unless the mycorrhizae can eventually be restored from residual forest in the environs (supposing any such is close enough), the forest community will not get a chance to re-colonize the cleared area. The ecosystem will remain disrupted, and a different assembly of plants, perhaps scrub or coarse grass, will take over.

Still other nutrient-conserving mechanisms operate in tropical forests. Evergreen trees produce leaves all through the year, rather than in a single season, which means that their nutrient demand remains constantly low. Moreover, the leaves of many trees are thick and leathery, preventing the leaching of their water-soluble elements by rainfall.[11]

All in all, the cycling of nutrients in a tropical forest is fast and efficient, occurring within a virtually closed ecosystem. So capable is the process that streams draining from some forests actually have smaller concentrations of certain nutrients than are found in the rainfall descending on the forest. Many plant communities store at least 75 percent, and sometimes as much as 90 percent, of their nutrient stocks in their vegetation. As a result, forest soils have virtually no exchangeable minerals such as calcium, magnesium, potassium, and phosphorus—precisely the reason they remain highly acidic and infertile and, hence, unsuitable for agriculture. Contrary to popular understanding, the forest thrives despite its soil rather than because of it.

Because the bulk of the forest's nutrients are held in the vegetation, burning the forest triggers a flood of minerals into the soil. After one or two years of heavy-rainfall leaching, this stock is washed deeply into the undersoil, where it lies beyond the reach of new plants (grasses, shrubs, etc.) that replace forest trees and that have only short roots. The nutrient cycle being broken, fertility quickly fades and can be restored only through increasing amounts of fertilizer.

Furthermore, when the forest cover is eliminated, and especially when the trees have been removed through heavy bulldozer-like equipment that compacts the soil, the exposed ground is left unprotected against the heavy beating of tropical thunderstorms. The result is massive water runoff, accompanied in some cases by soil erosion. In the main, a forest on undulating terrain allows erosion of no more than 1 ton of soil per hectare per year, whereas the same area can lose 20 to 160 tons if the forest is replaced with fairly dense vegetation such as that of a coffee plantation, 60 to 200 tons if replaced with man-established pasturelands, and 1,000 tons or more with field crops.[12]

These, then, are some of the risks associated with large-scale clearing of tropical forests. A small clearing, however, can generally recover, because key features of the ecosystem, such as nutrients, are available from nearby patches of residual forest as colonizing vegetation comes in to cover the denuded ground. Similarly available from the forested environs are tree seeds to restore the original plant growth—and these seeds, with their specialized patterns of reproductive biology, are so important that they deserve a section of this chapter to themselves.

SEEDS AND FRUITS

The paleoecological pedigree of tropical forests has been dominated, in the main, by the phenomenon of environmental stability. With only occasional breaks, there have been millions upon millions of years of ceaseless warmth and moisture; continuity unmatched. By contrast, temperate-zone forests, with their seasonal fluctuations, have had to cope more constantly with feast-and-famine conditions during their evolutionary past. The practical consequence for tropical forests is that when they are left undisturbed, they demonstrate exceptional stability and "survivorship." Equally to the point, they are little able to tolerate the abrupt and broadscale disruptions that modern man inflicts.

By way of illustrating how tropical forests adapt to conditions that vary only slightly, let us look at the regeneration systems of tropical trees.[13] Their seeds generally have little or no capacity to remain dormant: they don't need to. Whereas the seeds of a temperate-zone tree may well lie unchanged for ten years before finally sprouting, the seeds of trees in tropical forests often survive no more than twenty-five days. Having become finely tuned to unchanging conditions, they are acutely sensitive to perturbations in their environments. The dipterocarp family of trees in Southeast Asia produces seeds that germinate only so long as the microclimate varies hardly at all around the mean temperature of 73 to 80 degrees Fahrenheit. When a timber exploiter breaks open the canopy and allows the air to warm up to as much as 100 degrees F., many if not most of the seedlings die (though this is not invariably the case: in Central America, for example, seedlings of certain species respond vigorously to opening of the canopy[14]). By contrast, temperate-zone-tree seeds survive the heat of summer and the cold of winter year after year.

In addition, tropical trees do not all flower and seed continuously, or even regularly. Many tree flowers complete their cycle in a single day, or are receptive for only a few hours around the middle of each day. Many other trees seed heavily only every third year, some only every tenth year, and a few only every thirtieth year or so. What, we might wonder, is the reason for this curious circumstance? So far as we can tell, the tightness of the nutrient cycle does not encourage trees to produce large amounts of fruits at regular intervals.

But perhaps the most distinguishing feature of the self-regeneration systems of tropical trees is that very few of them depend on the wind to convey their pollen and seeds. This is in marked contrast to the pattern in temperate zones, where forests are more open, and there is more scope for the wind to blow through the plant community. Beneath the dense canopy of a tropical forest, there is little air movement. In the forest undergrowth of Barro Colorado Island, in Panama, the average wind speed measures only 1.6 kilometers per day, or 1/240th as much as in open fields. In this situation, there is next to no hope for pollen or seeds that would normally be carried by wind. In a survey of 40 hectares of forest in Brunei, scientists have found that only 1 out of 760 tree species relies on the wind for pollination. So most tree species and other plants depend on creatures such as insects, birds, and bats to transport pollen and seeds.

Furthermore, when the seeds of forest plants drop to the ground, they almost always have to germinate in very shady conditions. So a seedling must get a good start in life if it is ever to reach the sunlight. To survive in these unpromising conditions, it needs an immediate supply of food during its critical early phase—which helps to explain why many fruit seeds are large. We have only to think of the seed of an avocado, the size of an egg, to understand the appetite of the infant tree. But a large-sized seed needs a large-sized creature to carry it—one reason why we find, among the more regular residents of tropical forests, toucans, pigeons, monkeys, and other sizable creatures. To attract these seed carriers, the trees produce large, fleshy fruits—witness the splendid selection of outsize, juicy fruits found at stalls in marketplaces throughout the moist tropics. The forests of Southeast Asia produce well over 100 kinds of fruit that are widely consumed by humans, in contrast to Europe's forests, with less than 10 varieties.

In tropical forests, trees and other plants depend heavily on their dispersal agents. The difficulties of regeneration are illustrated by the fact that the islands of Borneo and Sulawesi are only 100 kilometers apart, yet no more than 2 of the 282 dipterocarp species on Borneo appear on Sulawesi as well.[15] Despite the crucial importance of large creatures in the self-regeneration systems of tropical forest plants, we generally encounter few such creatures in tropical forests. In fact the biomass of large mammals sometimes amounts to no more than the equivalent of two or three domestic cats per hectare.[16]

All of this has crucial implications for conservationists. Insofar as large animals exist at low densities, and with populations that are often patchy, we must anticipate that if tropical forests are ever reduced to a few parks and reserves, these protected areas will have to be very extensive if they are to safeguard not only the trees and other plants, but the rare creatures that play such a critical role in the plants' survival. On top of that, we must recognize that many fruit-eating animals turn to successional species in secondary forests for part of their food needs, on the grounds that only successional species provide the specialized types of fruits that the animals need for their year-round diets. So a protected area must take account of this further quirk. To safeguard a patch of primary forest may not be enough in itself, even though this is often the predominant goal of park planners. A protected area will often need to incorporate an associated patch of secondary forest, perhaps one-tenth of the total protected area, to supply all the needs of all members of the wildlife community.[17]

DURIAN AND BATS

As an illustration of the way that forest organisms make common ecological cause—and of the way in which the forest's fabric of life can be injured through marginal-seeming interventions on the part of man in the forest ecosystem—let us look at the saga of the durian fruit and its pollinator bat in the Malay Peninsula. The durian is considered a delicacy and may well have pleased the human palate for half a million years. The fruit, about the size of a small football, has a strong and distinctive taste, like a mixture of rancid garlic and best strawberries. Connoisseurs say there are few better fruits anywhere, even though the act of consuming it can be compared, because of the smell, to eating dessert in a run-down public toilet.

With the fruit selling for the equivalent of U.S. $3 a piece, the durian tree generates a crop worth as much as $35 million per year.[18] Whereas most of the fruits are harvested in the wild, a number are now raised on plantations. Cultivated durians cover some 12,500 hectares, with a value between $5,000 and $10,000 per hectare; throughout Southeast Asia, the crop is worth at least $120 million a year. The net annual return on investment for an orchard owner works out at between 15 and 20 percent during the twenty-five-year lifetime of a plantation—a rate of return matching that of a rubber or oilpalm plantation. For the peasantry of the Malay Peninsula, the durian represents big money.

The famous, or infamous, smell of the fruit plays a key part in the plant's life cycle. Like virtually all trees of tropical forests, the durian tree occurs only sparsely. Yet no matter how isolated the trees may be, the powerful smell attracts animals that disperse the fruit seeds. And herein lies a curious phenomenon. Together with a good number of other tree species, the durian is pollinated by a single species of bat. Conversely the bat, *Eonycteris spelaea*, appears to find its main source of nectar food in durian trees. The bat spends a good part of its day roosting in caves; and large numbers of this particular species occupy caves in the environs of Kuala Lumpur, known as the Batu Caves, about 40 km. from coastal mangrove swamps in which grows a particular flower that is also favored by the bats. The swamps are being reclaimed for building land, which reduces a significant food source of the bats. Moreover, the Malay Peninsula, like most other parts of the Third World, has developed a hearty appetite for concrete. So the Batu Caves have been steadily exploited for their limestone. In the wake of these two assaults upon their life-support systems, the bat populations have declined. Fortunately the blasting of the Batu Caves has recently been halted, in order to safeguard the bats' roosts, among other reasons. But as long as the mangrove swamplands continue to be eliminated, the bat populations steadily fade away. The economic backlash affects many hundreds of people who trade in the durian fruit of the area, and many thousands who enjoy that mealtime delicacy. Much the same applies wherever bat habitats in Southeast Asia are under human assault of whatever kind.

During the lengthy stretches of the year when the durian tree is not in flower, the bat sustains itself by feeding on the nectar of several other trees, not only the mangrove species, but certain trees that occur only in primary forest. So if durian enthusiasts are to conserve the ecological network that sustains the species, they must maintain extensive patches of forest in order to safeguard not only the pollinator bat but the entire ecosystems that sustain the durian economy—both the natural economy and the human economy.

This example of ecologic linkages within the forest community has come to light in just the past few years. If we were to trace the many dynamic relationships that constitute the full fabric of forest life, we would surely find many such instances. Meanwhile, we continue to intervene in the forest's workings, with next to no knowledge of what we are doing. We often start to learn only when we recognize the damage that we have unwittingly wrought.

For instance, we might ponder the lifestyles of other bats in tropical forests. Their popular image apart, bats are beneficial to humans in all kinds of unsuspected ways. Earth may support as many as 1,000 species of bats altogether, or almost one-quarter of all mammals; and we can reasonably reckon that a good half of them live in tropical forests. In Southeast Asia overall, at least one-third of all mammals are bats; in the Malay Peninsula, one-half; and in the La Selva Forest Reserve in Costa Rica, a full three-fifths. Bats range from the size of a bumblebee, making it the smallest mammal (weighing less than a penny) on Earth, living in limestone caves far up the notorious River Kwai in Thailand, to flying foxes with 2-meter wingspans. Because they dwell in localized refuges such as caves, and because their specialized foraging practices confine them to just a few sources of food, they are very vulnerable to man's disruption of their lifestyles. A single cave-dwelling colony can consist of millions of individuals, constituting the largest congregations of warm-blooded creatures on Earth—and leaving them susceptible to summary execution through dynamite blasting. Whereas it would take several weeks to dispatch the two million wildebeest in the Serengeti Plains, at least as many bats can be eliminated in a matter of seconds. Not such a spectacular demise, but perhaps more terminal in the long run and more insidious in its impact, is the destruction of forest ecosystems. In Thailand, for example, where 94 out of 263 mammal species are bats, forest cover has declined by well over half in the last twenty years, and much of the rest may disappear before the end of the century.

Together with the nectar-consuming bats, a good number of fruit-eating bats are likewise important to the workings of tropical forests. As we have seen above, the abundance and variety of forest fruits is complemented by an array of fruit-eaters that disperse the fruit seeds. In forests of Asia and Africa, fructivorous bat species are as numerous as all the birds, mammals, and other creatures that consume fruits. Moreover, fruit-seeking bats often visit flowers, thus pollinating them. All in all, bats pollinate hundreds of genera of tropical trees and shrubs, thereby fostering the prosperity, and often the very survival, of many plants that we encounter on our tables at mealtime, including guavas, avocados, bananas, breadfruits, mangoes, cashews, cloves, eugenol, and tequilla. In addition, we can view various other plants as "bat supported," notably the kapok tree and many timber and fiber trees.

The one-on-one link between the Batu bats and the durian tree is paralleled by mutually supportive relationships between fig trees and wasps. The fig genus, the most distinctive and widespread of plant genera in the tropics, comprises more than 900 species, *each* of which is pollinated by its own species of wasp. At the same time, the wasps depend on the ovaries of the figs as sites for their larvae to develop.[19] A further illustration lies with the Brazil nut tree, which is commonly known for its tasty nut, and is widely harvested in Amazonia as a source of cash. This huge tree is pollinated solely by an irridescent insect, a member of the euglossine group of bees (the busy euglossines occasionally cover as much as 20 kilometers in 65 minutes). Likewise the tree's nut depends for its germination on a sharp-toothed rodent, the agouti, which chews and softens the seed coat. So this towering tree requires, for its reproductive system, the services of a high-flying bee that pollinates flowers in its crown, and a forest-floor rodent that disperses its nuts. Of the two creatures, the euglossine bee appears to be the more important in that it also pollinates, among many other plants, orchids—and in turn, these plants often supply prime sources of food to sundry other insects, which pollinate further plants, and so on. In this crucial sense, we can view the euglossines as "mobile link" species, and their plant hosts, by virtue of supplying food to extensive associations of mobile links, as "keystone mutualists."[20]

Many other cash crops, after the pattern of the Brazil nut tree, are pollinated by obligate insects or bats or birds. Tiny midges and thrips pollinate rubber and cocoa; bees and others of the Hymenoptera order pollinate passion fruit and curcurbits; flies pollinate cashew, mango, and kola nut; nocturnal moths and bats pollinate callabash, kapok, and balsa trees; and hummingbirds pollinate wild pineapples. All these specific pollinator relationships form part of larger "food webs," often with their mobile links and keystone mutualists.

This key concept of pivotal linkages within tropical forest ecosystems can be extended to thousands of plants that through their nectar, pollen, and fruit supply critical support for multitudes of insects, mammals, and birds: for example, figs, with their several hundred species, and epiphytes, with their thousands of species. Further, if as a result of man's disturbance of the forest, a keystone mutualist is eliminated from its ecosystem, the loss may lead to that of several other species. Still more to the point, these additional losses may, in certain circumstances, trigger a cascade of linked extinctions. Eventually a series of the forest's food webs could become unraveled; developed as they are through the co-evolution of plants and animals that have sustained each other through ever-more complex relationships,[21] they can steadily become destabilized from start to finish of their workings, with a shatter effect throughout their ecosystems. Thus, the exceptional intricacy of the forest's fabric of life. Interdependency is all. When human incursion causes the severing of a few threads, the damage can ultimately lead to a rending of the fabric from top to bottom.

While many of the species, both plant and animal, live only in primary forest, certain of them are "successional" species that depend on disturbed patches of forest, with their specialized foods and creatures, for survival. These successional species include certain figs, epiphytes, orchids, and aroids, also many euglossine bees and butterflies. Conservationists must aim therefore to protect all sorts of forest formations with their multiple habitats, if they are to preserve multiple communities of species. Furthermore, they need to bear in mind that many pollinators of tropical forests, in common with a good number of other species, exist at very low densities.[22] This leaves them unusually susceptible to sudden elimination. Not only are there obvious threats such as forest burning and other disruptions at the hand of humans, but by virtue of their sparse numbers, these species are subject to stochastic extinction (meaning, roughly, "random" extinction). The phenomenon occurs when populations, fluctuating as they normally do between high and low points, descend a demographic curve to a point from which they cannot recover. Or a population may die out because of genetic quirks. Under natural circumstances, these localized events do not matter, since a population can be reestablished by colonizers from neighboring populations. But if a park were established in a tropical forest without space enough to account for small-scale disappearances of "fragile" species, the park might be unable to sustain a complete forest community, even though it were to total hundreds of square kilometers. In temperate-zone forests, a park can generally protect a sufficient spread of ecosystems while covering a smaller tract of forest. In the humid tropics, however, we need to consider that a park should usually cover at least 1,000 square kilometers, often much more, to do its job.

11. THE IMPACT

OF MODERN MAN

5

The Commercial Logger

I REMEMBER seeing a tree felled by a commercial logger in a forest of Borneo. It stood, I guessed, about 70 meters tall, with 50 meters of clear bole. A stately specimen. The chainsaw toppled it in ten minutes, or one-millionth part, I reckoned, of the tree's probable lifespan. Each day, many thousands of such trees are cut in Borneo. Logging contributes to a pattern of depletion that may leave little forest of any sort, except degraded remnants, in Southeast Asia by the start of the next century. These ecosystems, the most ancient on Earth, have been in existence for at least 50 million years, and they are being eliminated within a period of half a century or so, or one-millionth part of their history.

As I watched the tree totter and fall, it set off an outburst of rending sounds in the forest canopy. Its crown, I saw, was festooned with vines, lianas, and other climbing plants, linking it to neighboring trees. When the bole was all but severed, the tree shuddered but remained upright, its top being held aloft by all those supporting climbers. When it finally came crashing, the climbers ripped off entire branches from other trees. The tree had been selected by the logger by virtue of its stature, with its topmost sections reaching way above the main forest canopy. An "emergent" such as this benefits from the additional sunlight it enjoys, developing a crown that measures 15 meters across or more. When one of these giants is brought down, its crown damages the trees around it. These damaged trees are then vulnerable to attack by pathogens at places where large strips of bark are torn off. Even a seemingly minor injury can leave a tree harmed beyond recovery. As a result of harvesting this one tree, another ten trees would probably die within a year.

After being "lopped and topped," the trunk supplied two good-sized logs, each about 20 meters long and 1 meter in diameter. The logs would contain more than 10 cu. m. of timber each, saleable on international markets for about $2000 altogether. Whatever my sense of regret at seeing such a splendid plant die, I reflected on all those statistics about Third World countries containing the poorest of the poor, people with a cash income amounting to no more than I spend on alcohol or running shoes each year. Indonesia is one such country, and it can hardly be expected to neglect the basic welfare of its citizenry in response to overseas conservationists who cry "Cut not another tree." In any case, there was a good chance, I speculated, that at least one of the logs I saw lying on the ground would end up in the possession of a rich-world citizen. It would serve as a decorative panel for his home, or a quality desk for his office, or even planking for his weekend yacht. More than half of all tropical timber felled finds its way into the main market for specialist hardwoods of the tropics, that is, the advanced nations. This consumerist demand by affluent people many thousands of kilometers away from Borneo or Amazonia is a prime impulse behind the headlong rush of many nations that have tropical forests to harvest their hardwood timber at rates beyond which the forests can renew it. Well might we ask, then, "Whose hand is on the chainsaw?"

I still have a vivid image in my mind of that Borneo forest. I left the scene with conflicting reactions, part of me muttering "Regrettable but necessary," part of me asking, "Yet must it be so?" After I returned to the city life of Indonesia's capital, Jakarta, I had occasion to phone an architect friend back in Kenya. My wife and I were having a house built, and the architect was

A giant forest tree, such as this in Sumatra, can be felled with a chainsaw in a matter of minutes. A few hours later, it will have been "lopped and topped" and sawn into logs of a size that can be hauled away to timber trucks. Thus a tree several hundred years old can be dispatched during the course of a single day. A specimen of the size featured here may earn well over $1,000 in foreign exchange. *(A. Compost, World Wildlife Fund)*

overseeing its interior décor. This was the first new home my wife and I had designed for our family, so I urged him to let rip, buy the best, right from the front door onward. Lots of luxury paneling around the place, parquet floors in more rooms than not, fine furniture on every side. When I arrived in Nairobi several weeks later, I was very pleased with the immediate impression. Hardwood throughout. What else, my friend asked, could possibly do such a fine job?

Touché.

OUR APPETITE FOR WOOD

One of the main reasons why forests around the world, especially tropical forests, are increasingly exploited (and sometimes over-exploited) is that more people want more wood. We have a remarkable appetite for wood, wood of any sort. An average American consumes about one and a half tons per year, or as much by volume as would fill the space occupied by an average dining-room table. Wood plays a part in more activities of a modern economy than does any other commodity. Almost every major industry depends on forest products in one way or another. Wood is an economically competitive material, since substitutes such as steel, aluminum, cement, and plastics need more energy for their production. Not surprisingly, the total of industrial timber (i.e., wood other than fuelwood) that we consume has risen to 1,500 million cubic meters, an amount that exceeds, by weight, our consumption of steel and plastics combined. On the average, an American uses more than twice as much wood as all metals together.

The reader may wish to consider his or her own home, the building itself. Housing is the economic sector that uses most wood. A wooden wall requires about 20 percent less energy for heating and 30 percent less for cooling than does a house made of other construction materials. Three American homes out of four are made largely of wood. On average, each of the two million or so U.S. houses that will be started in 1984 will contain at least $12,000 worth of wood.

How much wood do we use altogether? Rough estimates suggest that we now consume 3,000 million cubic meters per year, enough to cover Manhattan to the height of a ten-story building.[1] Of this amount, roughly half is used as fuel, over four-fifths of it to meet the needs of the developing world (see the following chapter). Of the other half, one-fifth is manufactured into paperpulp products, seven-eighths of which are channeled to meet the needs of the developed world; the remaining four-fifths are used as timber for construction, panels, and wherever "solid wood" is required, at least three-quarters being consumed by developed nations. By the year 2000, we could be consuming 4,000 million cubic meters per year. Thereafter the total will continue to increase far into the next century—probably with the same fifty-fifty split between commercial products and fuelwood that has been the pattern of wood consumption since the start of this century, when we were using only about one-quarter as much in total as today.

INDUSTRIAL WOOD

Since this chapter is concerned with industrial wood as opposed to fuelwood, let us look further at the breakdown figures. Global consumption of industrial wood is likely to double by the year 2000, and to double again by the year 2025. While developed-world citizens now use at least three-quarters of all industrial wood, their share will fall to about three-fifths by the year 2000, because once people in the Third World achieve some economic advancement, they will cultivate their own appetites for commercial timber. For many people in developing countries, wood plays such a pivotal part in their daily lives—housing, furniture, fencing, carts, plows, etc.—that they can fairly be described as living in the Wood Age. During the period from 1960 to 1980, the Third World increased its consumption of timber three times over, and they seek to achieve the same during the period from 1980 to 2000.

In short, our overall demand for industrial wood is projected to increase by leaps and bounds during the foreseeable future. The trend will generate growing pressures to exploit those forests that have hitherto contributed relatively little to global timber needs, tropical forests. Although they contain about as much wood as their larger temperate counterparts, tropical forests now supply little more than one-tenth of wood consumed worldwide for purposes other than fuel.

HARDWOODS

There are two sorts of industrial wood, hardwood and softwood. More than nine-tenths of tropical forests are made up of broad-leaf trees, nonconiferous species, that produce hardwoods. Demand for these tropical hardwoods has increased in the developed nations during the past three decades because hardwoods meet specialized needs. For instance, they are highly resistant to termites, insect borers, fungi, and other causes of decay. Hardwoods simply last far longer than other types of timber.

Of the world's hardwood forests, the great majority are located in the humid tropics. Hardwood forests of temperate zones have been steadily depleted, or are coming under greater protection in order to safeguard supplies for the future, or in order to meet environmental interests (see below). As a result, exploitation is increasingly directed toward tropical forests.

In the United States a home may well contain a lengthy list of products with components drawn from the forests of Borneo or Amazonia: veneer on the dining table, the writing desk, the hi-fi cabinets, and the bedroom furniture; less costly items such as salad bowls and cheese boards; plus major parts of the house itself, sidings and deckings. In recent years, the United States has imported much tropical plywood for a boom in mobile homes; at least one-quarter of a million are constructed each year, or more than one in ten of all houses built. A further fast-growing application of tropical plywood lies with the house remodeling industry; with a value of $17 billion, it is growing faster than house construction overall. Remodeling is likely to use far more tropical hardwood than the original building.

Something similar applies in other parts of the developed world. In several countries of Western Europe, pleasure boats contain enough tropical timber to account for about 10 percent of the sale price. More important, sales of these boats have been growing at an annual rate of about 5 percent since 1980, making them a leading item of conspicuous consumerism. There is a good chance, furthermore, that some of the last raw material that a European citizen uses will derive from tropical forests. In Great Britain and Switzerland among other countries, most solid coffins are made of imported hardwoods, such as West African abachi wood and Brazilian mahogany.

After the economies of the developed world recovered from World War II, they began to import tropical hardwoods for the first time. Starting with only minor amounts, a little over 4 million cu. m. in 1950, the volume has grown to 70 million cu. m. and is projected to increase by more than half as much again by the year 2000. True, the populations in tropical regions use a lot of hardwood timber themselves, but the amount has increased only three times since 1950, whereas the developed world's imports have increased sixteen times, and the total has recently surpassed consumption by all tropical countries combined. The hardwood exports are worth a great deal to producer countries. Whereas they earned only about $250 million in the early 1950s, when the tropical hardwood boom was gathering momentum, they now pull in more than $8 billion a year, with a growth rate far faster than for trade in all forest products from whatever source. Tropical timber now ranks among the leading exports of the Third World, earning as much as cotton, twice as much as rubber, and almost three times as much as cocoa.

A number of developing countries depend on their hardwood

exports for no small share of their foreign exchange revenues. The front-runners include Indonesia, Malaysia, the Philippines, Papua New Guinea, Ivory Coast, Gabon, and Brazil. Let us bear in mind, however, that tropical hardwood exports constitute only one-seventh of global trade in all forest products. Whereas eight of the countries listed earn more than $100 million per year from their exports, the developed world of the temperate zones has twenty-two countries in the same position.

Tropical forest nations are led by the "big three" of Southeast Asia: Indonesia, Malaysia, and the Philippines. Together, they account for well over half of all tropical hardwood exports. When we add in the rest of Southeast Asia, the proportion rises to almost three-quarters.[2] In 1980, at the height of the global economic recession, Indonesia still earned $2.2 billion in foreign revenues from its timber, second only to oil, and more than all agricultural commodities combined.[3] Yet sizable and increasing as these export earnings are, trade in tropical timber still amounts to only a small share of overall trade in all forest products, now worth almost $60 billion per year. The Congo and Finland have land areas and forest estates of roughly the same size, yet the value of Finland's forest exports is around fifty times greater than that of the Congo's. The six main countries of the Amazon Basin, plus Guyana, Surinam, and French Guiana, possess 2.7 hectares of forest per person, compared with a world average of about 1 hectare, yet they pay more for their imports of forest products than they earn through their exports. Brazil's sector of Amazonia contains some 50 billion cu. m. of standing timber, yet loggers extract only 4 million cu. m. per year—a harvest that contributes less than one-tenth to Brazil's exports of tropical timber, which in turn amount to less than 3 percent of timber exports world-wide.

In light of the commercial importance tropical hardwoods have already demonstrated in the marketplace, many people in tropical forest countries believe, with some justification, that they possess a resource that can fuel their drive for development overall. The tiny state of Brunei, for instance, comprises only 5,500 sq. km., of which about 5,000 sq. km. are forested. If a timber harvester takes 25 marketable trees per hectare, out of the hundreds of trees available, and if each 5-ton tree (average weight) is worth $750, the country's total timber stock is potentially worth over $9 billion to the populace of one-quarter of a million people —or more than all trade in tropical timber from all sources.

Not surprisingly, then, many people in tropical forest countries look forward to the day when commercial logging will contribute much more to the welfare of their citizens. Given the impoverished state of most citizens in these countries, it is difficult to contest this goal. There are, however, two provisos. First, their efforts to meet immediate needs should not preclude opportunities to accomplish long-term goals of development overall (see, for example, the discussion of how forests support agriculture and several other economic sectors, described in chapter 14). Second, the desire to cash in on their hardwood forests should not undercut other legitimate considerations, such as humankind's heritage in the most exuberant expression of life ever known on the planet.

Regrettably, the tendency is for tropical forests to be logged on a once-and-for-all basis. After years of over-harvest, Thailand and Nigeria have become net importers of timber. Virtually the whole of West Africa could follow suit by the year 1990, and the Philippines and Peninsular Malaysia shortly thereafter. Indonesia will continue to export for a longer period—if only because its forests are larger, many of them are richer, and the logging began later. But at current rates of harvest, the downturn for Indonesia is only a matter of time, and it could well arrive by the end of the century.

The problem lies not only with mismanagement, widespread though that is. Because of their compulsive desire to lift themselves out of poverty, many Third World countries are inclined to treat their forests as an oil well, to be drained with all due dispatch. This is not because of ignorance or stupidity or worse.

It has to do with abstruse economic factors, notably discount rates and investment returns, that impel all modern economies, whether West or East, North or South, to emphasize today in preference to a long line of tomorrows. Third World nations treat their hardwood stocks in much the same way that Americans treat their remnant redwood forests, and Japanese and Russians treat whales: resources that could yield sustainable harvests in perpetuity, yet have been exploited to the brink of exhaustion. The reasons are complex; they reflect our inadequate understanding of modern economies, and they could fill a book in themselves.[4] Suffice it to note the comments of a Director of Forestry (who asked to remain anonymous) in a Southeast Asia nation: "It is not so much that we want to live today rather than invest in tomorrow. We want to survive today, then live a little tomorrow, and think about the future next week."

The largest single consumer of tropical hardwoods is Japan, accounting for over half of the developed world's imports.[5] Between 1950 and 1974, that is, until the onset of the recent global recession, Japan's imports of tropical hardwoods increased almost twenty times. Among all Japan's imports, wood now ranks a strong second to oil: some people in Japan foresee timber storages could eventually amount to "another oil crisis." Japan depends on overseas sources for two-thirds of its wood, a proportion that could rise as high as four-fifths by 1990. While most softwood imports are from North America, hardwood imports come almost entirely from the tropics. Although Japan has hitherto looked to Southeast Asia for three-quarters of its hardwood needs, it is increasingly turning an eye toward Latin America and West Africa.

Yet while Japan consumes ever-more hardwood timber from the tropics, on the grounds that supplies are both "cheap" and "abundant," it husbands its own hardwood forests against the day when tropical stocks could run out. Two-thirds of the country's land surface is covered with forests, a larger proportion than is the case for any other advanced nation; of the total growing stock, hardwoods account for well over two-fifths. Annual cuttings from the forests have actually been reduced by half during the past twenty years, until they now represent considerably less than annual growth (many of Japan's forests are in acute need of thinning). This management strategy applies particularly to the hardwood forests: year by year, more and more of Japan's hardwood forests are declared off bounds to commercial use.

The second-largest consumer of tropical hardwoods is the United States, where demand has been growing far faster than the country's growth rates for population and GNP.[6] There are two main reasons for this surge in demand. First of all, high-quality hardwoods from within the United States have become more expensive in the wake of environmentalists' concern for the recreational and esthetic values of hardwood forests (a highly legitimate concern in its own perspective). Yet without depleting an acceptable level of "non-consumptive" values, the United States could easily become self-sufficient in hardwood timber. It grows at least half as much again as it uses, and some foresters believe the sustainable output from the country's hardwood forests could be doubled quite readily, and tripled within five decades, through widespread application of proven management methods and intensive silvicultural techniques such as complete-tree utilization (see below).[7]

The second reason for the boom in American's demand for tropical hardwoods is that the timber, and particularly plywood and other paneling materials, can be obtained relatively cheaply from Southeast Asia, via East Asia. The United States buys about 70 percent of all tropical plywood and veneer entering world trade, the bulk of it coming from Japan, South Korea and Taiwan, often under the name of Philippine mahogany that actually originates from several countries of Southeast Asia. Between 1950 and 1973, that is, until the onset of the recent recession, U.S. imports of tropical hardwoods expanded nine times. By the year 2000, these imports are projected to double once more. While the imports make up only 2 percent of American's wood consumption measured by volume, their dollar value is high, amounting to about $1 billion. Moreover, while they make only a miniscule contribution to U.S. use of all timber, they represent a sizeable share of the harvest, or over-harvest, of hardwood currently being extracted from the region under greatest exploitation pressure to date, Southeast Asia. During the past three decades, Southeast Asia has expanded its exports more than 25 times, until the region now accounts for three-quarters of international trade in tropical hardwoods.

PAPER PULP

While global consumption of hardwood products is expected to expand steadily, we can look for yet greater growth in demand for paper pulp. With every year that goes by, the world needs more and more paper products. Demand has been increasing at a rate twice that of population growth—and is not explained merely by our reading more newspapers and magazines (massively true though this is of the Third World), since newsprint accounts for only about 15 percent of all paper products, and printing papers and other writing papers for only 20 percent. Almost two-thirds of paper pulp produced is manufactured into packaging materials, tissues, and convenience products such as paper cups and plates.

Tropical forests contain relatively little of the conventional source of paper pulp: softwoods. But as a result of recent changes in forestry technology, paper corporations increasingly turn to tropical forests for their raw materials. Until the mid-1970s it was thought that the multiplicity of hardwood species in tropical forests would prevent them from being pulped after the manner of softwoods. New techniques, however, can put every tree through the hopper, converting wood chips from 200 species at a time into pulp. Whereas tropical forests comprise well over half of the world's forests, they currently produce only about 7 percent of the world's paper and paperboard, a percentage that could soon start to grow rapidly.[8]

During the past three decades, world output of paper products has expanded from 40 million tons to 180 million tons. If long-term trends continue, demand could rise to 400 million tons by the year 2000, and to twice as much again within only another two decades. The developed world now consumes 155 million tons a year, the developing world only 26 million. A citizen of Great Britain or Japan uses just over 150 kg. per year, and an average American twice as much, whereas a citizen of the developing world is unlikely to use more than 5 kg. If consumption continues to grow as in recent years, the developing world is projected to expand its amount roughly three times by the end of the century, rather more than the increase on the part of the developed world. But by far the largest share of total consumption will remain with the developed world—again, the sector most responsible for growing pressure on forests around the world.

Many nations of North America and Western Europe supply their own paper pulp; some even produce a surplus. Japan, by contrast, is heavily dependent on foreign sources and seeks about half of its supplies overseas.[9] In the wake of recent price increases for U.S. softwood chips, Japan has turned to Southeast Asia and tropical America for hardwood chips; tropical forests already meet over half of Japan's chip needs.

As for the developing world, many citizens consume only a couple of kilograms of paper products per year, or about as much as a copy of the Sunday edition of the *New York Times.* Demand for paper is rising not only with growing human numbers, it is rising much more as a consequence of growing human aspirations. As more people become literate, they require far more newspapers, magazines, and books. Brazil's consumption of paper doubled during the 1960s, doubled again during the 1970s, and may well reveal a similar pattern for some while to come. Not all developing countries, however, can afford to follow Brazil's example. In all too many cases, sheer lack of paper causes serious problems in education and communications. Developing nations as a whole pay almost $4 billion per year for their paper pulp—

almost half of total earnings from tropical hardwood exports. Hence tropical forest countries perceive a strong incentive to turn their hardwood forests into paper pulp. They have now found a way to do so, in the form of the chipper.

CHIPPING

Long ago when loggers used to fell their trees with axes, they would take a whole day to topple a single forest giant. With a chainsaw, a single man can do the job in one-fiftieth of the time. Remarkable as this change has been, it is surpassed by the chipping machine, a manifestation of modern technology that can eliminate a complete patch of forest in super-swift fashion. The machine reduces an average-sized tree, branches and all, into chips, each about the size of a dollar coin, in about one minute. With a 2½-meter disc equipped with eight blades, revolving 320 times a minute, the machine consumes every bit of every tree, whatever the species. In just one day, it can eliminate five hectares of forest, as much as would cover half a city block.

The chips are utilized for various forms of paper goods, and for cardboard packing to encase industrial products such as cameras and calculators. The next time we unpack a Japanese electronic gadget, swathed in packing several centimeters thick, we might ask whether we really need all that packing material. Of course the Japanese throw in such a lot of wrapping because the chips are supposedly cheap at the source. The Jant-Honshu Paper Company, in Papua New Guinea, for example, chips every last twig in a forest concession of 830 square kilometers so that affluent-world citizens can enjoy the benefits of the company's 500 types of boxes. In 1974, Papua New Guinea exported no wood chips at all; today it dispatches almost half a million cu. m. per year. And that is just the beginning. Japanese forest corporations are gearing up to put forests through the chipper in such disparate places as the Philippines, Thailand, Cameroon, and Colombia.

What shall we make of this new phenomenon?[10] In theory,

it could be good news, in that it could enable timber harvesters to make much more intensive use of forests. If they obtained their many millions of tons of timber from smaller areas, it might relieve exploitation pressures on remaining primary forests. Since the chipper can not only be applied to take sound trees of any sort, but can be used to salvage defective trees and dead standing timber, it can expand the yield of a temperate-zone forest by as much as an additional 300 percent of usable wood, and that of a tropical forest by very much more.

Some observers look on this harvesting of any and all trees, also known as "full forest harvesting," as an innovation that could represent some sort of salvation for tropical forests—in certain places, under certain circumstances. But other observers believe that it could be a two-edged sword. Thanks to the chipper, a patch of forest is not exploited selectively, on a sustained-yield basis. Segments are exploited entirely, once and for all. If, of course, a harvested forest is then replaced by a man-made forest, this intensive use, both of the original forest and of the land on which it stood, may serve to confine timber harvesting to limited localities.

All too regrettably, however, the new technology is being put to work in a manner that could pose a threat to many forests. Whether or not a forest can re-establish itself on the cleared lands, we do not yet know. What we do know is that when forestland soils are exposed to the tropical sun and rain, they often become degraded beyond restoration unless new vegetation takes over within a matter of months—and this will almost certainly require a helping hand from man. As we have seen earlier in this book, most of a tropical forest's nutrients are in its vegetation. Logging of tree trunks alone removes one-tenth of nitrogen, one-fifth of calcium, almost one-fifth of potassium and phosphorus, and well over half of magnesium—amounts that are twice as much as those in most temperate-zone forests.[11] When entire trees are removed, branches and twigs and all, the nutrient drain is that much greater, especially insofar as many tree species concentrate their nutrient stocks in their extremities rather than their trunks. If a chipped-over patch of forestland is to grow even grasses and shrubs as a start toward new vegetation cover, it will probably need costly stacks of fertilizer to restore the nutrients. We are still waiting to hear from the Japanese how much they must invest in each deforested hectare to avoid leaving behind a cleared landscape that remains barren.

ECOLOGICAL IMPACT OF COMMERCIAL LOGGING

As we have seen, tropical forests contain a vast diversity of tree species. Yet international timber markets prefer to take only a small proportion of the wood types available (at least for industrial lumber). A commercial logger prefers to harvest as selectively as he can, taking a few choice specimens without regard for the rest —a "creaming" operation. In Southeast Asia, he focuses on less than 100 tree species, with exports consisting mainly of only two dozen or so. This means that he takes perhaps only 20 out of 400 trees per hectare. Despite this marginal harvest, however, the logger leaves many of the remaining trees damaged beyond recovery, far more than would be the case in a temperate-zone forest. When one giant is felled, it is likely to cause several others to be broken or pulled down.

Repeated surveys in Southeast Asia reveal that average logging leaves between one-third and two-thirds of residual trees injured beyond repair.[12] In addition, almost one-third of the ground may be left bare, often with the soil impacted by heavy machinery (in undisturbed forest, rainfall infiltrates the soil at a rate of up to 25 centimeters per hour, but in a bulldozed site, virtually none to start with, and only 3 centimeters per hour after one year). Hence the cynic's account of how a typical corporation practices selective logging: it selects a forest, and then logs it.

Yet with only moderate care, the damage can be greatly reduced. The climbing plants can be cut before felling, and the crown reduced in size. Through directional felling, a skilled logger can cause a tree to fall within a fairly narrow arc, at a point where it will cause least damage. A skilled logger can also leave aside those defective boles that are likely to split when they fall: on one logging concession in Borneo, I have seen dozens of shattered trunks left to rot on the forest floor. When skid trails and logging tracks are planned ahead of time, their lay-out causes far less permanent disruption. Finally, logs can be extracted from the forest by air: an expensive method, but helicopters are already used in steep terrain in central Borneo, so why not elsewhere?

According to experiments in Sabah and Sarawak, these mea-

For every tree harvested from the forest for commercial lumber, several others, sometimes dozens, are damaged beyond recovery. Still more injury is done to the forest ecosystem through extensive loading zones, such as this one in Malaysia. Experiments show that the unnecessary destruction of commercial logging can be reduced by at least one-half through simple techniques such as directional felling and better planning of timber tracts. *(Ken Scriven, World Wildlife Fund)*

sures enable loggers to reduce the damage by half. Of course, less destructive harvesting raises timber prices for the end-product consumer, a prospect that the main markets in developed nations are reluctant to accept on the grounds that it is inflationary. But at present, the price that a rich-world citizen pays for his tropical timber does not reflect all the costs that go into its production, notably the environmental costs. Nor need it be difficult to persuade the logger to implement these measures. All a government has to do is to levy a royalty not only on wood actually extracted from the forest, but also on trees unnecessarily damaged. Sarawak does it already.

FOREST DEPLETION

What shall we conclude about the actual amount of forest that is logged each year? According to official documentation by governments, the amount of forest that is newly logged each year in Southern and Southeast Asia is around 21,000 sq. km.; in tropical Africa, 7,500; and in Latin America, 15,500. This makes a total of 44,000 sq. km. for the biome each year. In light of growing demand by consumers in many parts of the world for tropical hardwoods, we can expect all these totals to increase steadily, faster for a while in Asia and Africa, with Latin America catching up toward the end of the century.

These figures, however, refer only to legal fellings. They are figures published by governments, based on annual returns from logging companies and other commercial concerns in the field. They take no account of illegal fellings. In several countries, timber poaching is a great and growing problem. In Ghana, foresters estimate that the country loses at least $50 million worth of timber each year, through smuggling to neighboring countries. In Thailand, official figures for legal felling in a recent year indicate a total of 2.8 million cu. m. cut, yet sawmill returns for the country indicate a total of 6.5 million cu. m. processed, meaning that 3.7 million cu. m. must have been taken from the forest without being recorded.[13] In addition, substantial quantities of

Thailand's teak among other prized species are smuggled over the border into Kampuchea and Laos by guerillas in order to raise foreign exchange for their clandestine activities. Similar fund raising is practiced by insurgents in peripheral forest zones of Burma. In the Philippines, customs figures reveal that export of logs to Japan in 1981 amounted to 365,441 cu. m., yet Japan's import records reveal that the country imported more than 1.4 million cu. m. of Philippine logs.[14] Incongruities in trade figures also indicate that Japan imports a good deal of illegal timber from Indonesia. In light of these activities by commercial loggers, it seems reasonable to upgrade the figure of 44,000 sq. km. of forests logged each year to at least 50,000 sq. km. per year.

Clearly the consequences of logging vary from area to area: in some places, only light modification of the forest; in others, gross degradation. But ultimately, the greatest damage done by the logger is often unwitting. By coincidence, those parts of the biome where logging is widespread are also, generally speaking, those where human numbers are highest. This means there is much population pressure on any lands that appear to become "freely available" to cultivators, and they speedily become occupied by landless peasants (see chapter 8). Timber tracks laid down by timber harvesters offer excellent access for hordes of itinerant farmers who can then penetrate deep into forest heartlands. In virtually all areas of Southeast Asia and West Africa—areas where the cropland squeeze hits hardest at land-hungry peasants—logging sites are subject to a slash-and-burn agriculture of such intensive form that the forest gets little chance, if any, to regenerate. As a result, the logger, through no fault of his own, encourages small-scale cultivators to damage the forest far more than the logger ever did.

SCOPE FOR ACTION

We have already looked at a few ways the logger could harvest timber in a less negligent fashion. We shall now explore some other alternatives.

Secondary Forests

Virtually all exploiters of tropical forests could derive the goods and services they require from secondary (i.e., disturbed) forests alone. They could then leave primary forests intact until we can determine, through systematic research, the ways in which we can put them to best long-term use. Secondary forests, covering as much as 2 million square kilometers (some experts believe 3 million), offer good commercial potential.[15]

When a patch of forest is in early successional stages of secondary growth following disruption, it is very likely to contain fewer species than a primary forest, yet a higher proportion of fast-growing species—certain of them with established economic value. Generally speaking, secondary forests represent an ecological phase when biomass production, notably wood growth, is greater than at subsequent stages. In addition, a secondary forest serves as a more readily renewable resource than a primary forest, insofar as it tends to be "aggressive," or quick to restore itself. While trees are smaller there than in a primary forest, they offer more uniform structure in their early stages. True, second-growth timber is not so strong as that of mature trees half a century old or more. But they are fine for paper pulp, thin-slice plywood, certain types of veneer, and for particleboard, chipboard, fiberboard, and laminated-wood products—all of which rank among the leading sectors of the timber trade. Finally, secondary forests are often located close to settled areas with roads, making marketing easier.

For these reasons, then, disturbance of a forest ecosystem may not mean, from the timber harvester's viewpoint at least, degradation. Secondary forests could prove to be the wave of the future. Some observers believe that already there are enough of these forests to produce all the industrial wood we need from tropical forests up to the year 2000, provided they are helped along through management techniques that are known collectively as silviculture (from the Latin word "silva" for forest), an approach that corresponds to agriculture. After all, our present approach to tropical forestry resembles that of agriculture some ten thousand years ago, that is, a hunter-gatherer activity (with a few primitive cultivars in plantations, see below). Under the general heading of silviculture, we can consider measures such as enrichment planting of desired species, selective girdling of unwanted species, and other steps to foster the sort of forest that a timber harvester prefers. While this may promote the emergence of an ever-more "artificial" forest, it helps to divert exploitation pressures from primary forests. Experiments at a forestry research institution in Turrialba, Costa Rica, show that when a forest patch is reduced, during the course of ten years, from 85 species of whatever sort to 17 species of commercial value, the forest produces several times more marketable timber than a similar-sized patch of primary forest.

The trouble with sivilculture is not that it has been tried and found wanting. It has not been given a fair trial at all. Less than one-twentieth of tropical forests are under any kind of management whatever, and over three-quarters of this is in India. Even the most rudimentary measures are generally disregarded. In eastern Borneo, the stronghold of Indonesia's timber boom, logging firms agree under their contract to implement a "select fell" system; it enables loggers to return to the site for a further harvest in thirty to fifty years. This system requires that only larger trees of desired species will be cut, and only in those areas where enough smaller specimens are established to replace the harvested trees. Yet according to a recent survey of logging companies, including giant timber corporations of international repute, not one leaves the required select crop trees to replenish the species' population, and indeed many companies choose to extract their harvest in areas where there are insufficient trees at the start to comply with the regulation.[16]

Plantations

Another way to reduce exploitation pressures on primary forests is to establish plantations. When we look at the rapid growth of harvesting rates for natural forests since about 1970, and when we take into account the deforestation rates due to forest farmers and other agents, we can reckon that soon after 1990, well over half of the dwindling natural forests may well be needed to meet prospective demand for hardwoods—unless we can shift an expanding part of the exploitation pressure onto manmade forests, that is, plantations.

We could conceivably plant enough "tree farms" to produce all the tropical timber and pulp we need by the year 2000. A plantation in Canada generates only between 3 and 5 cu. m. per hectare of wood per year, and in northwestern United States the average is between 5 and 8 cu. m. In the humid tropics, several fast-growing tree species produce wood at a rate of at least 20 cu. m. per year, a volume that can be obtained only from ten times as large an area of natural forest.

Regrettably, tropical plantations are not without problems. They prove costly to establish, as much as $1,000 per hectare, and since the plantation owner must wait at least eight years for a return on his investment, this type of forestry is hyper-sensitive to changes in investment trends and inflationary patterns. In addition, there are biological problems. In common with many monoculture crops, plantations attract diseases and pests that are numerous and varied. When the trees are harvested, they generally take a large stock of nutrients with them, requiring ever-growing amounts of fertilizer to maintain productivity.

Thus far, plantations are being established at only one-tenth of the rate needed to keep up with demand. Because of the scale of investment required, many tropical forest countries look increasingly toward the foreign entrepreneur. Half of Latin America's total sales of paper and paperboard have recently been accounted for by affiliates of U.S. timber corporations;[17] and Brazil has reached agreement with overseas corporations for a $1 billion plantation project. Yet foreign investors are not so interested in tropical plantations as they might be. They are not sufficiently assured about the security of prices for their products or the security of tenure for their holdings. Each inflationary upheaval in world currency systems, and each nationalization of a foreign enterprise in a tropical country, defers the day when enough plantations will be established to relieve exploitation pressures on natural forests. In fact, the plantation option depends as much on advances in the economic and political fields as on improvements in biological and forestry fields.

Processing

Tropical hardwoods exported as logs earn only one-third as much as when they are processed into plywood and other semi-manufactured products. Were tropical forest nations to undertake more processing of their timber, they could generate greater revenues for themselves while cutting fewer trees.[18] To date, however, the bulk of tropical timber exports still consists of raw logs.

Southeast Asia, for instance, now possesses the technological skills and the capital plant to process most of its hardwood output. Were the region to engage in more preliminary processing and thus to export sawn wood and panelboard materials for greater foreign exchange, it would be more ready to husband its dwindling forest stocks. Were the region to engage in advanced processing, producing semi-manufactured goods such as furniture assemblies, it would earn much more; and if it could eventually aim to turn out final products, it could earn as much as $4,000 per cubic meter for top-quality veneer. Were the three main forest countries of Southeast Asia (Malaysia, Indonesia, and the Philippines) to have processed all their log exports in the late 1970s, they could, according to John S. Spears, Forestry Adviser of the World Bank, have earned an additional $1 billion.

Furthermore, an advance along these lines would benefit the principal end-product consumer, the United States citizen, on the grounds that Southeast Asia nations can do the job more cheaply than the nations that seek to maintain a lockhold on profitable processing, that is, Japan, Taiwan, and South Korea. Part of the problem lies with the same giant timber corporations that we discussed earlier in this chapter, notably those from Japan. They prefer to export raw logs from, say, Borneo, so that they can supply processing plants in their own conglomerates in Japan. In part too, the problem lies with tariffs and other trade restrictions imposed by Japan and the others: on veneer sheets, levies can be as high as 37.5 percent, and on plywood as much as 50 percent.[19]

In order to counter this inequitable situation, the Southeast Asian Lumber Producers Association is trying to form a cartel of its timber trade, cutting back on log exports in order to take some of the processing action from the lands to the north. After years of frustration, they are finally achieving some success. By 1990 if not sooner, they should manage to reduce their log exports to a trickle, and to gain a greater share of the final market value of finished products by selling direct to the United States. A parallel endeavor is being attempted by the Organization of African Producers and Exporters of Wood, vis-à-vis their main markets in Western Europe. Whoever would have thought that cartelization could ever be good for us?

6

The Fuelwood Gatherer

WE HEAR MUCH about fuelwood problems in the Third World. For at least one and a half billion developing-world citizens, that is the real energy crisis, even the dominant factor in their daily lives. Yet in many tropical forest countries, it tends to be a moderate rather than a serious problem, not comparable with the impact of other major agents of forest depletion such as the timber logger and the forest farmer. The main problem occurs in tropical woodland and savannah countries.

In certain localities, however, fuelwood gathering ranks as one of the major pressures to which tropical forests are subjected. Areas in question include Nepal, northern India, Pakistan and Bangladesh, parts of Sri Lanka and Burma, much of Thailand, most of Vietnam, all of Java, northern Sumatra, southern Sulawesi, much of northern Philippines, eastern Madagascar, several sectors of West Africa, and portions of Central America. In these areas, albeit limited localities compared with the biome as a whole, patches of forest are being severely degraded, if not eliminated altogether, by the fuelwood gatherer.

Moreover, the situation today is likely to deteriorate. In many more areas, the pressure of fuelwood gatherers on tropical forests will steadily grow more pronounced, until a sustainable impact becomes a destructive impact.

SCALE OF THE FUELWOOD PROBLEM OVERALL

Whereas one-half of all wood cut worldwide each year is used as fuel, more than four-fifths of this amount is cut in the developing world, amounting to at least 1,200 million tons per year.[1] (This is a minimum estimate; much consumption goes unrecorded, and some experts believe the amount could be twice as large.) More than two billion people, or the great majority of developing-world citizens, rely on wood in its original role for humankind: as fuel to cook meals. In many tropical countries, as much as half of all energy is used for cooking and other basic needs such as heating homes. The amount consumed for these domestic purposes amounts to almost twice as much as is accounted for by agriculture and industry combined. In most developed countries, by contrast, cooking takes less than 5 percent of *all* energy used.

At least 1.5 billion people in the Third World have difficulty finding enough fuelwood. They meet their minimum needs only at the cost of extracting fuelwood faster than it is supplied by regrowth. Worse still, more than 110 million people simply cannot lay hands on enough fuelwood, or its derivatives such as charcoal, to meet even their minimal needs. For many families, it costs as much to heat the supper bowl as to fill it. To cook one kilogram of, say, rice, can require at least one kilogram of fuelwood, especially if the cooking stove or the fireplace is inefficient, as is usually the case. For people without sufficient fuelwood for cooking, there are health problems: under-cooked food is not only indigestible, it may well contain parasites.

The fuelwood crisis is not the "other" energy crisis. It is the original energy crisis that was exacerbating the worst poverty in the Third World for years before OPEC reared its head. For many Third World households, fuelwood costs as much as $200 per year, which can represent two-fifths of total cash income. So for many developing-world citizens, the energy crisis does not devolve into a debate on how to reduce the number of times the car is used each day, or on how to limit the electricity consumed in a gadget-oriented home. It is a problem that strikes directly at the struggle to keep body and soul together.

How much fuelwood is actually used? Good estimates are rare, since fuelwood consumption does not usually enter into national accounting figures. Figures vary from half a ton per person per year, to at least twice as much. Of course in an area where supplies have just about given out, people may get by with less than half a ton, even a quarter of a ton; and in areas where there seems to be plenty, people may use at least two tons, even more. But as a good working figure, we can reckon that the average often amounts to one ton per person per year. This may not sound like much, less than three kilograms per day, equivalent to a few thick sticks. But when we multiply this amount by the number of people who are short of sustainable supplies, we come up with a deficit figure of one billion tons per year.

As for the future, the outlook is bleak at best, unless we can make some fundamental changes. By the year 2000, the number of people suffering some scarcity of fuelwood may increase from 1.5 billion to 2.3 billion, and the number facing acute shortages may grow from 112 million to over 350 million. As for the further future, the year 2025 may find 3.5 billion people unable to keep themselves properly supplied with fuelwood, while the number of people who suffer severely through shortages may approach 1 billion.

IMPACT ON TROPICAL FORESTS

Fortunately for tropical forests, only a small proportion of total fuelwood is taken from forests of any sort, and a still smaller proportion from primary forests. In many parts of most countries, fuelwood is obtained mainly from savannah woodlands, scrub and brush patches, and local woodlots. In India, for example, people living in or near forests meet their entire fuelwood needs from the forest; people living within 10 km. of forest boundaries take about 70 percent of their fuelwood from the forest; but beyond 10 km., the use of fuelwood from the forest steadily diminishes, until at about 15 km. it is almost nil. In other countries of humid tropical Asia, those where the fuelwood problem is most marked, between one-third and one-half of all fuelwood is gathered from outside forests proper. Similar findings are reported from Africa. Fortunately in Latin America, Central America excepted, there is usually not much of a fuelwood problem.

Every day, however, the situation grows worse. While one person can, generally speaking, obtain a self-renewing supply of fuelwood from the equivalent of half a hectare of typical forest, it is more usual that, in order to exploit supplies close at hand, as many as fifteen people may be taking wood from one hectare, which amounts to an excessively concentrated harvest, one that the forest cannot sustain. This means that local supplies rapidly become exhausted, and people begin to direct their over-exploitative pressure at ever-larger tracts of forest.

Worse still, this trend is likely to accelerate. As human populations expand, they find that they can no longer depend on traditional sources of fuelwood. As remaining forests continue to be reduced (not only through exploitation for fuelwood and commercial timber, but to make way for agriculture, etc.), residual patches of forest become subject to increasing exploitation, and then to rapidly increasing over-exploitation. Within just a few years, a patch of forest that once amounted to a renewable resource is no longer being used, it is used up.

Furthermore, this pattern generates a compensatory response among people who suddenly find themselves without enough fuel. In several countries, large-scale commercialization of the fuelwood trade causes professional gatherers to seek supplies from far-distant sources. It quickly becomes financially worthwhile for charcoal manufacturers to transport their supplies from forests in one part of a country to markets in another part of the country, sometimes hundreds of kilometers away. This practice is already established in such countries as Thailand. The Bangkok conurbation, with its five million inhabitants, now derives much of its charcoal supplies from forests in remote parts of the country.[2]

Those sectors of tropical forest countries that fall within the humid tropics contain roughly one billion people. Of this total, some 300 million can be considered to live within the forests or on their fringes, or at least within their environs. So far as we can ascertain, each of these people consumes somewhere between 0.4 and 1.2 tons of fuelwood per year. In addition to this local use, substantial amounts are, as we have seen, taken to serve commercial markets in far-off urban areas. As a crude assessment, we can reckon that the total amount of fuelwood cut from tropical forests could well amount to some 180 million tons per year. Unfortunately we have insufficient information to say whether most of the forest areas in question can sustain an annual harvest at present levels of exploitation. We know that for a number of areas, such as those listed above, the harvest amounts to once-and-for-all use. Suffice it to say here that the areas listed may, by the year 2000, be joined by several others, namely the rest of Central America and West Africa, the rest of Thailand, the rest of Sumatra, and parts of Peninsular Malaysia.

CATTLE DUNG FOR FUEL

Fuelwood shortages cause problems not only for food cooking, but for food growing. When fuelwood supplies run out, farmers look for substitutes. A frequent option is cattle dung. In developing countries of Asia and Africa alone, at least 400 million tons of dung are reputed to be burned each year.[3] Each ton of dung that is burned instead of used as fertilizer on croplands means a loss of at least 50 kilograms of potential grain grown, that is, a total of 20 million tons of grain per year (some experts calculate the total at twice as much). To put this calculation in perspective, 20 million tons of grain can feed 100 million people for one year. To purchase the grain on world markets could cost as much as $3 billion—a figure to bear in mind when we consider the costs of establishing fuelwood plantations (see below).

SCOPE FOR ACTION

Fortunately there are several practical ways to attack the problem.

Better Stoves

Curiously enough, a lot could be achieved through better cooking stoves.[4] If the limited heat generated by fuelwood were to be put to more efficient use, people could cook more food with each stick of wood. A typical stove is dreadfully inefficient (and an open fireplace, used by people who have not yet made the advance to a cooking stove of any sort, is even worse). Most stoves deliver no more than 20 percent of the fuelwood's energy content to the food being cooked. Many achieve only 10 percent, some even less. In Java, with its 100 million people, a broadscale survey reveals that an average stove captures only 6 percent of the heat value of fuelwood. By way of comparison, a typical American household gas stove is almost 70 percent efficient. A stove with simple improvements in design, costing perhaps an extra ten dollars, cuts heat losses by two-fifths. In addition, a cooking pot of a style that can be partially sunk into the stove can save almost a further one-fifth. These measures alone would save millions of tons of fuelwood each year.

Plantations

In many countries, a start is being made to establish farm woodlots and local-community tree blocs.[5] If a typical plantation can supply ten tons of fuelwood per hectare per year, the aggregate area required to meet all Third World needs will amount to 600,000 square kilometers—an area almost the size of of Texas, or the whole of Central America. To establish a cut-price plantation, using village labor and land available free, costs at least $100 per hectare. So the total amount of investment required is $6 billion. The present pace of plantation establishment is not nearly enough to keep up with expanding demand. In the Third World as a whole, the planting rate needs to be stepped up five times immediately, and in the worst-hit areas, such as those of West Africa, between fifteen and fifty times. As for the needs of people living in the humid tropics, they amount to rather more than one-seventh of the total. So the funding required to relieve excessive exploitation on the part of fuelwood gatherers in tropical moist forests is just over $900 million.

Raising money, however, is only a partial solution to the problem. Local people in many communities realized the need for plantations years ago, yet they have not gotten on with the job. The reason lies with social and cultural factors, as well as economic crimps. A fuelwood plantation is almost always a community resource. A minimum-sized area will be too large for a single family, or even a group of families. Hence the need for "social forests," with the local community playing a big part in the planning of the plantation right from the start. In other words, a grass-roots activity—in contrast to the spirit of many development projects, which are "top down" exercises, with decisions imposed by remote bureaucrats who may have little sense of how people at the ground level view their situation. Fortunately, many international aid agencies now realize the need for a new-spirited strategy; of approximately 650 forestry projects in the pipeline, 400 are concerned with social forestry in one way or another (and of these, 125 are projects of the U.S. Agency for International Development).

As an example of grass-roots initiative, let us note the track record of the Kenya Energy Non-Governmental Organizations (KENGO), a coalition of citizen activists who have decided to take the business of tree planting into their own hands. The participants are a diverse collection. They include church groups, university departments, polytechnics, scouts and guides, and women's clubs, among many others. Whereas citizen activists are a familiar phenomenon in countries such as the United States (e.g., the Sierra Club, Friends of the Earth, National Audubon Society), they are rare in developing countries. A Third World citizen tends to accept that it is the government that plans activities, that takes initiatives, that generally takes care of the community's welfare. Unfortunately the Kenya government has been doing less than a complete job on the energy front. The nation spends more than one-third of its hard-currency earnings on imports of petroleum, and it has been doing all too little to promote home-grown energy. Worse still, Kenya's forests have been declining from about 5 percent of national territory in 1960 to only 2.5 percent today. If present land-use trends persist, Kenya could find itself by the year 2000 with virtually no forests left at all.

Representing large numbers of citizens, KENGO has climbed to its feet with a confidence that surprises many observers. During my travels around eighty countries of the Third World during the past two decades, I have seen nothing like it emerge in such a short space of time. KENGO has planted more fuelwood trees in its first two years of existence than the government has accomplished in the previous five years. Even though many millions of Kenyans chop down more and more trees each day in order to have a digestible supper, Kenya in 1985 may have as many trees as it did in 1980.

Equally good news is that Kenyans are experimenting with new tree species for fuelwood plantations. Hitherto, the usual approach has been to use species that have worked well in developed nations of the temperate zones, and that will supposedly do well in developing nations of the tropics. Fortunately many more potential species are available and deserve to be brought into action with all speed. A survey conducted by the U.S. National Academy of Sciences reveals that over 1,200 tree species can qualify, 700 of them characterized as front-rank choices—and of these 700, at least 250 are trees of the tropical forest zone.[6]

For example, there is the calliandra, a small bush of Central America that has spread around the tropics, and has become established as the favorite fuelwood crop with the 100 million energy-starved people of Java. An average crop reaches 4 meters tall or so after just one year, whereupon the trees can be cut at 50 cm. above the ground, producing a harvest of between 10 and 20 cu. m. per hectare. The stump coppices readily, allowing another harvest at the end of the second year, this time as much as 35–65 cu. m. The sudden spurt in productivity is due to the coppicing trait, whereby the cut-back parent responds by sending out a host of shoots. In one locality of Java, an experimental plantation of half a hectare was established in 1963, and within twelve years it had supplied offspring for 250 hectares of plantations in the vicinity.

In addition to its services as a fuelwood supplier, calliandra, by virtue of an extensive and deep root system, binds the soil and thus helps prevent erosion. At the same time, being a legume, calliandra rejuvenates soils from which nutrients have been exhausted. The dense foliage supplies excellent fodder for livestock, with 22 percent crude protein. When it is grown in conjunction with competitive species such as elephant grass, which also makes useful fodder, it can revegetate areas that have remained barren for years on end.

Densified Wood Biomass

Ironically, many of the countries where millions of citizens live with the fuelwood crisis as a daily burden are precisely those countries where, in remote sectors, other citizens burn vast amounts of wood each year. These people are forest farmers who clear patches of forest in order to plant their food crops. The amount of wood they burn each year amounts to at least two billion tons, possible much more. In addition, several sectors of Latin America feature cattle raisers who annually torch at least 20,000 square kilometers of forest, representing between a half and one billion tons of potential fuelwood wasted. Clearly, if the fuelwood seekers could make common cause with even some of the fuelwood wasters, they could achieve a sizable step forward. But the two groups often live too far apart. In several areas of the Third World, however, there could be scope for them to coordinate their activities.

One approach is to turn the felled forest trees into charcoal. Using an age-old technology, timber can be converted into a form of fuel that is much lighter than wood. A good 30 million tons of charcoal are made around the world each year, since there is growing demand for this high-quality fuel. Charcoal supplies twice as much heat per unit weight as wood, even after 50–70 percent of the heat value has been lost during conversion: 8 kg. of wood are needed to produce 1 kg. of charcoal. Because charcoal is so much lighter than wood, with twice as much energy on a weight-for-weight basis, it can be economically transported for hundreds of kilometers, in some cases, such as Thailand, as much as 1,000 km. Were the technology for charcoal making to be improved, as would not be difficult to accomplish, there would soon be economic incentive for the forest farmer and the cattle raiser to look upon his cleared patch of forest with an eye toward the charcoal kiln. This is all the more pertinent in light of rapid urbanization in Third World countries, at a rate between two and three times that of population growth overall: it is city dwellers who generate by far the biggest demand for charcoal.

Even more promising could be an emergent technology known as densification of wood biomass.[7] This consists of compressing dried-out wood into small-volume blocs, known as briquettes or pellets. The process not only makes the material more readily transportable, it improves combustion by densifying and agglomerating the wood volume, thus restricting access of air and slowing down the burning in a cooking stove. The material needs to adhere with a binder such as resin or tar, or certain plants such as algae, even clay or mud—or, in the absence of anything better, sewage or manure. When treated in this fashion, the bulk density of wood can be increased by between three and six times. According to the Indian Institute of Technology, in New Delhi—a pioneer in this field—the fuel is not only equivalent to coal in energy content, but it produces fewer pollutants, reducing the "charcoal smog" that fouls the air over New Delhi and Calcutta, also Jakarta, Bangkok, and many another Third World metropolis.

This technology could probably not be applied directly by the forest farmer, since he would not want to undertake the work himself. So we can foresee a network of commercial practitioners, each with his wood-densifying machinery, serving the needs of a community of farmers who wish to have their felled timber hauled away in marketable form. The backwoods expert could use a rudimentary lever press, such as a village smith can manufacture in a single day, exerting pressure of 5–15 kilograms per square centimeter, or up to 100 kilograms per average-sized briquette. Or he could use a screw-type press (rather like those used outside the tropics to extract wine from grapes), able to generate a pressure of about 40 kg. per sq. cm., or up to two tons per briquette. More sophisticated devices are available, depending on the scale of the operation. Were the commercial entrepreneur to go into business in a big way, he could use advanced machinery that handles huge volumes of wood through power-driven equipment. A large-scale facility incorporates a wood chipper of a scale to accommodate entire tree trunks, and a press that generates fifty times as much pressure as do the simpler devices.

Present technology for densifying wood is rudimentary. With intensive research, however, new devices can surely be developed to do the job in a much more efficient manner. The funds to advance the technology need not be great, in fact a trifle of the vast budgets required for fuelwood plantations.

As for potential demand, there is little doubt that the new technology would meet the needs of millions. In Indonesia, for example, the island of Java lies only 800 km. from those southern parts of Sumatra and Borneo where small-scale settlers burn huge amounts of forest timber each year. In northern Philippines, large throngs of people, in both rural and urban areas, pay soaring prices for paraffin, while in southern parts of the country the forests are often concealed beneath palls of smoke. In northern and central Nigeria, 50 million people lack enough fuelwood, while 15 million people in the southern forests have too much. Along the coast of West Africa, forest burning is so widespread that the fires can be discerned by satellite, while a few hundred kilometers inland live throngs of peasants whose fuelwood stocks are all but exhausted. In northeastern Brazil, 40 million people are acutely short of energy, while a few hundred kilometers to the west, cattle ranchers of Amazonia each account for more wasted wood than any other fire-setters in tropical forests.

7

The Cattle Raiser

MANY A READER OF this book will have munched his way through a hamburger in a fast-food restaurant. This convenience-style meal is popular partly because it constitutes good, solid meat, and partly because it is inexpensive, only a little over half the cost of table-cut beef. Being rational consumers, most citizens seek best-quality hamburger at the least cost, and may thereby be stimulating—albeit unwittingly, but effectively and increasingly—an international trade in beef that contributes to the decline of forests in Central America, if not Amazonia.

The number one factor in elimination of Latin America's tropical forests is cattle raising. Throughout Central America, also in the Brazilian and Colombian sectors of Amazonia, it plays the dominant role (here we take the term Central America to refer to the six countries of the political region, plus Panama—and many of the comments apply equally to the small sector of forest remaining in southern Mexico). In order to establish his pasture, the cattleman clears the forest completely, and he hopes permanently, by felling and burning it. True, he often converts a patch of forest that has already been subjected to small-holder cultivation or spontaneous settlement by peasant farmers who serve as a pioneer wave (see the following chapter). But it is the rancher, with his large-scale enterprise and operations, who is the prime agent in final elimination of the forest.[1]

Cattle raisers burn forest tracts to produce grazing land. Note charred tree stumps in the foreground. *(National Geographic Society)*

Roughly speaking, we can reckon that the cattle raiser is accounting for at least 20,000 square kilometers of forest in Latin America each year. This amount is far less than is accounted for by the commercial logger and the forest farmer in Africa and Asia. But the stockman's impact is expanding rapidly, primarily—in Central America at least—in response to demand from the international beef trade. The majority of additional beef produced does not find its way into the stomachs of local Latin Americans. Most of it is exported to the United States, with a little, though an expanding amount, going to Western Europe. To this degree, the cattle-raising phenomenon reveals another ecologic-economic linkage among the community of nations. It points up the way in which consumerist lifestyles of the affluent temperate zones contribute to the destruction of forests in the tropics.

THE FAST-FOOD TRADE IN THE UNITED STATES

Why, one might ask, should there be such a big demand for Central American beef in the United States? Well, beef produced within the United States has become increasingly expensive, partly because of declining supplies of home-grown beef, partly because of Americans' almost insatiable appetite for beef. The United States, with one-twentieth of the world's population, is not only the world's biggest producer and consumer of beef. It is the biggest importer too, accounting for about one-third of all beef in international trade. In 1960 an average American consumed well under 40 kilograms of beef each year, a total that rose by 1976 to well over 60 kilograms. Americans now eat rather less beef per person, but because of expanding population, the overall total continues to edge upward.

At the same time, few items in the shopping basket have increased more in price during the past few years than beef, with upswings far steeper than in the overall cost of living. Between 1975 and 1979, for instance, the price of a Montana steer at the packing plant rose from $0.63 per kilogram to $1.55, and during the same period retail prices rose by 30 percent per year, until by 1980 they topped $5.00 per kilogram. The cause of this inflationary climb in beef prices lies with changes in strategy on the part of American ranchers. In 1975 the United States contained 132 million head of cattle, but, because of poor prices at that time, ranchers allowed the total to decline to only 111 million head by late 1978, on the ground that it was becoming unprofitable for them to maintain their huge herds. Now that high beef prices encourage cattlemen to rebuild their herds, we shall have to wait several more years (due to the lengthy production cycle) before the national herd recovers to its mid-1970s size, and we shall have to wait for a further few years still before prices for U.S. beef reflect these changes. In short, it may be a good while before an American shopper can confront the hamburger counter in the supermarket without flinching.

Not surprisingly, then, Americans have been looking elsewhere for beef at non-inflationary prices. Central America has proved to be a ready supplier. In recent years, the average price of beef imported from Central America has been only half as much as the price of similar beef produced in the United States. Cattle in Latin America are raised on grass rather than grain, so the beef is very lean. This makes it suitable for just a single sector of the U.S. beef market, the fast-food trade. According to the Meat Importers Council of America, virtually all Central American beef makes its way into hamburgers, frankfurters, chili stew, frozen dinners, baby foods, luncheon meat, salami, and other processed meats for humans (also products for the pet trade). Of these items, hamburgers are easily the leading category, accounting for way over one-third of the fast-food industry.

At the same time, convenience foods, which account for one-quarter of all beef consumed, constitute the fastest-growing part of the entire food industry in the United States. During much of the period since 1970, the fast-food trade has grown at a rate of 20 percent each year, or two and a half times faster than the restaurant industry overall; the total now tops $20 billion per year. One-fifth of America's food budgets is spent on dining out, a proportion that could rise to one-half by 1990. Every other meal eaten in the United States is bought in a fast-food facility or at an office, factory, hospital, or school. Part of this phenomenon reflects the increasing number of married women who work and for whom time is more valuable than money. It also reflects the increasing number of older people in the population, for whom a meal eaten out represents a ready way to escape loneliness.

So the demand for cheap, lean beef is unlikely to slacken. Indeed we can anticipate that if grain prices continue to rise, U.S. cattle raising will shift toward more pastureland feeding, a change that may eventually lead to a decline of at least 50 percent in the amount of animal protein grown in the United States—a factor that in turn will stimulate further imports of foreign beef.[2]

Faced with one inflationary price surge after another during the past several years, the U.S. government has repeatedly stepped up beef imports. In 1960 the United States purchased virtually no beef abroad. Now the nation imports about 10 percent of its total consumption. Of these imports, 17 percent come from tropical Latin America, three-quarters of them, just over 100,000 tons, from Central America (most of the rest from Australia and New Zealand). Although the extra purchases from Central America contribute only about 1 percent to Americans' consumption of beef, the U.S. government estimates that they supply enough additional cheap meat to trim a nickel off the price of a hamburger. Food is the sector of the U.S. economy most susceptible to inflation, and the greatest price increases in recent years have been for meat, especially beef. The American government believes that increased imports of cheap beef from Central America have done more to stem food inflation than virtually any other initiative. Indeed the government claims that the additional imports save consumers at least $500 million each year.

Some of the major fast-food chains used imported beef, but others deny that they use anything other than domestic meat. Complicating matters, however, is the fact that meat imported into the United States is categorized, after it leaves its point of entry, as domestic beef. For wholesale and marketing networks, it becomes just beef. So when a chain asserts that it obtains its meat within the United States, it may be speaking the truth as it sees it, without comprehending that the beef may originate elsewhere.

CATTLE RAISING IN CENTRAL AMERICA

The seven countries of the region, although covering only a little over half a million square kilometers (smaller than Montana and Idaho), have contained extensive forests in the past, many of them rainforests, and some of them unusually diverse with rich stocks of species. But since about 1950, the area of man-established pastureland and the number of beef cattle in the region have more than doubled—an expansion that has occurred almost entirely at the expense of primary forests, of which two-thirds have now been cleared (Table 1). Still more to the point, beef exports have increased almost five times over.

TABLE 1. **Central America Beef Production and Exports**

	Country*			
	COSTA RICA	GUATEMALA	HONDURAS	NICARAGUA
Area (sq. km.)	50,700	108,890	112,090	130,000
No. of Cattle				
1960	937,000	1,050,000	1,285,000	1,176,000
1980	2,263,000	2,525,000	1,825,000	2,926,000
Beef Produced (tons)				
1960	21,350	33,800	17,150	26,800
1980	79,250	91,600	52,350	78,900
Net Beef Exports (tons)				
1960	6,900	1,420	3,100	8,150
1980	42,100	24,950	19,950	23,000[1]
Pasture (sq. km.)				
1960	9,540	9,850	19,920	16,900
1980	18,650	13,900	26,250	19,500
Forests & Woodlands (sq. km.)				
1960	29,650	86,000	71,000	65,300
1980	21,250	49,000	58,000	58,600

*No data included for Belize, El Salvador, and Panama, since their beef industries are too small to figure significantly in the overall picture for Central America.
[1]In 1978, the last year before the revolution, 46,200 tons were exported.

Sources: Information supplied by Dairy, Livestock and Poultry Division, Foreign Agricultural Service, and by the Economics and Statistics Division, U.S. Department of Agriculture, Washington D.C.

As an extreme example of the hamburger connection, let us consider Costa Rica. In 1950 cattle-raising areas accounted for only about one-eighth of the country, whereas they now account for over one-third. In 1960 the country's cattle herds totaled around 900,000 head, but by 1980 they had reached 2.2 million. Between 1960 and 1980, beef production almost quadrupled, yet during that period local consumption of beef actually declined by more than 40 percent, to a mere 15.7 kilograms per year (or way under one-third of what an American consumes—and much less than an American cat eats). Almost all the extra output has been exported, a total of around 42,000 tons per year, or well over half of total output (by contrast with an average of one-third for Central America as a whole).

We find a similar situation dominated the scene in Nicaragua until the 1979 revolution. Between 1960 and 1979, beef production in Nicaragua increased almost three times, and beef exports increased five and a half times. Of the country's farmlands, the Somoza family alone owned about one-quarter prior to the revolution—whereupon 10,000 square kilometers of their holdings were redistributed to landless peasants, with the welcome effect of relieving pressure on the country's Atlantic coast rainforest, the largest bloc of this rich type of forest in Latin America north of Amazonia. The Somoza family owned six beef-importing companies in Miami, which purchased beef worth $30 million per year.[4] Regrettably, the new Nicaraguan government is persisting with a plan to encourage agricultural settlements in extensive parts of the eastern region, including a sector of the uniquely diverse Mosquitia Forest.

The land tenure problem in Nicaragua is replicated in other parts of Central America, a problem that is grossly aggravated by cattle ranchers.[5] In Guatemala, for instance, 2.2 percent of the population owns 70 percent of the agricultural land, most in the form of cattle ranches. Of approximately 400,000 farms in Central America, 90 percent are smaller than 50 hectares; the remaining 10 percent account for three-quarters of all pasturelands and about two-thirds of all cattle. Well over half of Central America's rural families, some 35 million people altogether, either own no land or own too little to support themselves.

To relieve the problem, governments promote the migration of these impoverished peasants into lowland forest zones, with the rationale of using this settlement pattern to gain temporary release from political pressure for land reform. As they become established in forest areas, these migrants tend to become professional deforesters, pursuing a squatter lifestyle and cutting and burning the vegetation, before planting coarse grains and then selling the "improved" land for about $80 per hectare to leisure-class ranchers from the big cities. By the year 2000 the populations of the eight countries concerned (Mexico, southern sector only) are projected to reach almost 171 million people, almost double the 1980 total of 89 million. Despite rapid urbanization, at least three-quarters of the additional people are expected to live in the environs of forests, if not the forests themselves.

The inequitable circumstances of the land tenure systems are made worse by a cultural factor in virtually all these countries, where national oligarchies view cattle raising as a prestige activity. In Costa Rica, for example, the beef industry is dominated by only about 2,000 ranchers, holding an average of 750 hectares each and controlling over half of all agricultural land. To own land and cattle fosters both social standing and political power. Ranchers include many professionals, notably government officials, who retire to their country estates on weekends to ride horseback and enjoy an image as gentlemen stock raisers. By contrast with the small-scale peasant, who of necessity must make very intensive use of his smallholding, the prestige rancher is often content to use his land in an extensive fashion, which usually means an inefficient and wasteful fashion. Many of these large-scale landholders feel little incentive to intensify their beef-raising methods, especially when they believe there is still plenty of untouched forest left "out there." Thus those who control the largest amounts of agricultural lands are those with the least motivation to use it efficiently; and because of their privileged position in the political structure, they are unlikely to meet serious opposition.

All in all, cattle ranching boomed in Central America during the 1960s and 1970s. Beef exports soared from 20,000 tons per year to almost 150,000 tons, most making their way to the United States. Beef exports now represent the most dynamic sector of

Central America's trade with the world outside. The consequence is that while the region possessed at least 400,000 sq. km. of forest in 1960 (and 500,000 square km. prior to white settlement 400 years ago), the forests have declined to less than 200,000 sq. km. At present rates of conversion to grassland, virtually all remaining forests will be eliminated by 1990.

NEW BEEF IMPORTERS

How about other nations of the rich world? Do they not also import beef? Well, not nearly so much—at least, not yet. Most other citizens of the rich world do not share Americans' appetite for beef. Whereas an average American eats about 50 kilograms of beef each year, an average Western European is content with less than 25 kilograms, while a Japanese confines himself to a mere 5 kilograms. In all these countries, beef consumption is actually declining slightly, except for Japan, where the age-old diet of rice, fish, and tea is rapidly giving way to animal meats, wine, and beer. Whereas the consumption of meat of all kinds increased during the period 1961–1980 by only 42 percent in the United States, and by 45 percent in Western Europe, it soared in Japan by around 600 percent. Much of this Japanese meat consumption is made up of pork and mutton and other "inferior" meats (plus a good deal of marine mammal meat). The affluent fashionable Japanese now feels that a sound way for him to join the late twentieth century is by matching other fat-cat citizens' consumption of beef. Although Japan now produces about 450,000 tons of its own beef each year, this is not nearly enough to meet the Japanese desire for the good life at the steak table. Hence the skyhigh prices in Tokyo as far back as the mid-1970s: $33 a kilogram (with prime Kobe beef $66). Japan expects that by 1985 it will need a total of 625,000 tons of beef per year, of which it will produce only little more than 500,000 tons itself. So it will seek to import 125,000 tons, or at least as much as all U.S. imports from Central America. Were the Japanese ever to expand their consumption to match Western European levels, this would practically double global demand for beef in international trade.

In addition, the countries of Southern Europe are growing increasingly hungry for beef, as are the nations of the Soviet bloc, plus the oil-producing nations of the Middle East. All in all, we can anticipate that the recent demand for ever-more beef will persist for a good while to come. The Food and Agriculture Organization projects that until 1990 at least, overall demand for beef will rise more rapidly than for any other food category except fish. At the same time, we can note that whereas world beef production per global citizen expanded from 9.32 kilograms in 1960 to a peak of 11.58 kilograms in 1976, it has thereafter declined to about 10 kilograms today. In 1980 world cattle numbers declined for the fifth straight year, to about 950 million head. All major producers of beef, notably the United States with around 115 million head, the Soviet Union with 110 million, New Zealand 40 million and Australia 35 million (compare Central America, less than 12 million) have allowed their herds to decline. Ever-soaring prices for beef are the result.

NEW BEEF EXPORTERS

To meet this expanding appetite for beef, with its lucrative profits, several other nations of Latin America, especially Brazil and Colombia, are looking for ways to set fire to some of those "useless" forests and to step up the spread of pasturelands. Brazil already exports a sizable amount of beef, though its principal market lies with Western Europe rather than North America, four-fifths of its exports heading for Britain, West Germany, Italy, and others. In the long run, Brazil entertains ambitions of becoming one of the main beef exporters in the world, hoping to double its 1970 output of meat by 1990 at the latest. The nation has taken careful note of the upheavals in the global beef trade caused by the Falklands crisis in early 1982, which severed the sizable trade between Argentina and Britain. Brazil has leaped into the breach, and aims to dispatch an annual total of around 12,000 tons.

In order to foster its cattle-ranching sector, the Brazilian government offers many incentives to the prospective rancher in Amazonia. He can import certain forms of equipment into the country duty free. Until recently, he received a 50 percent rebate on his income tax on activities elsewhere in Brazil, though this proportion is now being reduced, perhaps to as low as 25 percent. A foreign investor can repatriate his profits and his capital. Moreover, clearing away the forest need not be costly; if the rancher employs hand labor to fell and burn the forest, expenses run to only $125 to $250 per hectare; and if he employs casual family labor, he need offer no more remuneration than opportunity for the family to take off one food crop from the forestland concession before leaving the area planted with grass for the new rancher's cattle. Thus there have been many entrepreneurs ready to try their hand at cattle raising in Brazil's sector of Amazonia, even though experience to date shows that it is a questionable form of land use.[6]

Between 1966 and 1983, well over 100,000 square kilometers of forestlands in Brazil's sector of Amazonia were converted to cattle pasturelands, in the form of almost 350 major ranching projects and a further 20,000 smaller projects, supporting a cattle population of around 10 million head. A good-sized ranch covers more than 2000 square kilometers, supporting 15,000 head of cattle or more. Virtually none of the timber felled has been disposed of as commercial wood, even though the marketable lumber could fetch at least $35 per cubic meter; the subsidized rancher does not find it worth his while to do other than douse herbicide on the area to be cleared, then set it on fire. Huge areas of usable timber thus go up in smoke, representing a total loss to date of $1 billion, more than twice the total investment realized by all livestock projects during its heyday phase of 1966–1978.[7]

An increasing number of ranching enterprises are foreign owned. While they make up less than 1 percent of total ranchlands, they are significant as a measure of Brazil's readiness to harness foreign capital to promote its development of Amazonia. A U.S. consortium of Brescan-Swift-Armour-King Ranch holds 720 square kilometers in eastern Amazonia, with an investment of some $10 million, while other U.S. corporations with at least part shares in ranching enterprises include Twin Agricultural and Industrial Developers, Caterpillar International, Beltec International, Dow Chemical, International Foods, Massey Ferguson, W.R. Grace, United Brands, Hublein and Sifco Industries, Anderson Clayton, Gulf and Western, and Goodyear. Further foreign enterprises, from countries other than the United States, include Mitsui, Tsuzuki, Marubeni, and Spinning-Nichimen, from Japan; Liquigas, from Italy; Volkswagen, from West Germany; DeBuis Roessingh, from Switzerland; and George Markhof, from Austria, among many other well-known companies.

Investment on the part of the twelve largest enterprises totaled $21 million by the end of 1977, except for Volkswagen with $35 million. Volkswagen holds a concession of 1,400 square kilometers in southeastern Para, of which half is eventually to be converted into pastureland, enough for a herd of 120,000 cattle. Large as this Volkswagen enterprise might sound, it is far from the biggest ranch.

As mentioned, experience to date indicates that cattle raising in tropical forestlands proves a doubtful venture. Stocking rates are generally low, only one head of cattle to one hectare immediately after the new pasture is established, and only one animal to every five hectares within just five years, due to decline in soil fertility and nutritional grasses. Steers require an average of four years to reach a weight of 450 kilograms for slaughter. This makes for a productivity rate only one-quarter or less of what it is in ranchlands in the United States. Formerly forested soils quickly become exhausted of nutrients, and pastures feature poorer and poorer grass unless they receive ever-greater amounts of fertilizer. Furthermore, in the warm humid climate (some localities receive over 3500 mm. of annual rainfall), there is a problem of toxic weeds that invade as the soil becomes compacted.

Some ranches quickly lose one-fifth of their cattle holdings. But to the wealthy stockman who obtains forestland for next to nothing (and may even receive tax incentives to develop it from its "unproductive" natural state), it matters little if he needs five or ten times as much space to raise his beef as would be the case if he were to manage his ranch properly, with improved breeds of livestock, prophylactic drugs to counter epizootics, and the like. True, many ranches have already been abandoned, and several hundred look likely to become unprofitable after a mere half dozen years. But when a ranch becomes worn out, the owner readily moves on to a fresh patch of undisturbed forest. Wasteful as may be this "shifting ranching," Brazil seems to have had little hesitation about fostering this kind of development—on the grounds that there is plenty of untouched forest to support the ranching industry for many years to come.

Ironically, Brazil could almost certainly derive much more animal protein through exploitation of its Amazon River system, which, with over 2,000 known species of fish, could supply the nation's entire requirements for animal protein and produce a surplus for export. In addition, the water buffalo, being native to the humid tropics, is much better adapted to Amazonia than are cattle. The beast is very resistant to parasites and diseases, and needs very little care. It produces as much meat as does a beef bullock, and a female yields between 5 and 20 liters of nutritious milk per day. Since the buffalo thrives in flooded lands, it could be raised in those selected sectors of the varzea alluvial floodplains that, proving fertile enough for at least two bumper crops of irrigated rice per year, could reasonably be converted from forest to farmlands.

Still further sources of protein could be found in certain forest animals such as the paca and the capybara. A large male paca weighs about 10 kilograms and yields at least 6 kilograms of first-rate meat. The capybara, looking like an outsize guinea pig, is the largest rodent known, and is already farmed on an increasing scale in Venezuela as an alternative to rare and expensive beef. A capybara weighs up to 50 kilograms and thrives off coarse swamp grass and weedy plants such as water hyacinth. Since it digests food three and a half times more efficiently than cattle, and out-performs cattle in reproduction to the extent of six times more offspring each year, and provides much-sought leather (ideal for gloves, because it stretches in only one direction), the capybara could prove an ideal domesticate for Amazonia.

SOME FURTHER OPTIONS FOR OUTSIDERS

In the case of Brazil, with its super-sensitive resistance to proposals, however constructive, from foreigners, there is little that distant observers can do. But Central America offers much more opportunity for leverage, on the part both of the United States and of international aid agencies.

First, the United States could opt to import no more beef from Central America. This would mean that the United States would have to reduce its beef consumption by only a very marginal amount (or import more from non-tropical Latin America, or from other suppliers such as Australia). This would make more beef available in Central America for local consumption, and it would encourage the region to follow an agricultural strategy that is inevitable in the long run, one involving greater diversification into other agricultural products. It would be no great hardship for the United States to simply eliminate 1 percent of its beef intake —a trifling gesture compared with the 5 percent reduction that the nation achieved in its gasoline consumption in the single year of 1980.

A second, and more productive, strategy lies with the United States using its influence to persuade Central American ranchers to become much more efficient in their operations. With only a moderate amount of effort in the form of better management, most established ranchers could double their beef output in fairly short order. This means there could be increasing amounts of beef available for both the foreign market and domestic needs, *without* felling a single additional forest tree. The United States' purchasing power is plainly large enough to be used as leverage: much as foreign tuna fishermen are now required to use nets that are supposedly not harmful to dolphins, Central American beef exporters could be obliged to demonstrate that they raise their cattle on already established pasturelands, not on newly cleared forestlands.

Third, both bilateral and multilateral aid agencies could do more to support conservation activities in the countries concerned. Costa Rica, for example, while presenting a dismal record in forestland stockraising, has nevertheless achieved several first-rate measures in conservation. During the mid-1970s, President Daniel Oduber formulated a systematic strategy for his country's forests; he set up a Natural Resources Institute to conduct ecologic evaluation studies and to produce a program for integrated land-use planning of the entire country. He also expanded the national parks network until it now contains a greater proportion of land than is the case for any other Latin American country. Under President Rodrigo Carazo, the parks network is being expanded still further, with the aim that it will eventually cover about 10 percent of national territory. At the same time, the country is implementing a natural resource conservation program, with $9 million support from the U.S. Agency for International Development (AID).

Fourth, along the lines of AID support for Costa Rica, and a further AID gesture in the form of a $10 million contribution toward a $17 million project for watershed reforestation in the Panama Canal zone, international aid agencies can do much to foster sustainable use of forests in Central America. To date, the tendency has been for development agencies, notably the World Bank and the Inter-American Development Bank, to promote cattle raising in forestlands (almost $4 billion in loans and grants during the 1970s), a trend that should generally be reversed.

In conclusion, let us note an irony inherent in the hamburger connection. If cattle raising in Central America causes the region's forests to disappear within the foreseeable future, not only local people will suffer by way of environmental degradation, decline of watershed services, and the like. (Already Costa Rica's hydropower dams are being silted up, and Honduras has undergone undue hurricane damage through loss of forest cover.) Other human communities will suffer, notably North Americans. I have already mentioned the exceptional variety of plants and animals in these forests. According to a botanist from South Carolina, Dr. Monie S. Hudson, who specializes in medicinal applications of phytochemicals, a screening program to evaluate 1,500 tree species in Costa Rica's forests has revealed that around 15 percent might have potential for treatment of cancer.

It is clear, then, that both Central Americans and North Americans are contributing to what we might call the "hamburgerization" of the forests, and that both will suffer from the loss of them. It is equally clear that both must cooperate if the problem is to be solved. Either all will lose together, or all could gain together: a paradigm of interdependent resource relationships within the international community.

8

The Forest Farmer

By FAR the number one factor in disruption and destruction of tropical forests is the small-scale farmer. In strong contrast to the next two most significant contributors to forest depletion, that is the timber logger and the cattle raiser, the farmer sees no alternative to what he is doing. He simply does not believe there is any other means available for him to gain a livelihood. Being much more numerous, moreover, than the timber logger, the cattle raiser, and the fuelwood cutter combined, he affects much larger areas of forest.

Amounting to roughly one in twenty of all people on Earth, these farmers represent the most pervasive form of environmental degradation overtaking the tropics. Despite all the problems associated with the forest farmer, however, there is hope that he can be helped—if we mobilize enough economic muscle and political will—to adopt a lifestyle that will supply him with food while being less harmful to the forest.

During my travels in all tropical forest regions, I have come across many types of these farmers. There is the shifting cultivator of traditional style. There is the itinerant smallholder of more recent kind. There are sundry types of landless migrants and other squatters. All these forest farmers practice a common technique in that they fell virtually all the trees in a patch of forest and then burn the lot. As an across-the-board term, then, we can call these farmers slash-and-burn cultivators. This label is frequently applied to forest farmers throughout the biome, and it is a more accurate designation than shifting cultivators—a term that should be limited to farmers with a centuries-old lifestyle that has persisted well enough as long as they have not become too numerous and there has been enough forest to support them.

TRADITIONAL PATTERN: SHIFTING CULTIVATION

Indeed, shifting cultivation has been the established form of forestland agriculture until just the last few decades.[1] Even though it is now a minority type of forestland farming, shifting cultivation has predominated for centuries; it can be regarded as one of the major agricultural systems developed by humankind. The cultivators practice a form of rotational agriculture: after clearing a patch of forest, they raise crops such as corn or cassava for a few years until the soil loses its fertility and weeds encroach, then they move on to repeat the cycle in another sector of forest, eventually returning, decades later, to the original site. As long as there have been only a few cultivators per square kilometer, generally ten or less, and provided the patch of farmed forestland could be left fallow for at least ten years and more often twenty-five years while it renewed itself, the system has worked.

Shifting cultivation has proved highly adaptable to a broad spectrum of conditions. When applied with an understanding of the forest's capabilities and limitations—an understanding that seems to have characterized shifting-cultivator communities of many different cultures—the system has turned out to be an

Primary Forest, in the Tai region of the Ivory Coast, being cleared for use by small-scale forest farmers (also known as "shifting cultivators"). *(Professor U. Rahm, World Wildlife Fund)*

acceptable way to use forestlands where poor soils and heavy rainfall make conventional farming difficult if not impossible.

Since about 1950, however, when the population explosion began to affect tropical lands, the numbers of shifting cultivators have increased to a point where there are often three times as many people as formerly. Result, they have less space for local migration. Further result, they start to make intensive as well as extensive demands on forest environments, leaving local ecosystems with insufficient time to recover. Soils rapidly become exhausted, and the area degenerates to scrub.

RECENT PATTERN: LANDLESS MIGRANTS

Still more important, these traditional farmers have recently been joined by large communities of subsistence peasants, who, due to lack of land elsewhere, move into forests where they adopt a variation of slash-and-burn agriculture that leaves even less scope for the forest to regenerate. These recent arrivals, possessing little cultural adaptation to forest environments, tend to advance upon the forest in waves. They operate on pioneer fronts, pushing ever-deeper into forest territories, leaving behind them a mosaic of degraded croplands and bush growth where there is next to no prospect of natural forest reestablishing itself, not even as impoverished secondary forest. Not to be mistaken for shifting cultivators, these are *shifted* cultivators.

Thailand serves as an illustration of what happens. During the course of my repeated visits to the country, I have found that most of the permanent deforestation is due, in the first instance, to the work of itinerant squatters. These are people who, finding themselves landless in other parts of the country, feel they have no alternative but to head for the only "free lands" left, the forests. Since the early 1960s, Thailand's established farmlands have expanded by at least one half, due mostly to an increase in rice growing.[2] Because much of this stepped-up output has been accomplished through highly commercialized farmers who cater for the rice export market, and smaller-scale cash croppers who cater for the cassava export market, a host of subsistence peasants have been squeezed out of their homelands." Mainly because of the impact of these "marginalized" peasants, Thailand's remaining forests have declined in extent from 53 percent of national territory in 1960 to only 23 percent at most in 1980.

The last decade has seen an outburst of cassava growing, a phenomenon that illustrates the economic-ecologic linkages that operate at international level, with adverse impact on tropical forests. Sometimes known as manioc, cassava has come into increasing demand as a feedstuff for livestock in Europe. The three main countries concerned, West Germany, Belgium and the Netherlands, absorb as much as one-tenth of the tropical crop, much of the exported stock being grown in Thailand. Being very rich in calories, and available at ostensibly cheap prices from Thailand, cassava makes an excellent candidate for Europe's live-stock feed supplements; in 1973, the European Common Market imported 1.5 million tons of cassava from Thailand, an amount that rapidly soared by 1978 to almost 6 million. Cassava growers in Thailand include a fast increasing number of smallscale farm-ers, who establish their crops at the expense of natural forest.

Until fairly recently, the majority of Thailand's forest farmers have been shifting cultivators of the traditional sort, causing little permanent damage to the forests. Now the majority are lowland-ers, who are not ecologically attuned to forest environments and have little awareness of how to make sustainable use of forest ecosystems. The lowlanders hack away at the forest fringe to plant patches of cassava and corn for a year or two, then they push further into the forest—whereupon their cleared patches are taken over by a wave of follow-on agriculturalists who establish themselves on the deforested territories with various forms of larger-scale commercial crops such as sugarcane. This means that the forest receives no chance to regenerate, whether in two or twenty years time. The original ecosystem is eliminated entirely and for good.

This experience in Thailand is repeated elsewhere. Time after time, I have talked with farmers in the Philippines, Bangladesh, Madagascar, and Colombia, plus a couple of dozen other coun-tries in between. The story I hear is, in essence, the same, with variations according to local circumstances. The impoverished peasants realize they are literally chopping away at the natural-resource base of their livelihoods. Prospects for the long run, they know, are growing bleaker. But when I press them about whether they truly perceive no alternative, they shrug their shoulders and talk about conditions in the remainder of their countries, in the savannahs and other areas that are generally more suited to agriculture. They do not spell out their plight in detail to me, because they do not understand its intricacies. But the heart of the situation amounts to this: At least 800 million people in rural zones of the Third World are either completely landless or so close to being landless that their basic security is too shaky for them to stay where they are. Of these 800 million, the majority live in tropical forest countries (Table 2). Not only do they have no land. They have no work, and no prospect of work; and they have next to no cash income, and a declining real income. At times of food shortages, the death rate among these disadvantaged millions can be three times as high as it is among people who own a mere one and a half hectares of land.

To put these generalities into perspective, let us look again at an individual country. Peru possesses 80,000 square kilometers of cultivable land, four-fifths of which features large farms.[3] Yet of all these arable territories, only one hectare in four is being used in worthwhile fashion, the rest being owned by absentee landlords who allow the land to lie idle or who indulge in grossly inefficient forms of management—and produce only half as much food per hectare as does a small-scale farmer. Because of this poor distribution of agricultural land, nine farms out of ten have become too small to support a family in the long run. By far the best way for the country to cope with its food problems would be to split the large farms into small holdings, and to offer much more support (investment credits, marketing networks, and the like) to peasant farmers. But because the government is dominated by a land-owning elite, it spends a great part of its agricultural budget on encouraging landless people to seek virgin territories in other parts of the country. In practice, this results in an overspill of impoverished peasantry from their traditional farmlands into the forests of Amazonia.

The migration phenomenon has been graphically described by a former Director of Forestry in Peru: "The populations overflowing from the Andes down to the Amazon plains do not settle there. They advance like a slow burning fire, concentrating along

TABLE 2. **Population Projections for Some Tropical Forest Countries**

COUNTRY	POPULATION IN 1980 (IN MILLIONS)	RURAL POPULATION (%)	POPULATION IN 2000 (IN MILLIONS)	YEAR OF NET REPRODUCTION RATE	YEAR OF STATIONARY POPULATION	TOTAL OF STATIONARY POPULATION (IN MILLIONS)
			Latin America			
Brazil	118.7	39	176.5	2010–15	2075	281
Colombia	26.7	40	39.4	2005–10	2065	60.2
Costa Rica	2.2	59	3.3	2000–05	2065	4.8
Ecuador	8	58	13.6	2020–25	2085	28.4
Guatemala	7.2	64	12	2020–25	2085	24.2
Guyana	0.8	60	1.2	2000–05	2065	1.8
Honduras	3.7	69	6.7	2025–30	2090	16
Mexico	69.8	36	115	2010–15	2075	202.6
Nicaragua	2.6	51	4.7	2025–30	2090	10.8
Panama	1.8	50	2.8	2005–10	2070	4.3
Peru	17.4	38	27.3	2015–20	2080	48.8
Surinam	0.4	34	0.6	2005–10	2070	0.9
Venezuela	14.9	25	23.8	2005–10	2075	38.6
			Southern/Southeast Asia			
Bangladesh	88.5	91	141	2030–35	2125	338.2
Burma	34.8	78	54	2025–30	2095	89.7
India	673.2	78	994.1	2015–20	2115	1,621.5
Indonesia	146.6	82	216	2015–20	2110	388.4
Malaysia	13.9	73	20.8	2000–10	2070	30
Papua New Guinea	3.1	87	4.4	2030–35	2125	9.4
Philippines	49.1	68	76.9	2010–15	2075	125
Thailand	46.9	87	68	2000–05	2070	102.5
Vietnam	54.2	78	87.9	2010–15	2075	153.4
			Africa			
Cameroon	8.5	71	14.2	2035–40	2110	40.8
Congo	1.5	60	2.7	2035–40	2100	10.5
Gabon	0.7	68	0.9	2035–40	2130	2
Ivory Coast	8.2	68	14.8	2035–40	2110	47.4
Madagascar	8.7	84	16.1	2035–40	2110	51.1
Zaire	28.3	70	51	2035–40	2110	156.1

Source: M.T. Vu and A. Elwan, 1982, "Short-Term Population Projection 1980–2000, and Long–Term Projection 2000 to Stationary Stage, by Age and Sex for All Countries of the World," The World Bank, Washington D.C.

a narrow margin between the land they are destroying and are about to leave behind, and the forests lying ahead of them."[4] During the past two decades or so, the population of Amazonian Peru has increased from just a few tens of thousands to 1.5 million.

A similar situation exists in much of Latin America. This is especially the case in tropical forest countries where land-tenure systems are dominated by huge private estates that tend to grow cash crops for export, such as bananas and coffee (and beef, which was discussed in the preceding chapter). In tropical Latin America overall, a mere 7 percent of land owners possess over 90 percent of arable land, by contrast with the United States, where the largest 7 percent of farms account for only 27 percent of farmlands.

In short, the forest farmer deserves to be regarded as an unwitting instrument, rather than a deliberate agent, of forest destruction. He is no more to be blamed for what happens to the forest than a soldier is to be blamed for starting a war. The root causes of his lifestyle lie in a set of circumstances often many horizons away from the forest zones. Far from being an enthusiastic pioneer of forest settlement, he finds himself pushed into the forest by circumstances beyond his control. As we have already seen in the previous chapter, small-scale farmers in Central America feel themselves driven into forests by large-scale cattle ranchers and others of the agricultural oligarchies. In Colombia, mini-farms produce three times as much food per hectare as do large holdings, yet smallholders have so little space left to pursue their farming that they head toward Amazonia, where there are now nine times more people than there were in the mid-1950s. In Latin America as a whole, 7 percent of land owners control 93 percent of arable lands, while the poorest one-third must manage with just 1 percent of the land; while in the Philippines, one-twenty fifth of farms cover more than one-third of the country's croplands. In Kenya the amount of land held by large farmers, now the African elite, is proportionately greater than it was during the regime of white settlers during Kenya's colonialist days. The largest farmers, three thousand of them, own more land than 750,000 smallholders; over half of Kenya's farmers hold just two hectares or less, accounting for way below 15 percent of all croplands. By the early 1980s, one Kenyan farmer in three could be accounted landless, with perhaps as many again ranking as near landless. Thus it is the gentleman farmers of Kenya's new elite who in effect light the fires in Kenya's few remaining forests. A similar tale can be told of many other tropical forest countries.

THE PHILIPPINES

To see how the detailed situation works out in still another country, let us consider the Philippines.[5] The main problem here lies with land-tenure pressures. At least one in ten of all farms amounts to well below one hectare in size, whereas one farm in five hundred amounts to more than 50 hectares—of which, generally speaking, only a small part is intensively used.[6] The result is that almost four farmers out of five are landless or almost landless. Equally to the point, the poorest 40 percent of farmers receive only one-seventh of all farming income, while the richest 40 percent receive almost three-quarters.

Not surprisingly, the number of forest farmers has grown apace, and the forest cover has declined. In 1950 forests extended across 55 percent of the country, but today they account for only 35 percent (the government believes that, for economic and environmental reasons, they should cover at least 47 percent). Forest farmers are officially estimated to total about two-thirds of a million people. But this is a minimum figure, for official needs only. During my visits to the country, I have talked with representatives of the Population Center Foundation of the Philippines and the National Environment Protection Council as well as the Bureau of Forest Development. I have found an informed consensus to the effect that forest farmers now total at least three million people, occupying more than 23,000 sq. km. of forests,

13,000 of them in critical watersheds.[7] The amount of forestland that they abandon permanently each year, after degrading it beyond recovery, is estimated at about 6000 sq. km. per year, three times as much as the official figure. In the monsoon forests of the large northern island of Luzon, forests become highly inflammable during the protracted November-May dry season, and wild fires in this area alone are believed to eliminate 1,500 sq. km. of forest.

So far as I can discern from my travels to all main parts of the Philippines, forest farming is widely practiced by both established hill tribes and newly arrived lowlanders. Those in the first category, who once predominated, are now less numerous than those in the second, who consist of two types. First, and expanding most rapidly in numbers, are "obligate cultivators," that is, persons who originate from outside the forests and who feel compelled, for lack of land elsewhere, to try their hand at forest farming, even though they know next to nothing about the ecological factors involved. Second, and becoming increasingly important, are "speculator cultivators," people who enjoy a sufficient source of livelihood elsewhere, but take up forest farming on a substantial scale as a commercial proposition.

These various categories of forest farmers encounter different responses from the government. Whereas the commercial operators are viewed positively, mainly because of their connections with the ruling oligarchy, the rest enjoy no such official favor. To cite the opinion of numerous officials with whom I have spoken, the government is inclined to look on all small-scale forest farmers as trouble makers—a label that is applied whether from the standpoint of forestry, economics, environment, or politics. They are considered to be people who operate outside the mainstream of national affairs. They are viewed as a messy problem that leaves officialdom baffled: how to come to grips with so many people who deliberately try to escape the eye of authority? Since the farmers maintain a low profile (otherwise they may get arrested for trespassing on public forestlands), the government returns the gesture by ignoring them, pretending they do not exist (as is reflected by demographic data, which appear to deliberately underestimate the numbers of forest farmers). As one minister in Manila observed to me, the government, wishing the forest farmers would simply disappear, hopes they will at least keep quiet—and anyway, if they start to riot in those remote forests, who is going to hear them? Meanwhile, if they insist on wrecking their living space, that is their affair.

Moreover, forest farmers in the southern sector of the country tend to make common cause with Muslim rebels who use the forest for the cover it offers to their clandestine activities. The rebels provide support and sympathy for the forest farmers' plight, which is a good deal more than the farmers receive from the government. Of course, the rebels find sympathizers among the displaced peasants from the lowlands. Because the government does not wish to recognize the insurgents' cause, and because it would be difficult to get them to cooperate with national census takers, nobody really knows how many rebels are living in the forests.

In short, and as the government admits, there is a good number of "non-persons" living in the Philippines' forests. No government leader with whom I have spoken is prepared to offer an accurate estimate, but insofar as the government must make some sort of official assessment of the problem's scope (if only to determine how to respond to the threat is represents), I have come across several best-judgment estimates to the effect that there could be at least 1.3 million such persons in the forests—in addition to the two-thirds of a million "acknowledged" cultivators.

Since this issue is so important for the future of the Philippines' forests, I have checked the figure of 1.3 million with numerous authorities. I have discussed it with foresters, scientists, development technocrats, land-use planners, resource economists, and many others who get their boots well and truly muddied through regular forays into the furthest reaches of these forests. While nobody cares to hazard a documented estimate, they virtually all agree that if the figure of 1.3 million is wrong, it is wrong on the low side.

I have encountered parallel situations, with a policy of neglect (to put it mildly) directed by governments toward forest farmers, in country after country in each of the three main regions. In Thailand the forest farmers in border zones of the country are joined by throngs of refugees and guerillas from neighboring Burma, Laos, and Kampuchea—people whom the government prefers not to recognize. Population totals are thus grossly underestimated. In Indonesia the Central Bureau of Statistics has acknowledged that the 1983 Agricultural Census would not make an estimate of numbers of "unofficial" forest farmers, on the grounds that they form no part of the government's development plans. In Liberia forest farmers derive from indigenous African stock, whereas the communities of the coastal plains, who dominate affairs in Liberia, tend to be descendants of persons returned from America: ethnic tensions cause the forest farmer situation to be played down, and officials claim it is much smaller than it is. In Ivory Coast, as in several other countries of West Africa, millions of immigrants flood in from the north to escape the ravages of the Sahel drought, and, sensing their interloper status, they frequently take refuge in forests: again, governments prefer to turn a bland eye to the intractable problems presented by these unwanted arrivals. In Guatemala many forest farmers are Indians, who receive a rough deal of many sorts from officialdom: more belittling of numbers. Throughout the biome, forest farmers do not fit in; they are a nuisance; best to look away. In short, the greatest single problem of tropical forests often fails to be recognized.

TOTAL NUMBERS INVOLVED

The crucial question is, how many forest farmers are there? Coming up with an answer is more than difficult, but we must try—since it is central to the survival outlook of tropical forests.

In the mid-1970s, the total number of forest farmers was roughly estimated at 140 million.[8] This was a very preliminary and rudimentary assessment, devised solely to enable forest planners to comprehend the challenge they were facing. (Moreover, it was a strictly minimum estimate; some experts believed that there were at least 200 million as far back as the late 1950s.) Roughly speaking, it was thought that about 50 million of these people were cultivating in primary forests, occupying at least 640,000 sq. km., an area the size of Texas. Another 90 million were believed to be operating in secondary forests, occupying twice as much land. Together, the two groups were accounting for almost 2 million square kilometers of forest, or at least one-fifth of the entire biome. In about one-half of this expanse, they were making sustainable use of forest environments, albeit with steady decline of soil fertility, biological diversity, and so forth: the forest tracts could still be classified as forests, though secondary forests of impoverished type. In the other half, forest environments were being so rapidly degraded that the tree cover was giving way to bush, scrub, bamboo, and coarse grasslands.

The greatest loss was thought to be occurring in Southeast Asia, where cultivators were clearing a minimum of 85,000 sq. km. per year, adding to the 1.2 million sq. km. of formerly forested croplands in the region.[9] Africa south of the Sahara was believed to have lost 1 million sq. km. of forest to these cultivators even before modern development trends started to gain momentum after World War II. In the mid-1970s, the region's loss was put at 40,000 sq. km. per year, in a zone of 400,000 sq. km. of forestlands then under that style of agriculture.[10] A similar story applied in Latin America, though fewer details were available.[11] All forms of expanding agriculture in Latin America, of which slash-and-burn cultivation was a major type, were thought to be accounting for 50,000 sq. km. each year.

How reliable are these estimates? Of all factors in tropical deforestation, this one is the most difficult to evaluate, while also the most crucial. But we can assert that the figures listed above appear to represent *minimum* estimates for the situation as it existed back in the mid-1970s. This is the strong impression I have derived while making repeated travels around the twenty-five countries which account for 85 percent of all forest farmers.

The situation is clearly growing worse with every year that goes by. To update the figures, I have made an adjustment for the under-counting of forest farmers that seems to be a common practice in many countries. The official estimates for the Philippines, Thailand, and several other countries could well be doubled or even tripled to make a realistic figure; and in Ivory Coast and at least half a dozen countries in West Africa and Central Africa, official estimates deserve to be expanded by at least one-half. In most of the remaining countries, a marginal increase of 10 percent appears to be reasonable. According to my field observations, these calculations step up the total from a base minimum figure of 140 million to 170 million. Then we must allow for population growth. In the countries concerned, national growth rates run at an average of 2.3 percent per year. But as we have seen, population increase for forest farmers is much higher, due to migration into the forests by throngs of peasants who find themselves landless. So far as I can determine after numerous consultations with colleagues, I put the farmers' growth rate at 3.5 percent per year.[12] Allowing for the compounding effect of population growth, this means a stepping-up of the 170 million by at least 30 million to up-date it to a 1980 level. This expands the overall total to 200 million (a minimum estimate, building out as it does from the minimum figure proposed for the mid 1970s of 140 million).

From this base-line figure of 200 million for 1980, I have gone on to use an average number of persons per family, eight, producing a total of 25 million families. How much forest does a family dispose of in a year? While there is much variation in clearing rates, in accord with the type of forest, population pressure, soil capacity, and so forth, an average appears to be one hectare per family per year.[13]

In rough terms, then, forest farmers can be reckoned to be clearing, whether permanently or temporarily, 250,000 sq. km. of forest each year. Many of these farmers exploit secondary rather than primary forest. Furthermore, population densities in certain sectors of primary forest, for example, in Central Africa, are still low enough to permit the forest to be used while sustaining its quality and composition, with the prospect of eventual regeneration of primary vegetation. Overall, however, it seems realistic to suppose that forest farmers are converting at least 80,000 sq. km. of primary forest to permanent cultivation, or to degraded scrub and grassland, each year. In addition, they are causing gross disruption to the "biological integrity" of another 80,000 sq. km. of primary forest each year, that is, they are modifying forest ecosystems to such a degree that the original biological community, with its full richness and variety, will not be restored before many decades—if ever.

FUTURE BUILDUP OF POPULATION NUMBERS

There is no sign that the rate of increase in farmer populations will slacken. Many of the countries in question are finding that the fastest buildup of population is in urban communities, which are often growing at a rate twice that of population increase overall. But many countries already possess large populations in relation to available cultivable land—generally countries where the problem of forest farmers is most pronounced, the Philippines, Indonesia, Vietnam, Thailand, Bangladesh, India, Madagascar, most countries of West Africa, Colombia, Peru, Ecuador, and all countries of Central America (Table 2). Unless these large-population countries proceed faster than hitherto with economic development that provides opportunities other than forest farming for landless people to make a living (i.e., through urban manufacturing and the like), the numbers of forest farmers will continue to grow at a rate far faster than that of national populations. Whereas population growth in the countries concerned is projected to produce an increase of almost 50 percent during the last two decades of this century, the number of forest farmers could double.

As for the longer-term future, we should note that none of these countries is projected to reach zero population growth, even with stepped-up family planning programs, until late in the next century, possibly not until early in the twenty-second century (Table 2). The problem lies with the "youthfulness" of their populations. When we look at the demographic profile of these countries, we often find that as much as one-half of the populace is aged fifteen or less, by contrast with United States and other developed nations, where little more than 20 percent is in that range. This means that a very large number of future parents have already been born. Even if these countries were immediately to reduce their family size to two children, their populations would continue to expand for another two generations, at least doubling before reaching stabilization at zero growth. So we must anticipate a future where the population of the Philippines grows (if they all find the wherewithal to support themselves) from its 1980 total of 49.1 million to 125 million; of Indonesia, from 146.6 million to 388.4 million; of Brazil, from 118.7 million to 281 million; and of coastal West Africa (not counting Nigeria, with its 85 million projected to reach 459 million), from 46 million to 263 million. At the same time there is a hopeful aspect: Zaire, Gabon, and Congo, with a tropical forest area of 1.2 million sq. km., are projected to grow from a 1980 total of only 30.5 million to an eventual total of "only" 168 million.

THREE COUNTRY PROFILES

Let us now take a look at three more countries to see how their populations may be affected.

Indonesia

Of Indonesia's present population of almost 160 million people, at least four out of five are rural inhabitants. Of the country's total area of roughly 2 million sq. km., equivalent to the United States west of the Rockies, the island of Java constitutes only 7 percent, making it half the size of Oregon. In the year 1800, the island is thought to have supported fewer than 5 million people, and in the year 1900 only 28 million. Today it contains around 100 million, and these dense throngs are exerting unsustainable pressures on the island's life-support systems. So the government is mounting a transmigration program to shift millions of families to the forestlands of the outer islands of Borneo, Sumatra, and Sulawesi. The government argues that the average population density in Java is more than 700 persons per sq. km., in contrast to Kalimantan (Indonesia's sector of Borneo), which has fewer than 10, and Irian Jaya (western New Guinea), with only 2 people per sq. km. The program aims to settle 2.5 million people from Java during the period 1980–84, at a cost of $2.5 billion—even though, during the same five years, Java will add about 9 million people to its population.[14] The long-term aim is to translocate 15 million people, occupying 100,000 sq. km.

Since the cost amounts to $1,000 per family, we might well ask whether the same money could not be better spent in upgrading farming techniques in Java. Over-crowded and deteriorating as it is, much of Java does not support the densities of neighboring Bali, where, with a different cultural background, farmers engage in some of the most intensive and sustainable agriculture ever known. In Java one sees many devastated landscapes, whereas Bali is a delightful garden isle where people live in apparent harmony with their environment: cropland terraces range across every hillside, fostering the best use of every square meter of land and every cubic meter of irrigation water. There is vast scope for the same caring approach throughout Java.

Brazil

Of Brazil's present population of more than 130 million people, only 40 percent are rural inhabitants. But in the northeastern region live 40 million famine-stricken farmers, constituting the most deprived community in the whole country. So the next-door territory of Amazonia is often regarded, to cite the words of former President Medici, as "a land without men for men without land." Not that there is not space for the northeasterners in other parts of Brazil: 1 percent of farmers own 42 percent of cultivated lands, while the 52 percent who possess just a few hectares occupy less than 5 percent of cultivated lands.

The TransAmazon Highway system was built by Brazil during the 1970s in order to open up its sector of Amazonia to "development." To its misfortune, Brazil finds that virtually all the modes of development promoted by the road network, notably small-scale cultivator settlements and large-scale cattle ranches, have proven unsustainable. Brazil is taking a lengthy second look at its options. *(Michael Freeman, World Wildlife Fund)*

Since the late 1960s, the government has made various efforts to promote migration into Amazonia.[15] Among its initiatives has been a colonist settlement project, intended to turn a 50-km.-wide belt of forestland along the famed TransAmazon Highway into farms of various sorts and sizes.[16] The strategy originally envisaged a migration rate from northeastern Brazil into Amazonia of about 100,000 persons per year (which would still amount to only one-tenth of the annual increase of the northeast's population; but fewer than half of the migrants eventually came from the northeast). Settlers were to be offered a piece of free land, plus crop seeds and other inducements. Yet only 1 out of 15 of the anticipated 100,000 families eventually arrived, and by 1975 the total of 50,000 colonists had almost stopped growing. Many homesteaders tried to grow unsuitable crops; many were immobilized by disease; and soils almost invariably proved poor. A good number of settlers abandoned the plots allocated to them and turned to various forms of rotational agriculture and informal forest farming—precisely the types of uncontrolled settlement that the government had hoped to discourage. During the period 1966–1980, agricultural settlement by smallholder cultivators is believed to have accounted for at least 40,000 sq. km. of forest.

All in all, it is now reckoned that less than 7 percent of the planned settlement can be called successful. Due to a plethora of problems, not only those mentioned above but widespread soil erosion and threats from hostile forest tribes, government support for the settlement program has fallen off, and officials are looking closely at adverse ecological repercussions of the initiative. The upshot is that development funding for this kind of colonist farming has dwindled, and construction of several of the Highway's subsidiary roads has been halted.

The Ivory Coast

With a present population of somewhat over eight million, of which 80 percent are rural inhabitants, the Ivory Coast has already lost half of its forests, due primarily to land-hunger pressures from its burgeoning populace.[17] Yet the country is projected to reach around 15 million persons by the end of the century, and at least 47 million by the time zero population growth is attained in the year 2110.

Worse still, these figures are based only on natural fertility rates. They take no account of immigration from outside the country, even though one-third of the male labor force is made up of immigrants from other countries, notably from Upper Volta and Mali, fleeing from the Sahel droughts. Not surprisingly, the population of the forest zone in the southern sector of the country has been growing twice as fast as the population in the rest of the country.[18] To make matters worse yet again, the government now encourages commercial farmers to clear forest patches for export crops such as coffee and tea. As an incentive, the government offers the commercial farmer a land title to his cleared land. Within just a few years, these small-scale entrepreneurs have come to operate on a larger scale in the forests than do subsistence cultivators.

TWO RESPONSES TO THE SITUATION

Despite the many problems presented by forest farmers, we can still hope that, in the long run, the farmers can be persuaded to pursue their lifestyles in disrupted forests, that is, in secondary and successional forests of various types, and in areas that have lost their forest cover. Within the not too distant future, we could enable forest farmers to enjoy a better livelihood, and to do it without cutting another single tree in primary forests. Herewith two proposals.

Agroforestry

Basically, we need an agricultural strategy that encourages farmers to plant trees as well as to cut them down. An innovative approach known as agroforestry seeks to offer the cultivator an incentive to establish a patch of fast-growing trees on his land at the same time as he plants food crops—an approach that can be pursued in cleared forestlands.[19]

The key to agroforestry lies with combinations of plants that support each other rather than compete with each other. As we have noted (and see below), leguminous tree species supply nitrogenous nutrients to the soil, thus assisting food crops at the same time that the trees produce domestic wood, industrial timber, and paper pulp. By the time the cultivator is ready to move on, the newly planted trees will be ready to close their canopy, restoring forest cover to the area. The tree plantation itself can be intercropped with food plants such as corn and bananas, and with cash crops such as coffee, tea, spices, fruit trees, and many other items, plus fish and giant snails in water channels—a regular Dagwood sandwich of products. In short, food growing can be combined with restoration of the forest, instead of leaving behind degraded scrubland.

Obviously, a particularly important contribution lies with legumes. To cite an expert in this field, Dr. James A. Duke, of the Agricultural Research Service in Beltsville, Maryland, cereals may receive the credit for the Green Revolution, but it is members of the legume family that underpin most established agricultures of the world. Legumes range from great trees such as acacias, to shrubs like pigeon pea, to herbs like vetches and clovers, and to climbers like the winged bean and lima bean. They mostly fertilize themselves by fixing nitrogen from the atmosphere, a characteristic that not only makes their seeds some of the richest in protein of all the plant kingdom, but helps them supply additional nitrogen to other plants.[20] Trial projects show that when wheat, for example, is interplanted with ipilipil trees, the legumes supply massive amounts of nitrogen, plus useful amounts of phosphorus and potash, all of which enable the wheat crop to produce much more than would be the case without the trees.[21] The same trees, together with other natural suppliers of fertilizer, help increase the harvest of dryland rice in Indonesia by two and a half times, while the overall output from each hectare is greater than it would be if the rice and trees were grown separately. When ground-level legumes are intercropped with rubber trees in Malaysia, they bring forward the first tap of latex by two years.

This chapter has been critical of Thailand and the Philippines. So let us now see how governments in these two countries are taking some steps, albeit tentative ones, to promote agroforestry. Under projects of Thailand's Forest Industry Organization, farmers are encouraged to plant timber and rubber trees in conjunction with their food crops.[22] After six years, during which the organization pays all expenses of raising the trees in return for the farmer's labor, the farmer starts to tap the latex, paying the organization 30 percent of his income. After thirty years, when the rubber trees are worn out, they will become the property of the organization, which will sell them as timber. This approach produces almost 250 percent more net income for the organization than would be the case if it planted the trees itself, while the farmer receives more than $1,000 per year in wages during his first six years, plus around $800 thereafter from rubber (1980 values). In addition, farmers are being helped to abandon their migratory lifestyles, and to try their hand at stabilized agriculture, that is, agriculture that will exploit permanent croplands. The farmers assemble in newly established villages, which the government furnishes with basic services such as water supplies and sanitation. Agricultural experts demonstrate that whatever the farmers have earned from their main cash crop to date, opium poppies, they can earn still more from rice, fruits such as peaches, vegetables such as beans, cabbages and sesame, and high-grade coffee and tea. As a result, tens of thousands of farmers have almost eliminated their poppy fields. They no longer make wasteful use of forestlands, and, still more important, they feel the government wants to help them confront their problems.

As for the Philippines, the government sponsors, in conjunction with the World Bank, a project that is principally the activity of a private enterprise firm. The Paper Industries Corporation of the Philippines (PICOP) has embarked on a resettlement project to encourage forest farmers to rent patches of deforestated land on the Company's 183,000-hectare concession.[23] A typical smallholder is required to utilize four-fifths of his holding to grow pulpwood trees, and the remaining one-fifth to produce his food crops and to raise his livestock. The farmer receives a loan to cover up to 75 percent of his tree plantation's costs, and of his first year's food growing; his own main contribution is his labor in planting the trees, then tending them until they become ready to cut. By 1980 almost 4,000 farmers, previously living as illegal squatters on the PICOP concession, had planted 10.8 million trees for pulpwood. The company buys pulpwood from the farmers, who also grow charcoal, firewood, and leafmeal (for cattle feed) for their own use. An average smallholder can hope to earn, within eight years of startup, at least $2,600 per hectare—big money for the farmer, considering his alternatives. Insofar as the Philippines now spends around $55 million per year on paper-pulp imports, the PICOP scheme represents sound financial sense, as well as knocking a solid dent in the local forest farmer problem. Eventually, when 7,400 smallholders will be looking after 28,600 hectares, the project should represent gross import savings to the Philippines of at least $40 million per year (at 1980 prices).

Multi-Species Gardens

A related strategy lies with attempts to make the farmer's croplands grow much more produce, in order to leave him with less cause to move into fresh forest areas every few years. In other words, an extensive and wasteful pattern of agriculture gives way to an intensive and efficient form of agriculture. The essence of the approach lies with the accumulated experience of "garden farmers" of tropical forests, who have developed a traditional type of agriculture that remains stable and productive for years on end. In several parts of the biome, a few local communities depend on gardens that feature multiple crops in an array that mimics the diversity of the natural forest to a highly elaborate extent—and thereby supplies the environmental stability that is fostered by diversity.[24] A typical garden covers no more than two hectares, often a good deal less, yet it supplies virtually all the farmer needs in the way of food, fiber, fuelwood, and fodder.

The secret of these gardens lies with the structure of the vegetation. It comprises several stories, with scope for crops both annual and perennial, plants both herbaceous and woody, including sundry shrubs and creepers, and numerous trees between 5 and 30 meters tall. At ground level there are several categories of crops: short-stemmed cereals such as rice, taro, and other staples; vegetables such as squashes, spinach, carrots, and lettuce; root crops and tubers such as yams and sweet potatoes; and weedy plants for green manure; plus some medicinal herbs, generally the only source of drugs. In the second layer, from 2 to 5 meters high, there are taller-standing crops such as corn, cassava, garlic, and curcuma; fruit trees such as papaya, banana, plantain, star-apple, jackfruit, citrus, mango, mangosteen, and rambutan; legume trees to supply nitrogen fertilizer; and a range of cash crops such as coffee, cocoa, nutmeg, tamarind, cinnamon, cloves, kapok, sugar-palm, and betel palm. These two main layers are often interlinked throughout their structures by climbers such as pepper and va-nilla. In a third story are emergent trees such as coconut palm, areca palm, and durian. Interspersed around the garden are bam-boo thickets for fuelwood and construction materials. The entire area supplies forage for chickens, perhaps goats and sheep too, while in a corner of the plot is a fish pond.

The main secret of the multi-storied structure is that it effi-ciently exploits the sun's energy through plants of many forms that absorb light and warmth in different ways. Just beneath ground level, a dense network of roots promotes cycling of nutri-ents throughout the garden ecosystem. Deeper down, the tap roots of trees serve to bring up nutrients that would otherwise remain buried beyond the reach of conventional farm crops.

The range of species can be astonishing. In northern Thai-land, the Lua tribe grows as many as 75 food crops, together with 21 medicinal crops, 20 plants for ceremonial and decorative pur-poses, and 7 plants for weaving and dyes.[25] In New Guinea, the Tsembaga people distinguish 36 plant species and 264 crop varie-ties, frequently growing as many as 50 different crops at one time.[26] In the Philippines, the Hanunoo people are acquainted with 430 crops, commonly growing 40 of them in a one-hectare plot.

In Latin America, we encounter a similar strategy. In the Tabasco state of southeastern Mexico, farmers use 285 plant species, 65 for food, 118 for medicine, 92 for ornamental needs, and 36 for building materials, the rest being grown for fuel, latex, animal feed, implements, stimulants, spices, tannins, and perfumes, also for ceremonies—among almost 80 sundry applications.[27] Farther south in Mexico, the Lacandon people cram as many as 80 plant species into a garden plot that is rarely as large as one hectare.[28] After harvesting a plot for five to seven years, they take the patch out of cultivation for a rest period of five to ten years. But rather than abandon the cultivated area, the Lacandones plant it with at least two dozen species of trees, such as citrus, cacao, avocado, and papaya in a system of mixed agriculture and forestry that perpetuates the agro-ecosystem.

At the same time, the Lacandones derive sustenance from the forests surrounding their garden farms, a practice in which they show much expertise. They gather about 100 species of fruits and other wild foods, plus protein in the form of 20 sorts of fish, 6 of turtle, 3 of frog, and 2 each of crab, crayfish, crocodile, and snail.[28] This sustainable use of the natural forest's bounty is paralleled in Southeast Asia by the Sawaran Kenyoh tribe, who make up almost two-thirds of their diets by gathering plants and hunting animals and fish.[29] In similar style, the Trans-Gogol people of New Guinea utilize 42 bird species, 16 of mammals, 8 of reptiles, and 2 of frogs, including such well-known creatures as wild pigs, wallabies, cuscuses, bandicoots, and wild fowls.[30] The Trans-Gogol people also draw on forest wildlife for materials that they use for garments and decorations, weapons and tools, food containers, cordage, house-building materials, musical instruments, medicines, narcotics, stimulants, and intoxicants.

These, then, are some of the inventive ways by which a few forest-dwelling peoples find the wherewithal to keep body and soul together. They rely hardly at all on the outside world to keep themselves going, supplying their own variations of fertilizers and pesticides. Fertilizers are generated through organic mulches, green manure, crop residues, nitrogen-fixing legumes, livestock wastes, farm debris, household refuse, and general garbage. They also reduce their need for fertilizer by using combinations of crops that help maintain soil fertility. Were they to farm after the manner of the outside world, through planting mainly annual crops plus a few semi-perennial crops, they would quickly exhaust the soil of its nutrients. But when they integrate short-term crops with at least as many perennial crops, they reduce or even prevent the impoverishment of the soil. In addition, some garden farmers exploit innovative sources of fertilizer. An obvious example is bird guano from the forest. More exotic, while equally effective, is the fertilizer used by farmers in the Oaxaca Valley in Mexico—organic debris ejected from the fungus gardens of atta ants, which, resembling fine sawdust, is extremely rich in nutrients.[31]

As for pesticides, the forestland gardener lacks that great pesticide of temperate zones—winter. But he does not need to buy expensive pesticides from outside, he employs his own biological controls. In multi-crop fields, the variety of plants serves to hold down pest populations by limiting their food and living space, as well as by fostering natural enemies such as insect predators and parasites. The farmer can also call upon the natural pesticides contained in the tissues of certain plants. If, for example, he plants corn and peanuts in alternate rows, he exploits the pest-inhibiting biocompounds of peanuts in order to reduce the depradations of burrowing insects that would otherwise infect his corn crop.

All in all, then, traditional garden farms represent a sound way for a subsistence farmer to make out with meager resources. If the gardens were to be expanded into larger units for many more people, they could constitute precisely the type of agro-ecosystems that we need to meet the challenge of stabilized farming in tropical forestlands.

Regrettably, a good number of the forest peoples that practice garden farming are being squeezed out of living space. Generally disregarded by the outside world, on the grounds that they do not conform with conventional standards of "modern" agriculture, the first that we often hear about these forest-dwelling peoples is when they shoot arrows at bulldozers. While their accumulated folklore represents perhaps the most valuable resource that exists in tropical forests, many of these groups are being steadily eliminated. The Lacandones of southeastern Mexico are losing their forest habitats to newly arrived cattle ranchers. The Trans-Gogol people of Papua New Guinea are being pressured out of their homelands by Japanese timber corporations, which wield a weapon at least as potent as a bulldozer, let alone a rifle: a government-backed legal document that cuts the land property rights from under their feet. The Dayaks of Borneo are being taught the virtues of a nine-to-five workaday world, with its cash economy, its jeans, and other perquisites of the global culture. So fast are these "progressive" processes sweeping over the tribals, that in Brazilian Amazonia, which featured 230 native groups with an estimated six million people (conceivably several times as many) only five centuries ago, and probably one million as recently as 1900, there are now only half as many such groups, with a total of about 100,000 persons (perhaps five times as many in the rest of the Amazon Basin).

Fortunately there is a growing awareness in some sectors of modern society that we should take another look at the situation. Survival International, based in London since its foundation in 1969, and with branch groups in several other nations, is helping tribal peoples to help themselves with legal weapons. Allied agencies include the Anthropology Resource Center and Cultural Survival Inc., both located in Boston, Massachusetts. But these private groups, with their shoestring budgets, cannot do more than make a dent in the problem. As the tribals disappear, with their long-enduring lifestyles, let us reflect that it is thanks to Indians of Latin American forests, together with earlier forestland communities such as the Mayas, that we ever became aware of corn, potato, tomato, peanuts, cassava, avocado, guava, cashewnut, vanilla, and sunflower, plus a host of drugs such as quinine, curare, mescaline, and cocaine.

9

Bottom-Line Reckoning

Question: How much tropical forest are we losing each year?
Answer: Depends on what you mean by "losing."

ONE OBVIOUS AND straightforward definition is that a forest is lost when exploitation is so heavy-handed that scarcely a single tree remains. The landscape is transformed into bamboo, scrub or brush, or even coarse grasses. Or the forestland is turned over to rice paddies, rubber plantations, cattle ranches, or other forms of agriculture. Or it becomes flooded by a dam or some other installation. Or it is given over to urban settlement. In all these instances, hardly anybody would disagree that the forest has disappeared, entirely and for good.

Where people differ is when they look at a forest tract that is converted from its original form, yet retains some trees. In the view of some foresters, this still constitutes a forest. Yet suppose it has been heavily logged, as in several parts of Southeast Asia, with one tree in ten being removed for timber, and at least half the rest being damaged beyond recovery. We then look out on a devastated landscape, which does not remotely resemble its former state. I recall many a scene in Borneo and the Philippines during my visits to logging concessions. Whereas in a primary forest I would catch mere glimpses of sky through the canopy overhead, in the logged forest I would see much more sky than tree crowns. In a primary forest I walk around in the gloom of the forest floor, but in a logged forest I feel the direct glare of the sun most of the time, and in those extensive sectors where logging winches and bulldozing skidders have done their work, the ground sports no more vegetation that Times Square. Of wildlife, I see virtually nothing. Instead of the chattering of monkeys in the treetops, the constant rustle of small animals in the undergrowth, the steady songs of birds, and the buzz-buzz of insects all around, the "forest" is silent. Yet because the area has not become permanent farmland, or an oil palm estate or a tea plantation, it cannot, in some observers' eyes, be designated as anything other than forest. In any case, they say, while it may look little like a forest right now, it will regenerate to a forest within a few years.

But what sort of regenerated forest? I recall other visits, to areas that had been logged ten years previously. In those areas that show vigorous regrowth, there are plenty of trees, sturdy specimens that reach tens of meters high, plus a host of saplings on every side. The canopy is virtually complete, even if not so thick as in a primary forest. There is much scurrying of wild creatures at all levels of the forest, I hear a bedlam of birds, and I do not need to look for insects since they are all over me. But on looking about, I notice that many trees are the same. In a primary forest, I have sometimes identified one member of a tree species, then searched in vain for another member. Colleagues tell me that among the hundreds of individual trees they have surveyed in one hectare of primary forest, they find that most species are represented by but single specimens. In a logged-over patch, by contrast, I find lots of trees of the same species. Only a few are not replicated within just a few dozen meters. So the richness of the primary forest, the abundance of species, and the diversity of plant communities is missing. A secondary forest is a very much different form of forest. Vigorous as it may be, with its massive plant growth that may surpass in sheer weight the growth of a primary forest, a young secondary forest is often a poor substitute for what was once there.

After another few decades, it will become a better substitute. In outward appearance, it will even become a fair replica. Physiognomically it will match a primary forest. During my forays into older secondary forests, I have noticed that the structure of the vegetation is almost the same as in undisturbed forests, with stately pillars of tree trunks, distinctive stories to the forest "architecture," tree crowns interlinked with lianas, and sparse growth on the forest floor. Yet for sheer bulk of vegetation, it still does not match primary forest. According to comparative measurements by Dr. Christopher Uhl in Venezuela, a 10-to-15-year-old forest may feature around 40 tons of biomass above the ground, and by 50 to 60 years the total is likely to surge to 150 tons or so.[1] But an undisturbed forest can feature 375 tons. Successional processes often need well over 100 years to restore the sheer stature and bulk of the original forest.

Furthermore, I find, during my explorations in older secondary forests, that the species makeup is generally far different from what it is in a primary forest. Still more significant, the integrative variety of the primary forest is not the chief attribute of this secondary forest, which is still making its way toward the ecological complexity, as well as the biological richness, of the primary forest. If I were to take a few months to draw up an inventory of the forest's main components *and* of their interactions, I would find it remains a depauperate version of primary forest proper. It is not yet true "forest" in the sense understood by those scientists who study tropical forest ecosystems as a kind of climax of evolution's creativity. The floristic array, the animal community, the ecosystem's workings are simply different, much different.

Similar reservations apply to "fallowed forest," the forest growths that spring up in the wake of subsistence farmers who make rotational use of forest ecosystems. These are variations of the secondary forest theme and are so extensive that they cover at least 2 million square kilometers. As we have seen in chapter 8, the traditional type of forestland agriculture—shifting cultivation—could formerly make sustainable use of forest ecosystems, because the populations of these farmers were low enough to allow abandoned croplands a lengthy period to recover. A forest clearing could generally regain most of its original ecological character within a matter of decades, since it could become recolonized by seeds from a surrounding sea of primary forest, often just a few dozen meters distant. Today, the pattern is basically altered. Cropland patches are bigger and often aggregated into extensive farmlands, supporting a much larger number of farmers who exist in denser communities. A cleared patch thus takes far longer before it can be recolonized by primary forest plants, the nearest stocks of which are hundreds of meters away at best, possibly several kilometers away. On top of all this, an abandoned patch of cropland will be allowed only a few years to recover before it is brought back into cultivation. Ever-more frequently, as farmer populations soar in numbers, croplands are not abandoned at all for a period of recovery.

So a patch of forest that is being worked over by a community of farmers does not often get a chance nowadays to qualify as restored forest, even as degraded secondary forest. Of course, this is a broad generalization, and "often" is a relative term. Nonetheless, if the forest area has not been taken over by permanent agriculture (or is on its way to that new state), it is likely to degenerate into scrubby woodland or other forms of impoverished "forest." In certain instances, notably in Southeast Asia, the soil has become so exhausted of its nutrients that it can sustain nothing but a weedy type of grass cover, known as alangalang (described in chapter 8), with hardly a tree to be seen from one horizon to another. These coarse grasslands already account for 200,000 sq. km. of formerly forested lands in Indonesia, while similar impoverished vegetation occupies at least another 200,000 sq. km. in other countries of the region, plus a good deal more as we move westward toward Burma. Entire territories that are still colored rich green on maps, designated forest in the usual sense of the word, deserve to be re-colored pale green, if not brown.

Equally distant from natural forest—in the proper, broad sense of the term—is the tree plantation. About 110,000 sq. km. in the humid tropics are now covered with monocultural crops of eucalyptus, pine, and a handful of other species, almost all of them alien to the humid tropics. Certain foresters proclaim them as forest just as much as any other patch of trees—and in a strict sense of the term, they are right. But as an ecosystem with attributes that are supposedly kindred to those of primary forests, plantations are less than a travesty. In Madagascar's eucalyptus plantations, there is hardly a single wild creature among many millions of trees. Lemurs are more endangered through plantations that have replaced indigenous forest during the past few years than through illicit hunting of the past several decades.

In short, we need to be clear about what we mean when we talk about forests and deforestation. As understood in this book, a tropical forest is more than a bunch of trees. A whole world more.

CONVERSION OF TROPICAL FORESTS

It is within this perspective, then, that we shall now go on to consider the rate at which we are losing tropical forests. For purposes of this chapter, we shall use a catch-all term, conversion, to include degradation, impoverishment, destruction, and whatever other change merits our attention. Plainly, conversion can thus range from marginal modification to fundamental transformation. In essence, it refers here to depletion of biological integrity of forests, especially insofar as tropical forests contain uniquely large congregations of species and constitute a vast storehouse of biotic materials (for example, the genetic resources inherent in species) that, as we shall see in Part III, can serve humankind's long-term welfare in multitudes of ways that we only dream of as yet.

To spell out some specifics, modification can be construed as the result of human intervention whereby the outward appearance, the biological makeup, and the ecological dynamics of an original forest undergo discernible change. In turn, this change can be slight, substantial, or severe. In its lightest form, for example highly selective timber extraction, modification can entail merely some alteration in relative densities of tree populations and in the quantity of timber stocks; it need not necessarily cause a qualitative change in the species complement. A more significant form of modification lies with subsistence cultivation and other types of forest farming that induce secondary succession in the forest. Transformation, by contrast, amounts to a basically different category of conversion. In order to make way for permanent agriculture, plantations, or pasturelands, the forest is entirely eliminated, being replaced by a man-established ecosystem.

In between these extremes, all manner of variations are possible. So "conversion" can mean a host of different things. It can imply floristic impoverishment, a diminution of standing biomass, a temporary disruption of successional processes, and a gross disturbance of the species array (the change can be either quantitative or qualitative or both), together with a basic shift entailed in alteration of primary forest to secondary forest—plus of course many other variations, including gradients in between. Some changes may be of short duration, and the former state restored within just a few years. Other changes are more profound and may disturb the natural continuity for several decades. Yet other changes may persist, for whatever reason, for still longer periods. In the course of some of these processes, there may be threshold stages at which, in a relatively short space of time, deep-seated and irreversible alterations ensue.

REVIEW OF FOREST CONVERSION

In chapter 5, we found that the commercial logger damages some 45,000 sq. km. of primary forest each year. His impact is not significant in the broad picture, however, because most logged forests are soon occupied by small-scale cultivators and other farmers. In chapter 6 we saw that the fuelwood gatherer may well use 25,000 sq. km. of primary forest each year. In chapter 7, we read that the cattle raiser (confined to Latin America) is responsible for destroying at least 20,000 sq. km. of these forests. In chapter 8, we saw that the forest farmer is converting as much as 160,000 sq. km. of forest per year, possibly a good deal more. These figures make a total of 205,000 sq. km. per year, or, for the sake of a round figure, 200,000 sq. km. It means that about 2 percent of the biome is affected each year.

I regard this aggregate figure as conservative. It takes no account of other agents of conversion, such as the plantation man who replaces primary forest with monocultures of tree species such as eucalyptus and oil palms, plus others who convert forests for various reasons but whose overall impact is not on a scale to match the four main categories.

Of course—and to reiterate a central factor—these calculations represent no more than informed estimates. They are certainly not presented as a concise account of what is overtaking tropical forests. As the previous chapters indicate, the statistical details present a great range of reliability. Many quantifications should be qualified with phrases such as "So far as can be ascertained . . . ," or "Best-judgment assessments suggest that. . . ." In other words, I have tried on the one hand to avoid bogus accuracy, and on the other hand to describe semi-documented situations with an appropriate degree of "precise imprecision." To put it yet another way, my bottom-line reckoning should be amplified to read: "To the extent that we can establish the situation, on the basis of limited appraisals and with a less-than-satisfactory data base of variable reliability for parts of the biome, tropical forests appear to be undergoing conversion at a significant rate that can be put, for want of any better assessment, at somewhere in the region of 200,000 sq. km."

Some sectors of the biome are obviously being harmed more than others, and conversion patterns turn out to be highly differentiated. Certain sectors of the biome are undergoing widespread conversion at rapid rates, other sectors are experiencing moderate conversion at intermediate rates, while still other areas are encountering little change. We can conclude that virtually all lowland forests of the Philippines and Peninsular Malaysia seem likely to become logged by 1990, or very shortly thereafter. Much the same applies to most parts of West Africa. Little could remain of Central America's forests by 1990. Almost all of Indonesia's lowland forests have been scheduled for timber exploitation by the year 2000, and at least half by 1990. Extensive portions of Amazonia in Colombia and Peru could be claimed for cattle ranching and various forms of cultivator settlement by the end of the century; and something similar is true for southern and eastern sectors of Brazilian Amazonia.

By contrast, Central Africa is sparsely inhabited and possesses abundant minerals. This reduces the incentive for governments to liquidate their forest capital in order to supply funding for various forms of economic development. Hence there could well remain large expanses of little-disturbed forest in Central Africa at the turn of the century. Similarly, the western portion of Brazil's Amazonia, because of its remoteness and perhumid climate, could undergo only moderate change.

In sum, the overall outcome is likely to be extremely patchy, both in terms of geographic areas and in degrees of conversion. Readers who wish to assess the situation in further detail can consult the country-by-country review in the Appendix of this book.

If tropical forests are actually being depleted at the rate indicated, what will happen to the multitudes of species that exist in these forests? How many of them are becoming extinct through loss of habitat? Again, we can do no better than attempt some informed assessments. As indicated in chapter 4, many species in tropical forests are unusually susceptible to summary extinction, by virtue of their limited distributions, their low densities, or their ecological specializations. Roughly speaking, we can assume that even if conversion rates do not accelerate (as we can prudently expect they will), one-third of species in Latin America's tropical forests will disappear by the turn of the century, almost one-seventh in Africa's, and well over two-fifths in Southern/Southeast Asia's (by contrast, the rest of the world may lose only one-tenth of its species).[2] The total of species lost in tropical forests could be anywhere from half a million upward: we cannot rule out the prospect of losing one million species.

THE FAO / UNEP REPORT

So much for my conclusions, based on twenty years of travel and research, about conversion of tropical forests. How do my findings compare with those of other surveys? The report that I presented to the U.S. National Academy of Sciences in 1979, presenting figures along the lines of those set out above (they are here updated to 1983), accord with a number of earlier analyses.[3] A couple of years after my National Academy report appeared, a further review was published, prepared by the Food and Agriculture Organization and the United Nations Environment Programme.[4] This publication presents findings that, at first glance, are far different from my own conclusions, suggesting that the deforestation rate in 1980 was only—"only"—76,000 sq. km. per year. There is a basic reason for such a divergence between the two findings.

The apparent discrepancy lies with the two sets of criteria used. In my own case, the National Academy of Sciences required that I look at rates at which primary forests are being converted from their original state into some other state. I was to take, as my base line, primary forest rather than forest of any sort: the Academy made this stipulation because it was interested in biological values at stake, rather than other factors such as forestry (timber) values and the exploitation potential of forestland for agriculture. By contrast, the FAO/UNEP project confined itself in the main to forestry values. Its brief did not include biological values, such as repercussions for almost half of Earth's stock of species, and the many other non-timber values such as have been described in "pure science" terms in Part I, and will be analyzed in "applied science" terms in Part III. The United Nations team considered that a disturbed forest remains a forest, even if it has been subjected to heavy exploitation through timber harvesting, which, as we have seen, can sometimes mean that only one-quarter of the above-ground biomass remains alive in the wake of the harvester's activities. After all, say FAO and UNEP, if the forestland has not been given over to rice cultivation or a rubber plantation or some basically different type of land use, the forest is still a forest. While incapable of returning to a primary state, with all the exceptional biotic richness that implies, for a period of decades or even centuries, the forest still constitutes forest.

Of course it is no criticism of FAO/UNEP to observe that their report, as required by their terms of reference, views "tropical forest resources" (the wording of the report's title) in a restrictive commercial sense, that is, as so many billion cubic meters of commercial timber. This is FAO's principal purpose in life.

In short, my National Academy study looked at *significant conversion* of primary forests, that is, destruction plus degradation. The FAO/UNEP study focused instead on *outright elimination* of forests, that is, destruction alone. Hence the difference in the two sets of figures presented as the main conclusion of the two reports. While preparing this book, I worked with certain American colleagues to adjust the two sets of figures to take account of differences in criteria, definitions, and the like. We found that my own documentation of outright elimination amounts to a total of 92,000 square kilometers per year—which, while higher than the FAO/UNEP figure, is in much the same ballpark.[5]

On the following two pages satellite photos reveal the extent of forest burning by small-scale farmers in the coastal zone of West Africa and in much of the Philippines. *(Meteorological Satellite Program of U.S. Air Force. Courtesy of Dr. Thomas A. Croft)*

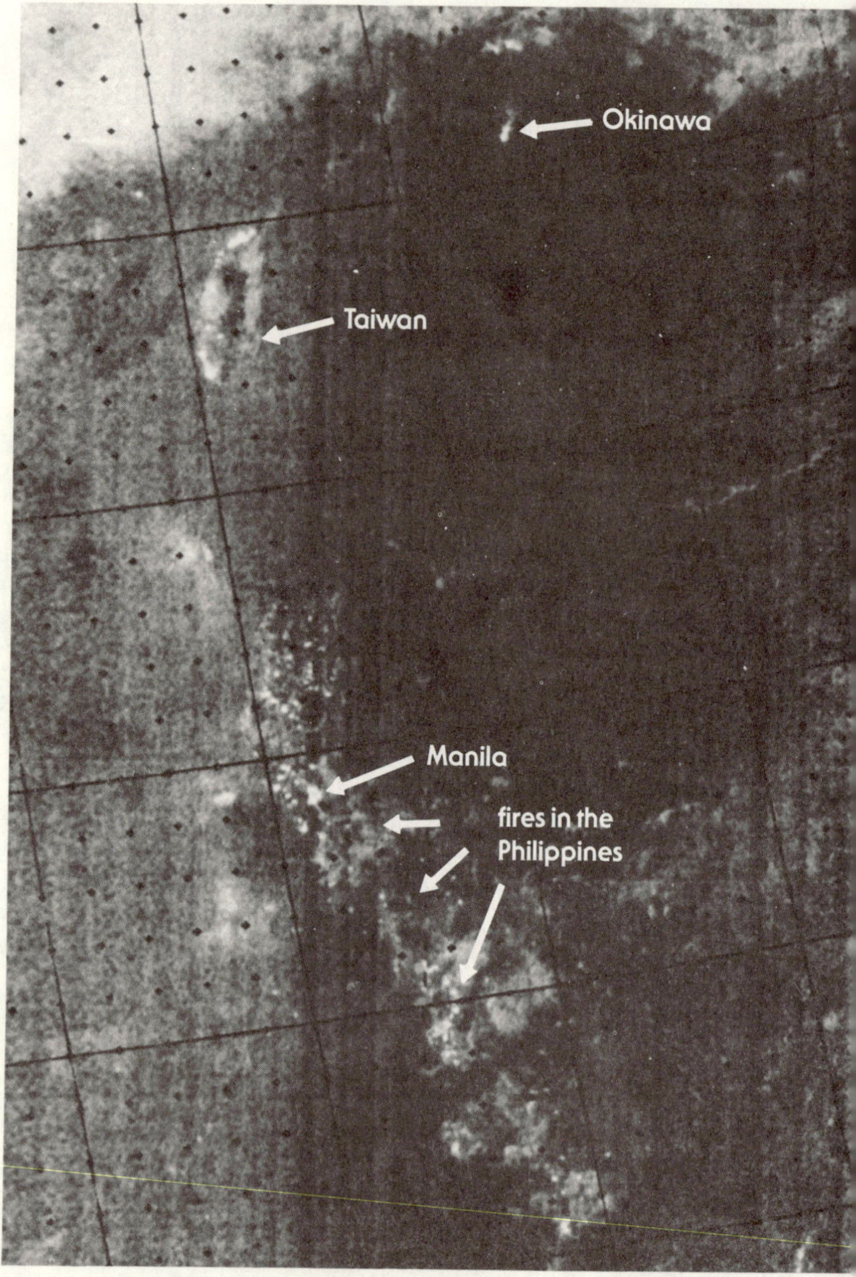
Okinawa
Taiwan
Manila
fires in the
Philippines

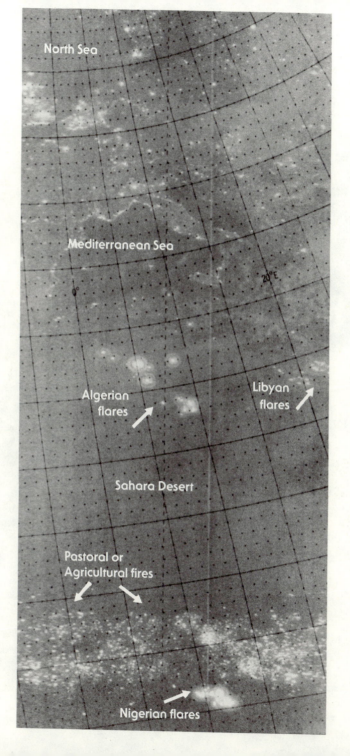

1976

1978

These satellite photographs show the state of Rondônia, in the southern part of Brazilian Amazonia, an area with an exceptional concentration of species. The settlement pattern depicted on the satellite images, covering an area of 30 kms. by 30 kms. in central Rondônia, reveals a network of roads established by the government at 5-kilometer intervals. Each settler is allocated a patch of forest measuring 2,000 m. by 500 m. The images, taken in 1976, 1978, and 1981, indicate the speed and scale of forest clearing; the last image reveals that about 30 percent of the area lost its forest within seven years. If the present accelerating rate of deforestation persists, Rondônia may well lose most of its forest within just another decade. *(NASA Goddard Space Flight Center, Greenbelt, Maryland, courtesy of Dr. C. J. Tucker)*

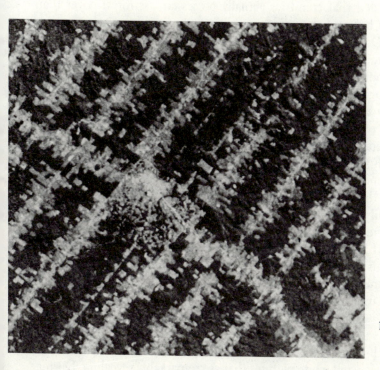

1981

REMOTE SENSING

In a situation where information is of varying quality, we are fortunate to call on a new form of technology that helps us in assessing our natural resource stocks. It is remote sensing. The merits of this technology are that it is systematic and objective. Through satellite imagery, we can make inventories of large portions of the Earth's surface in a short period, often at remarkably low cost. At the time of completing my National Academy study in 1979, I drew on remote-sensing findings for Thailand, the Philippines, Brazil, Nigeria, and several other countries, totaling almost half the biome. In more cases than not, satellite imagery revealed that countries actually possessed less forest cover than they believed they did.

During those days, however, I had to depend on the limited capacities of Landsat 3, which produced "images" covering blocs of land 185 kilometers on each side. This did not offer nearly enough resolution to indicate just what is going on at ground level. Worse still, it often could not differentiate between the vegetation of a forest and the vegetation of a cropland patch, hence it generally could not discern the impact of forest farmers.

Fortunately the United States has recently lofted a more advanced satellite, Landsat 4. Through one of its sensors, known as a thematic mapper, this satellite has two and a half times the detail documentation power of its predecessor, so it can engage in fine-grain mapping of forest features. About the size of a saloon car, Landsat 4 is sensitive enough to take pictures the size of a tennis court 650 km. below, and to do so through some cloud and darkness. Instruments on board pick up signals from the ground and interpret them to indicate whether they derive from rock, soil, water, or vegetation—and, most importantly, vegetation of different sorts. The images are then digitized and transmitted to Earth-based receiving stations where they are converted to color or black-and-white pictures that are remarkably revealing. During the course of a photographic cycle of 18 days, with 14 circuits of the globe per day, the satellite can report on virtually every last corner of the Earth.

Landsat 4 will mark a quantum advance in our ability to assess the tropical forest situation. Landsat 3 was not able to pick out small plots of forest clearings, which further helps to explain why FAO and UNEP, with their emphasis on remote-sensing information, from the earlier satellite, gave little emphasis to forest farmers. We can look forward to these problems being resolved as we accumulate information from Landsat 4. Within just a few years we should have solid evidence of what is truly happening to tropical forests.

iii. CONTRIBUTIONS

TO OUR WELFARE

A Cornucopia of Foods

OUR DAILY wake-up cup of coffee owes its far-past origin, and its currently acceptable price, to tropical forests. Perhaps we shall shortly sweeten that cup of coffee with a natural sweetener that contains next to no calories; if so, the plant from which the sweetener is prepared will almost certainly come from tropical forests. At breakfast, we may well enjoy a slice of papaya, a banana, a mango, an avocado, or any of at least two dozen fruits that derive from tropical forests. As a measure of our daily delight in these tasty offerings of tropical forests, a typical citizen of North America and Western Europe consumes about 10 kilograms per year of bananas; worldwide, people eat their way through 40 million tons of bananas each year, plus 13 million tons of mangoes, and about 1.5 million tons each of papaya and avocado. We can assert much the same about many other items that turn up in our daily diets, right through to the late-evening cup of hot chocolate.

Of course these items do not come to us directly from tropical forests. They come from plantations. But as is the case with many established crops around the world, tropical fruits, plus beverages such as coffee and drinking chocolate, cannot keep on flourishing without season-by-season infusions of new genetic material from their wild relatives, among other sources of "support germ-plasm."[1] However much we may suppose that a banana planta-tion remains superbly productive because the farmer throws masses of fertilizer and pesticides at it, the key lies rather with its genetic underpinnings. Indeed, all modern crops, being the refined products of selective breeding, constantly require new genetic material in order to maintain and even expand their productivity, to enhance their nutritive content, to improve their taste (or to restore it), and to resist emergent types of diseases and pests, as well as environmental stresses such as cold and drought.

During the past several decades, genetic resources from tropi-cal forests have saved a number of important crops, including cocoa, banana, and coffee. Wild germplasm for cocoa is found in the species' native habitats in western Amazonia and in relict patches of forest in the Pacific coast zone of Ecuador—where one particular variety of the cocoa plant has now been reduced to just a few surviving individuals in the 1.8 square-kilometer Biological Reserve at Rio Palenque. It is a type of cocoa with better taste and other virtues than almost all other gene pools of wild cocoa.[2] Another example is the sugarcane crop. During the mid-1920s, sugarcane growers in the Deep South of the United States ran into trouble from a mosaic virus transmitted by aphids, which brought the crop crashing from over 180,000 tons per year to only 43,000 tons. Fortunately, mosaic-tolerant varieties of sugarcane were found in a wild species that grows in secondary forests of Java, saving the U.S. sugarcane industry from bankruptcy.[3] Since that time, further wild types of sugarcane have supplied resistance to red rot, gummosis, and other pathogens that plague sugarcane growers.

More recently, coffee growers faced a major setback in 1970 when a rust disease appeared in southern Brazil. This was no ordinary type of disease. It revealed at least thirty different races and types. The rust soon spread to Central America, threatening the economies of several countries where coffee exports are the premier source of foreign exchange, earning a total of $3 billion a year. Coffee growers tried conventional methods of fighting the rust, principally by spraying of fungicides. But the cost ran to $200 per hectare, more than many farmers could afford. Again, it was crop geneticists who saved the day. A rust-resistant strain became available from germplasm collected in forests of Ethiopia —the original source of genetic variability for coffee. Even though at least four-fifths of Ethiopia's original forests have been eliminated, and the rest are facing imminent threat, germplasm collectors reached the wild gene reservoirs in time—whereupon coffee growers in Latin America were saved from catastrophe, and coffee drinkers around the world were saved from the one-dollar cup of coffee.[4] Even health addicts benefit from wild sources of coffee germplasm. A new variety of wild coffee, containing not a trace of caffeine, has been discovered in such tiny patches of forest as still survive in the Comoros Islands off eastern Africa.

Significant as these instances have been, the most remarkable contributions from wild germplasm almost certainly lie ahead of us, provided we safeguard the wild gene reservoirs in time. An illustration of the wild gene support yet to come involves the recent discovery, in a small patch of montane forest in south-central Mexico, of a weedy-looking form of wild teosinte, the closest relative of corn.[5] Curiously enough, this proved to be the first perennial type of teosinte with the same chromosome makeup as corn, allowing it to be crossbred with conventional, that is, annual, varieties of corn. A hybrid strain might eventually eliminate the heavy year-by-year costs of ploughing and sowing, since the crop would spring up by itself with every new season, just like daffodils. The wild corn also offers resistance to several diseases, including at least four out of eight major viruses and mycoplasmas that are more than troublesome to corn growers and to corn consumers worldwide.[6] Corn appears not only in cornflakes and popcorn, but in preserves, salad dressing, catsup, soft drinks, beer, and bourbon. Moreover, since the wild species has been discovered at elevations between 2,500 and 3,250 meters, where its cool mountainous habitats are often damp, it may well thrive in wet soils that have hitherto been beyond the survival capacities of conventional corn. The wild species could thus expand the cultivation range of corn by as much as one-tenth, or many millions of hectares. Overall benefits for the global corn industry could eventually be measured in billions of dollars per year.[7]

ENTIRELY NEW FOODS

Tropical forests not only support modern agriculture through their genetic contributions to established crops, they offer hosts of opportunities for entirely new crops, in the form of foods that we may scarcely have thought about thus far. Unaware as we may be, forest plants of the humid tropics have supplied us with many staple foods, including cereals such as rice and millet; pulses such as peanut and mung bean; roots and tubers such as yam and taro; and other well-known items such as cassava, and pineapple, to name but the leading foods.[8] But the fact that these crops are so widely grown does not necessarily mean that they are the best crops. We can argue that today's crops are accidents of history. While we may well believe that rice and other staple products of the tropical forest zone—together with wheat, potatoes, and two dozen other basic foods from other parts of the world—are eaten by the great majority of humankind each day on the grounds that they constitute the world's best food plants, we actually grow these plants because it turned out they were suited to cultivation by Neolithic man. The plants that were selected as most appropriate 10,000 years ago still supply virtually all items on our meal tables today. Many new crops could be awaiting our attention, capable of becoming front-rank crops in many lands if they were given a chance.

As an example of a plant with unusual promise, let us look at the winged bean, a plant from the forests of New Guinea.[9] Because of the shape of its pods, it is also known as the four-angled bean and the asparagus pea; and it has long been known to forest tribes of its native New Guinea as a crop with outstanding nutritional content. The vinelike plant contains far more protein than occurs in potatoes, cassava, and several other crops that serve as principal sources of food to many millions of people each day in the tropics. The winged bean offers nutritional value equivalent to soybean, with 40 percent protein and 17 percent edible oil, plus vitamins and other nutrients. It is not to be decried as a poor man's crop, a vegetable to be dismissed as a second-rate product for third-rate communities. Its capacity to match the soybean's nutritional value might remind us that the United States used to produce only scattered patches of soybean, for at least one century before the plant was finally upgraded to a widespread crop. Today the soybean is the principal protein crop in the world, flourishing in dozens of temperate-zone countries. Could not a similar prospect be in store for the winged bean, scheduled to become the long-sought "soybean of the tropics"? The bean has received a crash program of development and improvement during the last few years, until it now helps enhance human diets in more than fifty countries of the developing tropics.

Probably fruits offer the greatest promise for new foods. Temperate-zone forests have yielded only about 20 major fruits, whereas tropical forests feature at least 250 fruits that please human palates in their millions, and many more are enjoyed locally. In New Guinea alone, 251 tree species bear edible fruits, most of them consumed, at one time or another, by local communities—yet a mere 43 have become established as cultivated crops, and only about one dozen reach the marketplace, indicating the scope for future development.[10] There could well be 2,500 fruit species in tropical forests for human consumption—and of these, perhaps 250 are widespread, 50 are well known, and 15 rank as major commercial species.[11]

To get an idea of what these fruits might consist of, let us look at a couple that deserve to become as widely established as bananas and mangoes. First, the pummelo, a citrus fruit that appears splendidly suited to the warm, moist areas of the lowland tropics. Virtually all other citrus fruits require subtropical environments, with cool, damp winters followed by warm, drier summers. But the pummelo thrives in year-round warmth and moisture. Native to Southeast Asia, it prefers lowland areas, and it can even prosper in shoreline soils, revealing a high tolerance for saline conditions —a trait that would allow it to be grown in brackish, marshy areas. The pummelo is the largest of the citrus fruits, larger even than a grapefruit. As a measure of our global appetite for citrus fruits, we eat 55 million tons per year, and a U.S. citizen consumes an average of 10 kg. per year, probably enjoying the sharp flavor of oranges and so forth virtually every day. Second, a fruit from tropical America, the soursop. A large fleshy fruit that can weigh 4 kg. or more, with a white juicy interior, the soursop's smell is akin to that of a pineapple, but its taste is a striking mixture of musky and acidic flavors. The fruit can be eaten raw, or it can be pulped for use in ice cream and soft drinks.

Similar accounts could be presented, without further field research, for many dozens of other tropical forest fruits; for instance, the mangosteen of Malaysia, described as "perhaps the world's best tasting fruit."[12] So when we visit the fruit section of our local supermarket and marvel, as we rightly may during certain seasons of the year, at the bountiful variety of fruits shipped in from the tropics, let us reflect that these represent but a tiny fraction of the array that we could enjoy if botanists and agriculturists were to exploit the far greater variety of fruit trees available in tropical forests—and that steadily decline toward extinction before the axe and the chainsaw.

As for vegetables, we find that nutritious leaves and other greenery of tropical forests are equally diverse. At least 1,650 plants of tropical forests offer vegetablelike materials of various sorts.[13] Having looked at tropical forest fruits in some detail, we shall not linger with an instance-by-instance account of vegetables —except to note a distinctive way of deriving high-quality food from forest greenery, in the form of leaf protein.[14] This type of food is not yet something that we can purchase in cans in our local supermarket, yet it could become, in just a few more years, a significant factor in our diets—and even more importantly, a common item among protein-deficient areas of the world. Through a process that is sometimes known as green-crop fractionation, plant leaves and stems can be ruptured to release juices and saps that contain several proteins, also sugars, salts, lipids, and vitamins. The process is relatively simple. Leaves are passed through a pulper, the juice is drained and then heated to 70–80 degrees Fahrenheit, causing the protein to coagulate, whereupon it can be separated by means of a filter press. The protein material, dark green and cheeselike in appearance, is mixed with established foods as a protein reinforcer.

A key question is, Which plants best lend themselves to supply of protein through this technique, that is, which offer the greatest quantity of best-quality protein in their tissues, which grow fastest, which best tolerate less-than-ideal conditions, and so forth? Thus far, scientists have been inclined to focus on a series of legumes, with a protein content between 12 and 36 percent. Given present technology for green-crop fractionation, we can generate leaf protein from legumes at a cost of only one-fifth as much as soybean protein, and one-seventeenth as much as protein in the form of milk or meat. Via experimental projects, protein of this sort has been extracted from alfalfa, clover, oats, spinach, Chinese cabbage, water hyacinth, and numerous grasses. Now that we know that green-crop fractionation can be applied to a wide variety of leafy plants, the challenge is to find the most suitable candidates. Plainly, the best bet for investigators lies with the zone that features almost as many plant forms as in all the rest of the Earth.

NATURAL SWEETENERS

In the wake of the cyclamate and saccharin controversies, there is urgent need for an alternative non-nutritive sweetening agent, that is, one that does not add calories to our diet and centimeters to our waistlines. Although sweetness is only one of four basic tastes of humans, it commands far more attention than all the others combined—to the detriment of our health and our appearance, since a single cola drink contains as many calories, in the form of sugar, as half a kilogram of potatoes. We must break ourselves of our sugar habit, since we already consume far too much—100 million tons worldwide each year, or almost 20 kg. for each global citizen. The average American consumes at least 45 kg. of sugar per year, making it a key factor in overweight problems and associated diseases such as heart disorders and tooth decay.

Many plant pigments, such as carotenoids in sweet-tasting fruits, attract birds, insects, and other herbivores. Almost certainly, then, they are not toxic to mammals, including humans. The problem with the natural sweeteners of fruits that we consume is that they are nutritive sugars—glucose, fructose, and sucrose—and we already consume far more of them than is good for us. Fortunately we are finding that a few natural sweeteners in fruits are made up of protein compounds; and these materials, identified only during the last ten years, make up an entirely new class of natural sweeteners.[15] They are all 1,000 times sweeter than sucrose, and at least 300 times sweeter than saccharin.

A leading example of a source of protein sweeteners is the so-called miracle fruit *Synsepalum dulcificum*, a berry from West African forests. It causes sour foods to taste splendidly sweet, as anyone can tell by merely chewing on a berry at the same time he or she is eating lemons, limes, rhubarb, or grapefruit. Another fruit from West Africa's forests, known as the serendipity berry, *Dioscoreophyllum cumminsii*, has a sweetness 3,000 times greater than that of sucrose. The red, grapelike berry derives its curious name from an occasion in 1965 when an American scientist, Dr. George Inglett, was exploring West African forests for sweet fruits. Coming across the plant by chance, he was surprised to find that its berries tasted so sweet; he had gotten no indication from scientific reports.

Still more important than these two is a third fruit from the forests of West Africa. The katemfe, *Thaumatococcus danielli*, contains two sweet-tasting proteins, thaumatin I and II, both of which are 1,600 times sweeter than sucrose. Thaumatin is now widely marketed by the noted sugar corporation in Great Britain, Tate and Lyle Limited, under the trade name Talin. It is becoming strongly established in Japan, where it is used as a sweetener in such diverse products as candies, chewing gum, salad dressing, coffee drinks, soups, jellies, pickles, frozen desserts, fish and meat products, and table-top sachets. It should soon be become available in the United States after surviving tests by the Food and Drug Administration.

CONTROL OF AGRICULTURAL PESTS

Tropical forests can further support agriculture by supplying materials to help keep down the many insect pests that account for the loss of 40 percent of all food grown around the world each year. A sound way to control insect pests is to exploit chemicals from plants that have developed mechanisms to resist insects. The finest source of such plants lies with tropical forests and their exceptional variety of plant forms that have co-evolved in equilibrium with associated insects.[16] Tropical forest plants constitute a vast storehouse of chemical substances for defense against insects—not only biocompounds that serve as insect repellents and toxicants, but feeding deterrents of various sorts, inhibitors of insect growth and development, and the like. Since multitudes of plants and insects have evolved symbiotic relationships within their tropical forest ecosystems, we can surmise that there must be many other insect-resisting substances available in the forests, not only of the types listed, but of novel and unrecognized forms as well.

Moreover, all these compounds are biodegradable. This means that they do not accumulate in organisms and thus do not contribute to the environmental problems associated with synthetic chemical insecticides. Perhaps most important of all, they generally cause little, if any, harm to higher animals, such as birds and mammals, including humans.

By way of illustration, let us note two main categories of toxic compounds—the pyrethrins, from chrysanthemum-type plants, and the rotenoids, from roots of tropical forest legumes. Rotenoids are the more widely distributed, known to occur in the roots and seeds of at least 67 plant species found across Amazonia and Southeast Asia. Especially important as a source of rotenoids is the Derris group of woody climbing plants in Southeast Asia; their roots contain powerful toxins that have long been used by forest tribes as fish poisons. So powerful are these root compounds, that 1 part of plant to 300,000 parts of water is sufficient to kill fish. It was the use of these rotenoids as fish poisons that enabled their toxic characteristics to be identified by Western scientists, and today a series of Derris-based insecticides are used in the form of plant sprays for field crops, and as dips and dusting powders for livestock.

In the American tropics, a major agricultural pest consists of leaf-cutting ants. These ants make their living by stripping leaves from plants, including crop plants. Fortunately, at least one species of tropical forest tree produces a chemical compound that actively repels the ant—thus opening up the prospect that the compound could be used to produce an insect repellant. Better still, of course, would be to identify the genes that enable the tree to produce its own internal insect repellant, and then to splice the chemical-generating genes into certain of our food crops. The same approach could apply to many other plant-produced materials, for example the anti-feedant compound found in bark extracts of forest trees in Papua New Guinea.[17]

But while a gene-splicing strategy will represent by far the most efficient way to use anti-insect compounds produced by plants, the strategy will depend on the greatest possible stocks of genetic variability among wild plants. As the burgeoning industry of genetic engineering reaches a stage where inter-species crosses become not only possible but relatively straightforward, the industry will seek a maximum array of plants with insect-repelling and insect-killing capacity (as they will seek other prized traits, such as resistance to diseases, saline soils, drought, etc.).

At the same time, we should bear in mind that insect pests include variations that can multiply in numbers to overcome plant defenses in as little as ten years, sometimes a mere three years. Hence there is all the greater need to derive further genetic combinations of plants to enable farmers to "stay ahead of the game"—otherwise, they will start to encounter increasing numbers of immune insect strains.

Not all anti-insect defenses in plants are chemicals that make the insect feel sick to its stomach, or worse. A number of chemicals cause insects to moult at the wrong phase of their life cycle; or they inhibit the growth of an infant insect into an adult, leaving it as a perpetual juvenile of harmless scope; or they suppress the reproductive mechanisms of insects; or they make life more than difficult for insects in dozens of other ways.

PEST CONTROL THROUGH
NATURAL ENEMIES

Yet another weapon for the farmer to mobilize against insect pests lies with natural enemies. Certain insect species in the wild operate as predators or parasites to hold down the numbers of their host species—and it is precisely these hosts that frequently proliferate to become pests in agriculture. Predators and parasites attack the eggs, larvae, pupae, and adults of their target species.[18] Since many predators and parasites are highly specific in their choice of prey, they are inclined to target only certain species without doing damage to others. By contrast, the broad-scale use of persistent toxic chemicals tends to kill off far more insects than is necessary.

In any case, many insect pests are growing resistant to chemical insecticides. Among certain species, as few as fifteen generations are needed to build up resistance. According to the Food and Agriculture Organization of the United Nations, more than 300 species of insects, mites, and ticks throughout the world are known to have become resistant to one or more pesticides, and dozens more are suspected of becoming resistant. In the United States, entomologists in the Department of Agriculture estimate that approximately 700 insect species do significant damage to crops within the continental limits (not counting Hawaii, where the problem is still more acute); the loss is calculated at $5 billion per year. Of these 700 insect species, around half have developed resistance to at least one pesticide, some to two or more. American farmers apply 400,000 tons of chemical insecticides to their crops each year, ten times as much as in 1950, yet they lose twice as much food to insects. Hence the rationale for the natural enemies strategy. The track record of this approach is encouraging. Entomologists have recorded at least 250 cases of pest control accomplished through introduction of predators and parasites that attack troublesome species. For example, citrus growers in Florida have been able to save their industry $40 million each year through a one-time outlay of $35,000 in 1973 for the introduction from the tropics of parasitic insects that attack citrus-tree pests.[19]

Furthermore, natural enemies can be used against another form of pest, weed plants. In the United States, alligator weed has proved a notorious and costly problem for rivers and lakes, irrigation canals, waterways for boats, etc. No herbicide proved equal to the task. Fortunately a solution has been found in the form of a flea beetle from South American forests, which regards the alligator weed as its preferred host.[20] Another prominent weed in the United States, as elsewhere in the world, is the water hyacinth; a mere 10 plants can multiply to 600,000 in just eight months, forming a carpet a quarter of a hectare in size, and a carpet thick enough for a person to walk across. A promising response to the water hyacinth problem lies with herbivorous insects from tropical forests of South America.[21]

The economics of the natural enemies strategy are often

highly positive. Across-the-board performance indicates that there is an average of a thirty-dollar return for each dollar expended on importation of beneficial organisms (in contrast to the 4:1 benefit-to-cost ratio for chemical pesticides). In California alone, during the period 1928–1979, a series of natural-enemy projects has reduced crop losses to insects, and has reduced the need for pesticidal chemicals, for savings worth just short of $1 billion (at 1979 prices).[22]

At least 250 insect pests in the United States, accounting for around one-half of all crop losses to insects, are species of foreign origin. It is precisely against these imported pests that alien introductions of natural enemies offer the greatest promise.[23] According to preliminary estimates by the U.S. Department of Agriculture, a full 1,000 species of foreign organisms—not only insects, but also mites and pathogens—could be profitably introduced into the United States, at a cost of less than $40 million over a period of twenty years. Much potential is available in the form of, for example, wasps that act as parasites or predators, notably the chalcid wasps, among others that are numerous in tropical forests. At least as much potential appears to lie with the ichneumonid or braconid wasps, and with certain groups of beetles. To date, American pest experts have called on the services of natural enemies from abroad to overcome many dozens of foreign insect pests: the best times still lie ahead.

FOREST ANIMALS FOR FOOD

Thus far, this chapter has emphasized tropical forest plants that are new sources of food. Let us now take a look at animals.[24]

Within the forests of the Thailand-Kampuchea border lives a secretive, cowlike creature with resplendent horns: the kouprey. A bull kouprey sports horns that are longer and wider than those of any other living wild cattle except the buffalo; a female's are lyre-shaped, corkscrewing upward in a manner akin to those of the lesser kudu in Africa. The kouprey is believed to have been one of the wild ancestors of the humped zebu cattle of southern Asia, which means that further cross breeding between the two bovids could boost cattle raising throughout the entire region. In particular, the kouprey appears immune to rinderpest, a widespread disease that is fatal to cattle. Regrettably, the kouprey's very survival is doubtful, due to military activity in its habitats during the past several decades. Perhaps only a few individuals remain.

Other wild bovids of Southeast Asia's forests could likewise help cattle husbandry, notably the so-called dwarf water buffaloes, more correctly termed the selatang, the tamarau, and the anoa. But as with the kouprey, the numbers of all these wild cattle have been severely reduced through human disruption of the forest ecosystems. In addition, the forests of Southeast Asia support the babirusa, a distant relative of the pig that seems to be a rudimentary ruminant and, through its distinctive approach to the challenge of converting rough forage into good meat, could upgrade some of the half-billion pigs around the world.

When speculating on the potential contributions of these wild creatures to modern livestock, let us bear in mind that the ancestral stock of the domesticated water buffalo includes the wild *Bubalis arnee* of southern Asia; that among the progenitors of Indian cattle is the gaur, *Bos gaurus;* and that livestock people in Indonesia already use a hybrid animal, known as the madura—a cross between conventional cattle and the banteng, *Bos banteng.*

As for birds, we might reflect, the next time we enjoy a chicken supper, that this most numerous bird on Earth, numbering at least five billion, originated from a pheasantlike creature of India's forests, known as the red jungle fowl, together with some genetic support from the guinea fowl of West Africa.

WILD MEAT

In a number of tropical forest countries, wild meat counts as an important item in people's diets.[25] In Nigeria, for example, where the forest cover has been reduced by at least 90 percent, local people still derive a renewable harvest of almost 100,000 tons of good, solid meat per year from animals including grass-cutters (giant rats), small antelopes including bushbuck and duiker, and sundry monkeys. Four people out of five enjoy game meat as a regular item in the cooking pot. On average, it constitutes one-fifth of all animal protein of whatever sort for people in Nigeria's forest zone. In Zaire it rises to almost 27 percent, and in Cameroon, Ivory Coast, and Liberia, to a massive 70 percent.

Wild meat comes not only from mammals. In Amazonia, seven species of river turtles could, if properly managed, become sustainable sources of high-grade meat.[26] Turtles feed readily on aquatic plants of all sorts and conditions; they survive temporary food shortages at the end of the dry season without adverse effect; they have low metabolic rates, so they do not become nearly so hungry as warm-blooded creatures; and they appear to need far less living space than higher vertebrates. According to some theoretical calculations, a one-hectare lake of turtles could produce well over two tons of meat each year, in contrast to the 50 kg. of beef that could come from one hectare of average cattle pasture in the humid tropics. Other creatures of South American forests provide abundant meat, notably the tapir, agouti, and paca, all rodents of Amazonia, plus peccaries (wild pigs), monkeys, and other mammals, also snakes and lizards. In two separate areas of Peru's sector of Amazonia, people living outside townships depend on wild meat for 80 to 85 percent of their animal protein.

To date, the offtake of wildlife in Amazonia has tended to range from next to no harvest at all, to severe over-exploitation. If, however, a controlled-cropping system were to be mounted, the result could be encouraging, according to some preliminary figures worked out by Dr. Angel Paucar, of the Wildlife Division in Ecuador, and Dr. Alfred L. Gardner, of the U.S. Fish and Wildlife Service.[27] Their calculations represent no more than a tentative and speculative attempt to illuminate a situation that remains almost entirely unexplored. So the numbers represent best-judgment guesstimates—no more and no less.

In Ecuador's sector of Amazonia, local people have long taken a self-renewing harvest of wildlife products, partly in the form of meat for consumption and partly in the form of skins and hides for export. In some localities the offtake has been excessive in the case of certain species. But by and large, wildlife has proved itself to be a strongly renewable resource, given the low levels of exploitation to which it has generally been subjected. So few people live in this part of Amazonia, that only a moderate harvest can make a sizable contribution. As much as 85 percent of animal protein consumed by local people comes from wild animals, notably peccaries, deer, tapirs, pacas, and agoutis, among some forty species of mammals in all. A sustainable harvest of wild meat can, according to Paucar and Gardner, amount to 240 kg. per sq. km., with a market value of about $1.8 per kg., or a total of almost $440 overall. (Were the harvest to be systematized, and expanded to include birds, turtles, and fish, the minimum potential value could be increased as much as ten times.) Furthermore, since almost one-third of Amazonian Ecuador consists of swamps and rivers, there is plenty of scope for harvest of caimans, at least two per hectare per year. A caiman measuring 1.5 meters in length is worth $145. Thus 1 square kilometer of such habitats could yield $14,550 per year for hides alone, rising to $16,370 when we include meat from the caimans. In addition, each square kilometer of forest can renewably produce twenty primates each year for biomedical research, an individual being worth between $200 and $300. So a sustainable harvest of wild primates could generate a minimum of $4,000 per year. Many readers may object to the prospect of using wild primates to foster human health. In the past, the wild primate stocks have almost invariably been overexploited, sometimes critically so. But if this crucial factor can be taken care of, need any objection remain, particularly when a sustainable harvest of primates helps to promote the survival of their forest habitats?

This all means that a forest tract of 500 square kilometers could, under scientific management, produce a self-renewing crop of wildlife with a potential value of at least $10 million per year, or slightly more than $200 per hectare. These revenues are to be contrasted with a return from commercial logging of only little over $150 per hectare—and hardwood timber, as we have seen elsewhere in this book, tends to be harvested as a once-and-for-all product, leaving little prospect that a further harvest can be taken within several decades at least.

In conclusion, then, we can accept that tropical forests, while making significant contributions to modern agriculture already, play only a trifling part in the life of the farmer compared with what they could supply. Were we to undertake a systematic exploration and selective extraction of whatever "agricultural" materials are available from tropical forests, we could surely look for a steady stream of new products. Were this chapter to be re-written in the year 2000, who can guess at the host of exotic articles that it might enumerate? The problem surely does not lie with tropical forests themselves, with their extreme abundance of resources waiting to be brought into the mainstream of our lives. Rather, the deficiency lies with our imaginations: we constantly fail to grasp the scale of potential products from tropical forests that could enrich our lives.

II

Pharmaceutical Factories

Now THAT I have been away from my beloved Kenya for over one year, I sometimes speculate on returning for a brief sojourn. I don't have to think long before getting a clear idea about which locality would draw me most.

It would not be Amboseli or any of the great wildlife parks, much as I treasure my memories of over one thousand nights in camp among the elephants and the giraffes and the lions. Nor would it be the Kenya coast, with its palm-ringed lagoons, and the best coral-reef snorkeling I have found anywhere in the tropics. It would not be a Masai encampment, even though I was once made an honorary elder of the tribe for sundry services and I know I would still receive a welcome from them. It would not be the Rift Valley or the northern deserts or Lake Nakuru, much as I would rather visit any one of these areas than scores of other scenic places that I know around the world. It would be Mount Kenya.

Dear old Mount Kenya, almost 5,600 meters high, and 800

meters higher than the highest mountain in the United States outside Alaska: what times we have had together. It was to Mount Kenya that I first took a safari party of boys from the Nairobi school where I formerly taught; it was a try-out expedition for what became a series of 30-odd such trips to mountains of East Africa. It was Mount Kenya that my wife and I chose for our first mountaineering venture together; and it was Mount Kenya that our family selected for our farewell safari (whereupon our two daughters led their parents a merry romp up the moorlands toward the ice). All in all, I have made twenty-two treks to the snows and glaciers of Mount Kenya, sometimes in company of crowds of others, sometimes alone.

During my latter years in Kenya, after I took up marathon running, I specially enjoyed a series of single-day roamings on Mount Kenya. I would jog up to the head of the largest glacier and down again in just a dozen hours, an excursion that would leave me feeling as if I had done more living in a short span than I would normally accomplish in several weeks down on the plains. I would drive from Nairobi to the foot of the mountain on, say, a Friday evening and spend the night in a safari cabin located at 3,000 meters, in the thick forest that encircles the mountain's base. Next morning I would set off at first light in running gear, following a trail that eventually emerged from the forest onto the open slopes that led me to the screes and the iced ridges of the main peaks. By late morning I would reach a point higher than the Alps, looking out toward the glistening dome of Kilimanjaro 400 kilometers to the southeast. Then I would wander around the shoulders of the main massif, perhaps making a circular tour of the entire peak zone before heading off down the mountain in late afternoon.

All the while, I would talk with Mount Kenya. We would plan new explorations together; we would share thoughts on the cloud patterns in the immense sky; I would listen to the mountain's reflections on its life of stark solitude. Eventually I would plunge down into the forest again, sensing a sharp contrast between the harsh landscapes above and the reassuring environment of the forest's warm interior. Mount Kenya's forest was always softer-seeming than a regular rainforest of the lowlands, many of its trees being covered with moss and festooned with filmy creepers. It appeared a benign community, secure and supportive to me after all those hours I had spent around the near-lifeless summit. For all their bewildering variety, the plants of the forest would offer a single message to me as I returned from the world of shale and ice above: "You are welcome back with us."

So I keep a secret place within me reserved for my memories of that magic mountain and its forest. And I was not surprised to hear, shortly before I left Kenya, that Mount Kenya's forest supplies something special to local people too. It is a prime source of medicinal plants for tribal herbalists and other traditional healers. Scientific skepticism apart, it turns out that a number of these "witchdoctor" preparations do in fact work. I remember a consultation with Dr. John Kokwaro, a botanist at the Kenya Herbarium and an expert on plant medicines whose reknown has spread to North America and United Nations agencies. "Our local medicine men use a combination of plants from Mount Kenya to prepare a remedy for asthma," Dr. Kokwaro explained to me. "Their treatment works at least as well as anything they could get from a modern hospital in Nairobi. We don't know why the plants are so successful; we can't pin down a strictly scientific explanation. But we know that they do work, and we are having them checked out to see if they can be developed into a commercial drug for broadscale use throughout East Africa."

I have learned similar insights from another local guru on tribal medicine, Professor Njuguna Ngwaro, a biochemist at the University of Nairobi. He has told me of a drug concocted from seeds gathered on Mount Kenya, capable of easing labor during childbirth and more potent in some respects than anything available from a modern pharmacopeia. Colleagues of Professor Ngwaro have described to me how local people use plant-derived materials for their contraceptive needs. So effective are these plants, that they have been selected for investigation by the World Health Organization, a U.N. agency that is searching for anti-fertility materials that are safer and more efficient than the conventional pill.

When I think back, then, to the wonders of Mount Kenya's forest, and to the strange memories of hyper-experience that I used to sense as I came down from the mountain heights, I feel it is altogether fitting that the same forest should supply materials in support of human well-being of other sorts.

TRADITIONAL MEDICINES FROM TROPICAL FORESTS

Many other tribal peoples look to the plants and animals of tropical forests for their medicinal needs. Through long experience, they have learned that tropical forests can supply a welter of products, not only drugs and pharmaceuticals, but also stimulants, narcotics, and hallucinogens among diverse other products that make life more livable in remote territories. In northwestern Amazonia, for instance, Professor Richard E. Schultes, Director of the Botanical Museum at Harvard University, has found that forest-dwelling Indians employ at least 1,300 plant species for medicines and related purposes of one sort and another.[1] In Southeast Asia, traditional healers use some 6,500 plants in treatments for malaria, stomach ulcers, syphilis, and assorted other disorders.[2] In the 4,500-square-kilometer island of Siberut, off the west coast of Sumatra in Southeast Asia, medicine men draw on dozens of plant species, many of them found nowhere else, for herbal remedies against a host of illnesses, also for astringents, disinfectant materials, purgatives, emetics, and sedatives.[3]

So promising are many of these ethnobotanical materials when evaluated by scientists, that a good number are being screened and developed in medicinal research laboratories of North America and Europe.[4] Little as we may realize it when we visit our neighborhood pharmacy for a medicinal preparation or a pharmaceutical, there is roughly one chance in four that the product we purchase, whether by prescription or not, owes its origin, in some way or another, to plants and animals of tropical forests. The commercial value of these products worldwide now tops $20 billion per year.[5]

TROPICAL FORESTS AS VAST PHARMACEUTICAL FACTORIES

Tropical forests represent nature's main storehouse of raw materials for modern medicine. Plants alone offer a host of analgesics, antibiotics, heart drugs, enzymes, hormones, diuretics, anti-parasite compounds, ulcer treatments, dentifrices, laxatives, dysentery treatments, and anti-coagulants, among many others. The total number of plant-derived products in modern pharmacopoeias amounts to several thousand, including such well-known trade products as emetine, scopolamine, and pilocarpine.

Among the most important materials with which pharmacologists manufacture drugs are alkaloids—complex biocompounds produced by many categories of plants. Of all biomes, tropical forests contain the highest proportion of alkaloid-bearing plants, and the alkaloid yield of these plants is higher than elsewhere.[6] Tropical forests provide a greater abundance not only of alkaloids, but of antibiotic and antibacterial compounds. Plant alkaloids have many medicinal applications, including cocaine, reserpine, quinine, ipecac, ephedrine, caffeine, and nicotine.[7] Moreover, the capacity of plants to produce alkaloids is under strict genetic control, which means that provided we can safeguard sufficient genetic variability in the wild, genetic engineers may soon be able to devise improved varieties of alkaloid-producing plants in order to better serve the needs of modern medicine.[8]

Probably most famous of all alkaloidal drugs are vincristine and vinblastine, developed by Eli Lilly and Company of Indianapolis, after following up clues provided by tribal herbalists and other traditional healers in Madagascar and Jamaica. The

Forest fishermen in Amazonia use plant-derived poisons on their spears and arrows. It was one of these poisons that first gave Western pharmacologists the clue to curare, and we can thank tribal peoples, such as the man featured here, for one of the major supports of modern surgery. *(World Health Organization)*

Vinca alkaloids derive from a dainty plant of drier tropical forests, the rosy periwinkle, a flower so small that you might overlook it unless you notice its pretty appearance. Thanks to the *Vinca* alkaloids yielded by this plant, a child suffering from leukemia, who would have faced only one chance in five of remission in 1960, now enjoys four chances in five.[9] The two drugs are also used against Hodgkin's disease and other malignant lymphomas, breast, cervical, and testicular cancers, choricarcinoma and certain tumor-type cancers.[10] Not surprisingly, worldwide sales of these two drugs totaled $90 million per year in the late 1970s, and the commercial turnover continues to expand at a vigorous 15 percent each year.[11]

At least another 1,400 plant species of tropical forests are

The rosy periwinkle, of drier tropical forests, supplies alkaloidal materials for two potent drugs used against Hodgkin's disease, leukemia, and several other cancers, with commercial sales worth around $100 million worldwide each year. If, as is all too likely, we lose fifty tropical forest plants each year, who is to say that we are not losing more sources of anti-cancer products, plus many other medicines and pharmaceuticals? *(P. Steele, World Wildlife Fund)*

believed to offer potential against cancer.[12] And while ruminating on these sizable figures, let us bear in mind that scientists have conducted cursory screening of only 1 in 10 of tropical forest plants, and intensive screening of only 1 in 100 plants. In Latin America alone, with its estimated 90,000 species, we have tested no more than 10,000 for their anti-cancer properties; and our experience to date suggests that of the remaining 80,000 at least 8,000, should reveal some form of activity against laboratory cancers, and of these, 3 could eventually rank as "superstar" drugs.[13]

Apart from alkaloids, there are many other complex organic chemicals, sometimes known as secondary plant metabolites, that can serve the cause of modern medicine. Hitherto we have defined, in chemical terms, only about 10,000 of them, even though we can surmise that there could be as many as 400,000 or more on Earth. Of the 10,000, some 4,500 are alkaloids; 1,200 are flavonoids; and 1,100 are terpenes. As with alkaloids, we find that no other biome contains such concentrations of these metabolites as the tropical forest biome.[14]

So much for plant sources of medicinals. How about animal sources?[15] Scientists are finding that many insects, notably butterflies, offer potential for many medicinal products, including anti-cancer compounds.[16] As we have already seen, tropical forests support between 1.5 and 3.5 million insect species (possibly many times more). So much potential lies ahead of us: if we have so far looked, however superficially, at about 1 plant species in 10, we have hardly glanced at 1 animal species in 1,000. Let us consider the uses of bats, for example. These diminutive creatures use highly sophisticated sonar equipment; they live for an unusually long time; they are resistant to many diseases; they can become torpid at almost any time; and they possess virtually transparent wing membranes. Thus they are much sought for research, having contributed to navigational aids for blind people, to studies of aging processes, to development of vaccines, to testing of drugs, and to studies of creatures in space.

"BIOLOGICAL WARFARE" IN
TROPICAL FORESTS

Why, we might ask, do tropical forests appear to contain proportionately more of these biologically active compounds than is the case with other biomes? A leading answer lies with the compressed nature of many assemblies of plants and animals, accommodating a far greater number of species per unit area than we would find elsewhere (with the possible exception of tropical coral reefs). A concentration of species in a limited locality makes for competition, lots of it, as organisms jostle each other for food, light, and living space, generally. They compete constantly, and in every last which way: they generate "biological warfare" of the most intense kind we know. In response to this competition, tropical forest species throw up specialist lifestyles, often in the form of unique biocompounds in organisms' tissues, used for defense, attack, whatever purpose that helps the owner to stay alive in circumstances where survival is the principal preoccupation.

Furthermore, many species in tropical forests are either completely immobile, as is the case with trees, shrubs, and the like, or they are only moderately mobile, as is the case with ants, termites, beetles, and so forth. While ants and other insects move around busily within their habitats, a habitat may not extend beyond a few square meters. These conservative lifestyles, "tying" the organisms to a single small sector of the forest, mean that the creatures are little inclined or able to quit the scene when a predator or a parasite arrives. Should the most usual form of threat, that is, a bacterium or a virus, happen along, the targeted victim cannot escape by moving on elsewhere. It must stay where it is, risking infection—a threat to which it responds by producing antibacterial and antiviral defenses. In addition, the stationary lifestyle can even mean that an organism becomes overwhelmed, whether suddenly or slowly, by neighboring animals and plants: again, this is a situation that is best resisted through production of defensive organic chemicals that pass a loud-and-clear message to the encroacher, "Keep your distance."

All in all, many tropical forest organisms produce biocompounds that have unusual types of form and function. Such enormous reserves of strange substances occur in tropical forests, that this realm could ultimately prove to be a major source of start-point materials for medicines, drugs, and pharmaceuticals of the future. To help us gain a grasp of this potential, let us look at the forests' capacity to serve our daily health needs.

THREE MAIN APPLICATIONS FOR MODERN MEDICINE

In practical terms, tropical forest species serve modern medicine in three main ways.[17] First, extracts from organisms can be used directly as drugs; and they can serve in semi-synthesis of drug derivatives and active therapeutic agents. Second, the chemical structures of forest organisms offer templates, or blueprints, to enable researchers to chemically synthesize drug compounds. Third, forest organisms serve as research aids for the development and testing of drugs and pharmaceuticals, also for the needs of biomedical research generally.

1. Direct or Semi-Direct Use of Organisms as Drugs

First off, let us briefly review those biocompound extracts from plants and animals that can be used as drugs with little more ado. A vast number of tropical forests species yield substances that are pharmacologically important in the manufacture of ethical drugs. A survey of U.S. prescriptions in 1973 showed that the plant kingdom was already supplying 76 drug compounds, a total that has grown still higher today. Of these 76, only 7 could be commercially produced through chemical synthesis from start to finish.[18] Most of the drug compounds in question have been successfully synthesized in the laboratory, but it is generally more costly to synthesize them than to extract them directly from wild plants (or occasionally from cultivated, i.e., domesticated, forms of wild plants). While the sales volume of prescription medicines has continued to expand year by year, thus encouraging pharmacologists to cut corners in their manufacturing process whenever possible by synthesizing their materials, they have made no dent at all in the proportion of natural plant products used in modern medicine. In short, we benefit from the plant kingdom more and more as the years go by.

A prime example of this thesis is rauwolfia, a material from the so-called snakeroot plant of monsoon forests in India. Rauwolfia was actually used for more than four thousand years by Hindu seers to treat nervous disorders and mental illnesses, also dysentery, cholera, and fever, before its potential became known to Western-world scientists in the late 1940s—whereupon its alkaloid reserpine began to form the base of tranquilizer products and of materials for treatment of hypertension, anxiety, and schizophrenia, also for menstrual tension and menopausal disturbances.[19] Before that time, high blood pressure strongly disposed a patient toward stroke, heart failure, or kidney failure. Today, however, this one plant helps many millions of people to lead a reasonably normal and healthy life, partially freed from a set of ailments, that is, hypertension, that constitute the single greatest and fastest-growing source of mortality in advanced societies. As long ago as the early 1960s, this first modern tranquilizer generated sales in the United States alone worth $30 million per year. By the late 1960s, at least four out of five of all hypotensive drugs prescribed in the United States, amounting to almost 6 percent of all community drug prescriptions, contained alkaloidal extracts from rauwolfia species.[20] In recent years, retail sales of reserpine-based products in the United States have exceeded $260 million per year.[21] Although reserpine can be commercially synthesized, the process is a complicated and multi-stage affair, yielding the drug at a price of around $2 per gram, whereas reserpine can be extracted from natural sources at about half that cost.

In addition, the biochemical compounds of many tropical forest organisms serve as building blocks for the manufacture of semi-synthetic drug materials. A case in point is the glycoside saponin from the *Dioscorea* species of the Mexican yam, which yields steroidal sapogenins—including the well-known product diosgenin, a compound that came into wide use as an anti-inflammatory preparation after World War II, before finding very widespread application in the form of birth-control pills. Of all steroidal drugs on the market today, 95 percent derive from diosgenin as a raw material. These drugs contribute to our everyday welfare in that at least one out of seven prescriptions from community pharmacies contains steroids. Not only are diosgenin-derived steroids used in the manufacture of oral contraceptives, but they serve parallel purposes in sex hormones (progesterone, estrogens, androgens, etc.) and in cortisone and hydrocortisone used against rheumatoid arthritis, rheumatic fever, Addison's disease, several allergies, sciatica, and a number of skin diseases including contact dermatitis.

All in all, at least 125 *Dioscorea* species have been examined in detail to see if they could be suitable sources of diosgenin. Of these, only 2 have proven to be first-rate sources for commercial needs—both are climbing vines known as the Mexican yam.[22] Until just the last few years, the Mexican yam has yielded virtually the world's entire supply of diosgenin. By the mid 1970s, the world was using up to 180 tons of diosgenin per year; by 1985 the amount could rise to as much as 500 tons; and by 1995, to 3,000 tons, if the contraceptive needs of all women at risk are to be met. In 1980 some 80 million pills were being used each day; if total needs worldwide were to be recognized and supplied, the figure would probably be at least two and a half times as great. At the peak of production of diosgenin from the Mexican yam, at a price of $152.20 per kilogram, the wholesale value per year must have been around $25 million. When we look at further stages in the manufacturing process, we can reckon that chemical composition would have pushed up that figure to $70 million per year, and cross-counter sales for final products could have totaled some $700 million.

In view of this end-product turnover in the commercial marketplace, and in light of the fact that the yam could scarcely be successfully grown outside its native habitats in tropical forests, the Mexican government decided to sieze a greater part of the action through jacking up export prices. In other words, the Mexicans decided to "cartelize" their yam product. Regrettably, they misjudged the situation, and the bottom dropped out of the yam market, with the price for diosgenin rapidly falling to $95 per kilogram, as manufacturers looked around for alternative sources, both natural and synthetic. Today the contraceptive industry uses many other plant-derived steroidal compounds, notably from the soybean, the Calabar bean which flourishes in West Africa's forests from Sierra Leone to the Congo People's Republic, and from plants in Himalayan forests and other parts of tropical Asia.[23] Let us note, however, that the demand for contraceptive materials is likely to double between the late 1970s and the mid-1980s, leading to a shortfall of almost half total requirements—a gap that diosgenin from the Mexican yam can once again fill.

Many other plant species, around half of them from tropical forests, have been identified as containing antifertility compounds: at least 4,000 such species altogether.[24] Of these 4,000, at least 370 have been shown to offer special promise for safer and more effective contraceptive pills, suitable for both males and females. A leading candidate is the greenheart tree that flourishes in the forests of Guyana, where women of several tribes use the tree's nut as a traditionally reliable contraceptive. Among other promising plants are forest growths in Haiti, Bangladesh, Papua New Guinea, Fiji, Cameroon, Madagascar, Colombia, and Brazil. In addition, more than 600 plant species appear to offer potential as abortifacients, a good number of them from tropical forests.[25] A notable instance is a pretty ornamental plant from the forests of India, Thailand, and Malaysia, widely used locally to induce abortions.[26]

2. Chemical Structures as Templates and Blueprints

We have noted that, for all our sophisticated technology, we often cannot manufacture synthesized substitutes for natural materials at competitive costs. In the case of steroids, scientists have tried for twenty-five years to develop synthesizing techniques, but we still obtain 95 percent of our startpoint materials for steroids from natural sources.

Even if medicinal technology cannot always stand in for raw extracts from wild species, however, it can benefit from models supplied by biocompounds from plants and animals in tropical forests.[27] Working from these models, pharmacologists determine chemical formulas of natural compounds, and then go on to synthesize mimic compounds in the laboratory. Without model compounds derived from nature, many of our most common synthetic drugs could not have been devised. One century ago, an aspirinlike substance was extracted from the leaves of willow trees; today aspirin is produced artifically, and a lot more cheaply, through replication of the chemical formula that owes its origin to the willow trees' extracts. A notable instance from tropical forests is the ipecac plant of Brazil and Bolivia, yielding the alkaloid emetine, used to treat acute bronchitis, croup, bilharziasis, and guinea worm. In particular, emetine is used against amoebic dysentery: a concentration of 1 part in 100,000 is enough to overcome the amoeba. Emetine is now synthesized commercially, an advance that would not have been practical without the blueprint supplied by the wild plant.

Similarly, the coca bush contains cocaine in its leaves, which, together with other derivatives, serves as a local anesthetic. The chemical structure of cocaine supplies us with chemical information for synthesis of various other local anesthetics, notably procaine and other related drug compounds.[28]

Some observers may suppose that the template insights supplied by biocompounds in tropical forest plants may not persist as a long-term benefit. After all, they say, if pharmacological researchers were to stretch their imaginations a bit, they should surely be able to visualize the chemical structures without waiting for nature to provide the start-off clues. Our experience shows, however, that the scientists' creative capacities fall way short of the challenge. The biochemical makeup of natural drug compounds are simply too esoteric for the most talented white-coated researcher in his laboratory to dream up a fraction of the intricate formulations already designed for us by nature—existing in their most ingenious and sophisticated forms in tropical forests.

3. Organisms as Research Aids

Pharmacological enterprises find much help in their exploitation of plant chemical compounds as aids for research. By way of example, certain plant compounds promote tumors and thus serve to promote understanding of what makes cancer cells work. They also help to alert us to possible sources of tumor-causing agents in our drugs and foods.[29] Salient instances include the betel nut (a masticatory used in Southeast Asia), many tropical cycads, and plants that produce tannins, resins, and rotenone—all being plants that produce potentially oncogenic compounds.[30]

Similarly, neurologists have been assisted by a toxic compound, tetrodotoxin, derived from certain frogs of Central America's forests (also from some tropical coral reef fishes), by virtue of the fact that the compound possesses 160,000 times the potency of cocaine for blocking nervous impulses. As a result, it is much used in studies of nerve impulse transmissions and nervous excitation. Tetrodotoxin is also employed directly as a drug in Japanese clinics, where it serves as a pain killer, a local anesthetic, and a muscle relaxant, notably for persons suffering from neurogenic leprosy and terminal cancer.

Especially illuminating for psychiatrists are hallucinogenic plants of Latin America, where they play a part in religious rituals by virtue of their psychoactive principles.[31] The same compounds that yield these psychoactive principles can be employed for research into the central nervous system in general, and psychiatric disorders in particular. Of course a number of psychoactive drugs have long been familiar to us, such as analgesics and euphorics (e.g., cocaine and opium), sedatives and tranquilizers (e.g., reserpine), and hypnotics. But while other psychoactive substances are merely mood modifiers, hallucinogens produce, to cite a leading pharmacognosy expert and ethnobotanist, Professor Richard E. Schultes, "deep changes in the sphere of experience, in perception of reality, space and time, and in consciousness of self."[32] This means that hallucinogens can throw light on research into our perception of the external world, insofar as there seems to be a similarity between psychoactive plant compounds and naturally occurring brain hormones. To date, the principal source of these hallucinogens is Amazonia, but other tropical forest zones may prove promising, as witness the drug hyoscine (scopolamine) from the corkwood tree, a forest plant in eastern Australia, used against several forms of mental illness.

Exotic as they may sound, hallucinogens will almost certainly become more important in the years ahead. Being complex compounds, they serve a variety of purposes, not only in the treatment of mental disorders, but also in experimental psychiatry (and they bring relief to patients with terminal cancer). To date, pharmacologists have tracked down at least 150 plant species with hallucinogenic properties, and several times that number surely await our attention if the scientific researcher can get to the forest habitats before the sawman.

Most important as "research aids" are tropical forest animals, especially primates. The closest relatives of humankind, primates manifest reactions to drugs that all but replicate those of humans. Whatever we may believe about the ethics of using some of the most advanced representatives of the animal kingdom to save us from risks associated with new drugs, we can recall the thalidomide tragedy of the early 1960s as a cautionary tale. Too late we found that we could have alerted ourselves to the disaster if we had undertaken thorough testing with creatures that turned out to display identical birth defects when they were dosed with thalidomide: at least two species of macaque monkeys from the Indian sub-continent could have served as ideal "testing tools" (an abominable but accurate term).

The role of primates in evaluating drugs for pregnant women is plain. Not so apparent but equally important is their capacity for checking on additives in drugs and foods, and for general research in fields such as reproductive physiology and fetal pharmacology, drug metabolism, production of human vaccines, drug abuse, mental health, malnutrition, studies of cancer, the central nervous system, arteriosclerosis, and chronic degenerative diseases of organ systems.[33] Among the most common species in question are the squirrel monkey from forests of tropical Latin America, for research into cardiovascular diseases; the common marmosets from forests of eastern Brazil, for research into cancer and hepatitis; the so-called Celebes macaque, for study of diabetes; and the chimpanzee from tropical Africa, for research into human gonococcal urethritis in particular and psychobiological studies in general.[34]

SOME OTHER MEDICINAL WINNERS

Along with these three main categories of medicinal applications of tropical forest species, including some major products such as vincristine, reserpine, and emetine, are several other products with pharmacological value. Of obvious importance is curare, derived from a plant of western Amazonia. The compound induces paralysis of skeletal muscles during a number of critical conditions for humans, including abdominal surgery, tetanus convulsions, shock therapy, and spastic cerebral palsy. Without this key compound—originally brought to the notice of modern science through its use by forest-dwelling Indians as an arrow-tip concoction to paralyze prey—today's surgeons would often be severely restricted in their work. A somewhat similar product is picrotoxin, from seeds of the so-called Levant berry plant of forests in Southern and Southeast Asia. The compound helps to relieve schizophrenic convulsions, and is used to restore breathing in persons who have taken an over-dose of barbiturates. The papaya of Central America supplies us with papain, being a proteolytic enzyme that helps with dyspepsia and chronic diarrhea (it also serves to tenderize meat). Various balsams from tropical Latin America are used as antiseptics, as cough preparations, and as suppositories, also as dental cements. The Calabar bean, from forests of West Africa, already mentioned above as a new source of diosgenin, assists with eye disorders such as glaucoma; it also promotes contraction of the pupil and thus assists ophthalmologists in their work. We have already noted that the corkwood tree of eastern Australian forests alleviates several forms of mental illness; this product, together with atropine from the same plant, relieves goiter, and even helps those people who become sick when they travel in ships or airplanes. Finally, the benzoin tree in Malaysia and Indonesia yields a tincture used as an antiseptic and protective coating; it can also be taken internally for bronchitis.

The interested reader can find many more accounts in the scientific literature.[35] We can see that some sectors of the tropical forest biome rank higher than others as major sources of medicinal products. Let us also note that in India, with its 18,000 flowering plants, a full 2,500 are used for medicine; in the Philippines, with 9,000 flowering species (of which 2 out of 5 are found nowhere else), almost 900 are widely used for health needs; and in Java, with its 4,598 native species, at least 200 are used for medicine. Let us further reflect, then, that even were modern Western-world medicine to eventually find all its drugs and related materials in the synthesizing processes of the laboratory, there would still be a vast need for plant-derived compounds. For many years to come, citizens of the Third World are not likely to enjoy the benefits of costly health systems such as are within the reach of Americans and Europeans. So they must depend, as they have for centuries, on the "green medicine" that they devise from their native plants. The most fortunate of these Third Worlders are those living within the tropical forest zone.

BASE-LINE CALCULATIONS

On the basis of the figures given above for India, the Philippines, and Java, we can reasonably calculate that a "responsible minimum" figure of 5 percent of all plant species serve the cause of medicine in one way or another. Furthermore, we can presume that at least one-half of these plants occur in lowland primary forests. If we apply these reckonings to Southeast Asia, and to the richest of the region's forests (also those undergoing the most rapid exploitation), we can infer that the Malay Peninsula, Borneo, and New Guinea each probably support some 200 important medicinal species.[36] As we have also noted, it is not unusual for a plant-derived drug to achieve commercial sales worldwide worth many millions of dollars per year, ranging from $30 million to around $250 million. With $50 million as a safe low average, the three sectors of Southeast Asia could well generate plant-derived medicinals worth $10 billion per year.

While this is a very "iffy" speculation, it is advanced solely to cast a little light on a murky issue—and a little light is better than none. There is surely much in the situation that we do not know. Yet even if the figure were too high by a factor of ten, Borneo would still represent a sound bet for humankind's hopes for improved health. What we certainly know is that we already enjoy a sizable number of successful drugs from tropical forests that assist the lifestyles, if not the survival, of many millions of Americans, as of Britishers, Japanese, and Australians—and of Brazilians, Nigerians, and Indonesians. Moreover, we enjoy these naturally occurring medicinals after only the most haphazard examination of just a few species in tropical forests.

So, working with these calculations, crude and approximate as they are, we can hope that better times lie ahead, supposing we can exploit the tremendous pharmacopoeias of tropical forests while they remain in existence. All this is plain to me now as I sit at my Oxford writing desk, conducting my back-of-the-envelope calculations about the potential value of tropical forest drugs.

I did not realize all this in the days when I was running up and down Mount Kenya, otherwise I might have delighted even more in the welcoming embrace of that beautiful forest as I descended off the moorlands. In point of fact, my recent dashes up Mount Kenya, as part of the training of a runner who likes to finish in the leading 5 percent of the giant fields at the New York and Boston Marathons, have not been my fastest efforts on that mountain. My speediest ascent was back in the early 1960s, when, as a member of the Kenya Mountain Club rescue team and as a record holder for the up-and-down trip on Kilimanjaro, I used to act as a two-legged medicine mule to rush relief supplies to injured climbers on the topmost cliffs of Mount Kenya. Many of these fallen climbers quickly contracted a disease that is a major threat to injured mountaineers at high altitude—a lung ailment known as pulmonary edema, which causes the lung cavities to fill with fluid, whereupon the sufferer drowns. Little did I know then that the drug I administered, and that appeared to save several climbers' lives, was a heart stimulant known as ouabain, manufactured from the seeds of a strophanthus plant of tropical forests in West Africa. Curious, that a refined product of an obscure plant in the coastal zone of the far side of Africa should serve to keep alive the spark of life in broken humans at lofty altitude in East Africa.

12

Forest Industrial Complexes

I HAVE A DREAM. It is to industrialize tropical forests. Some of them. To help them survive in more or less undisturbed form.

How can this be, given the present over-use of tropical forests? I ask you to bear with me in this apparent contradiction, and let us explore the prospects of exploiting tropical forests for the hosts of industrial raw materials they contain. For many of these materials are speciality items, high in value while low in volume, and hence can be harvested with next to no disruption of forest ecosystems.

Tropical forests offer a variety of little-known industrial materials. Broad and diverse, it is in marked contrast to the very limited number of industrial products that we already derive from the forests, notably commercial timber. These little-known industrial materials are often categorized by foresters as "minor forest products," yet they can turn out to be not so minor. More important still, the spectrum of industrial materials can be greatly expanded through systematic investigation of all that the forests offer.[1]

What are the main categories of these so-called minor products? Fibers and canes from tropical forests appear in our lives in the form of wickerwork furniture made from rattans, and of kapok as an insulating and sound-proofing material and as stuffing for lifebuoys. Essential oils from tropical forest plants turn up in dozens of everyday products: in mouthwash and deodorant, in many beverages from instant coffee to absinthe, and in cough drops among a host of other medicinals. Edible oils—used not as food additives but for industrial applications—turn up in detergents, candles, emollients, lubricants, cellophane, and explosives. Exudates, notably gums, resins, and latexes, serve us each time we seal an envelope and affix a postage stamp, apply varnish to our furniture or our fingernails, whack a golf ball, or chew gum. Waxes support our lifestyles each time we read a glossy magazine, take a carbon copy of a credit-card slip, or feed the dog. Tannins and dyes make our Gucci items last longer and appear more elegant, and they also add a subtle hue to many a cocktail drink.

In addition to these better-known products from tropical forests, we use others such as terpenes, pectins, sterols, esters, phenols and polyphenols, elastomers and polystyrenes. We use plastics in dozens of ways that are immediately obvious and in some that are less so, such as ultrasuedes and as football-field turf. We use acids from tropical forest plants, and especially fatty acids, whenever we apply anti-corrosive to our automobiles, and each time we use surfactants, more usually known as soaps; and we benefit from the same esoteric materials whenever we exploit modern machinery that, working at higher speeds, higher pressures, and higher temperatures than ever before, requires supersophisticated kinds of lubricants. Equally unwittingly, we make use of tropical forest materials every time we brush our teeth or ride in a car, apply a shampoo or a sunscreen lotion, clean a rug or pull on our jogging shoes, buy goods that contain tinplate or glycerine, take a ride in a jet aircraft, swing a tennis racquet, put on a top-quality shirt or blouse, take an analgesic or an antiseptic,

swallow a sedative or a stimulant, touch up with lipstick or mascara or many another cosmetic, or polish the silverware.

Furthermore, we enjoy these many raw materials from tropical forests after only very random and limited exploration of the more than 90,000 plant species found there. To reiterate a central point, we have taken a passing look at a mere 1 species in 10 to assess its economic usefulness, and we have cast a longer look at only 1 species in 100. As for animal species, of which there are between 2 million and 4 million at a minimum and possibly 30 million (mostly invertebrates), in tropical forests, we have scarcely investigated more than a few hundred for their material contributions to our lifestyles. Thus present harvesting patterns of industrial raw materials from tropical forests can be said to correspond, in their haphazard form, to hunter-gatherer types of agriculture: rudimentary in the extreme. We could increase the total number of products many times over if we were to set about a systematized and comprehensive examination of all materials available— provided, of course, that the industrial chemist can get to the forests before the sawman.

Equally to the point, these varied materials, generally occurring in concentrated form while commanding high prices in the marketplace, can very often be extracted with only trifling disruption of forest ecosystems. Hence, an industrialization strategy that seeks to intensify our harvest of specialist materials from tropical forests stands in strong contrast with established modes of exploitation, such as timber harvesting. In sum, there is fine prospect that tropical forests can be industrialized in the sense that they can eventually feature nodes of economic development, where extractive and processing industries use raw materials drawn from a surrounding sea of relatively undisturbed forest.

Rattan, a climbing palm of Southeast Asia's forests, is a tough and resilient fiber that can be harvested without damage to the residual forest—in contrast to exploitation of hardwood timber. We know of 550 rattan species, plus many more that await discovery, plus many other fiber plants that could offer arrays of products to match those of rattans. *(Timothy C. Whitmore, World Wildlife Fund)*

GLOBAL CHEMICALS INDUSTRY

Before we go on to assess the capacity of tropical forests to supply us with array after array of industrial goods, let us put the issue in perspective by looking at the scale of the global chemicals industry. Already large at $400 billion per year (compare the global armaments industry, $650 billion per year), the industry is growing at an average annual rate of about 5 percent. In one way or another, it touches as much as one-third of Gross Global Product.

Particularly important is the synthetics sector of the chemicals industry. The sector is now worth about $150 billion per year, and expanding annually at a vigorous 9 percent. The leading category of synthetics is plastics, which enter our daily lives in literally hundreds of ways, not only in the form of packaging materials, appliances, fabrics, and artificial houseplants (no, the cheaper breads and breakfast cereals are not made of plastic—yet). But the synthetics sector of the chemical industry derives almost all its feedstocks from fossil petroleum. One whole decade after OPEC set up shop, American chemical corporations still derive only 3 percent of their raw materials from alternative sources such as vegetable fats and oils. An American citizen consumes 250 kg. of petrochemical products each year, out of his or her total petroleum consumption of more than 3,500 kg. But times are changing. Whereas many petrochemicals in 1973 cost a mere $0.07 per kg., they now cost, in the wake of OPEC, at least half a dollar per kg. By contrast, vegetables and oils, which cost around $0.22 in 1973, still remain way below half a dollar per kg.

Thus the chemicals industry now senses a strong incentive to move beyond its petroleum-dominated era. Fortunately many plant materials, together with a few animal products, can help us get off our "petroleum kick." Technology is available to manufacture 95 percent of synthetics products from selected plant materials. Salient instances include plastics, fibers, adhesives, polyisoprenes, and surface coatings. As the synthetics industry moves to replace ever-more costly petrochemicals, it is starting to seek alternative feedstocks among the abundant phytochemicals and other materials of the plant kingdom, especially in tropical forests. Already we use many vegetable oils for such well-known industrial products as castor, linseed, and tung oil. In addition, several oils that are familiar to us as edibles contribute to industry as well: for example, soybean oil, coconut oil, palm oil, and peanut oil. Within just a few years we may find that many other plant materials join the list, not only additional fats and oils, but new waxes, exudates, essential oils, and a lengthy list of other items that exist in greatest number and most novel forms in tropical forests.

So much for an introductory review of the main categories of materials available. Let us now look at some major items, group by group.

FIBERS AND CANES

Of all tropical forest plants, none does so much to conjure up an image of "jungle" as the long, climbing plants that festoon trees and seem to be waiting for Tarzan to happen along. Being woody and ropelike, they serve many purposes in the wickerwork trade, especially in furniture and woven goods. There is hardly a home in the United States that does not have a basket, a chair seat, a decorative screen, or a blind made from these split-stem climbers from Southeast Asia's forests.

The best known of these plants are rattans, which are the spiny stems of certain palms.[2] To date, we know of 550 species of rattans, mostly in Malaysia and Borneo, while a good number more almost certainly await discovery. Of all rattans that we are acquainted with, about 90 rank as commercial items, 20 of them accounting for the bulk of the market. Significantly, the most sought-after species occur rarely in the forests, even in favorable environments.

Rattans vary a good deal in form and size. Their stem diameter ranges from 3 milimeters to more than 10 centimeters. Growing as fast as 5 meters each year, they can reach a length of 100 meters, occasionally twice as long. They offer a broad range of uses, from obvious ones such as cordage, binding, utensils, and materials for thatch, to less apparent uses such as varnishes, dyes, and medicines.

Southeast Asia's export of rattans has now reached a value of around $60 million, with Indonesia accounting for 90 percent of the trade. These economic revenues are complemented by social benefits in that the crop assists local communities far more than does the harvesting of timber, by virtue of the fact that the gathering of rattans requires much labor. So diverse are rattans' applications in manufacturing that the value of their end-products now approaches $4 billion, making them Southeast Asia's most valuable export after timber. Because there is so much profit in the processing of rattans, Southeast Asian nations are starting to undertake much of the task themselves, which could rapidly make the crop far more valuable to them. At the same time, and in light of rattans' expanding use as substitutes for costly synthetic fibers, world demand for rattans is growing more rapidly than supplier nations can put materials onto the market.

Only about one-sixth of rattans are grown in plantations. Shade-loving plants, rattans can be planted alongside trees in fruit orchards. But in the main, domestication of rattans does not seem successful so far, and for the time being the principal source remains the primary forest. Fortunately rattans appear to flourish in secondary forests, opening up the prospect that they could become an important crop in forestlands that have been heavily logged.

Many other fibers and canes of tropical forests have the potential to match the success of rattans.[3] How many more fibers await scientific investigation, with the marketplace muscle of, for example, jute, roselle, and kenaf? As a measure of the scope that lies ahead, let us recall that bamboos, with their multitudinous uses, total 1,250 species in the humid tropics. They are woody perenni-

als, some of them capable of reaching 18 meters in height—
occasionally after only two months of growth. They offer a versa-
tility of end-uses that is hard to match, at least 100 commercial
applications, extending from cooking ingredients and chopsticks
to building poles and irrigation pipes, together with many hundreds of further applications among subsistence communities.

As a special form of fiber, let us consider the leaves of the
"tendu" tree in India, scientifically known as *Diospyros melanoxylon*. (There are another 500 species in the genus around the
tropics, many of them in moist forests, hardly investigated at all.)
The tendu's leaves are in great demand for cigarette wrappers, on
the grounds that they burn at a suitably slow pace and impart a
unique flavor to the cigarettes. Gathered entirely from natural
forests, the leaves, worth $1.42 million to the casual workers who
collect them, support an industry that now exceeds $400 million
per year.[4]

ESSENTIAL OILS

Essential oils are not "necessary" oils. Rather, they contain an
aroma, or an essence, that stems from an important group of
organic compounds known as terpenoids. Unlike fixed oils with
their basically different chemical makeup, essential oils are volatile, that is, they are ethereal substances. This trait makes them
very suitable for perfumery and cosmetics in the pharmaceuticals
sectors, and for scenting and flavoring agents in food manufacturing. Each type of essential oil generally contains several chemical
compounds, occasionaly more than a hundred.

Little though we may know it, we encounter essential oils on
dozens of occasions in our daily lives. Almost certainly there is
flavoring material from an essential oil in our coffee. We benefit
again when we apply skin creams, colognes, shaving preparations,
shampoos, pomades, antiperspirants, and cosmetics generally,
among many other toiletries, for both women and men. At the
breakfast table, numerous tasty items, especially sauces and con-

fections, owe their unique flavoring to essential oils—and the same at lunch and dinner, whenever we enjoy curries, puddings, ice creams, cheeses, and other desserts, plus condiments and carbonated drinks, also scented tobacco, attars, and incenses. A host of products that offer pleasant aromas or flavors owe their manufacture to essential oils from tropical forests.

As a measure of the economic importance of essential oils, we can note that global trade in these products, including their derived terpenes and related chemicals, now exceeds $1 billion per year.[5] The United States imports about 10,000 tons of essential oils each year, for a value of well over $100 million. So great is global demand for these oils, and so rapidly is it growing, that the essential-oil industry worldwide looks likely to multiply several times over before the end of the century.[6]

Of course, not all essential oils derive from tropical forests. But this single ecological zone supplies as much as the rest of the world combined. Notable sources of essential oils in tropical forests include such plants as camphor, citronella, sandalwood, palmrose, cassia, cinnamon, patchouli, clove, nutmeg, cardamom, and ginger, to name but a few of the better-known products.

Herewith a few examples of the oils themselves. The staranise, so called because of the delicate starlike shape of the fruit, occurs in Southeast Asia. The oil distilled from its fruit is a major source of anethole, used as a flavoring in oral pharmaceuticals, also as a mild expectorant in cough preparations. Another essential oil is from the cascarilla, a shrub that originated in the Bahamas and has a bark oil that yields a spicy scent akin to that of thyme, clove, and nutmeg. Because of its exquisitely delicate flavor, cascarilla oil is very expensive; it is used only in luxury items such as top-quality confections, aromatic bitters, and Vermouth. Sandalwood, associated with perfume manufacture for at least 4,000 years, is a small tree of Southern and Southeast Asia. When wood chips from the tree are distilled, they produce a pale yellow liquid with a woody aroma—and, since the oil gives off a soft, sweet smell of unique type, it is highly prized throughout the perfume industry.

In much of Southern and Southeast Asia, also in islands of the Indian Ocean, grows a tall tree known as the ylang-ylang. Its flowers, which must be picked in the early hours of the morning before the sun's heat volatilizes its oils, contribute prime ingredients to some of the finest French perfumes.

Given their varied and esoteric uses, what are essential oils worth in the marketplace? Citronella oil and cloves sell for about $5 per kg., star-anise oil about $12, patchouli oil around $30, and sandalwood oil for around $70. When we consider essential oils in conjunction with a few other perfumery compounds and associated resinoids, we find that in 1979 Madagascar exported materials worth $6,274,000; India, $9,137,000; Indonesia, $21,916,000; and Brazil, $21,528,000. A large proportion of these exports made their way to the United States.

Sizable as these sums are, they are likely to increase fast, following a steady decline during the third quarter of this century. In the year 1900, plant-derived raw materials constituted about 99 percent of ingredients of perfumery products. As a result, however, of the rapid emergence of petrochemical industries from 1950 onward, the proportion declined to only 15 percent, with the prospect that it would eventually taper off to zero. But the establishment of OPEC, with its upsurge in petroleum prices, has meant that the proportion of plant materials has already increased to 25 percent, with a rate of annual growth that could push the proportion way beyond 50 percent by the end of the century.

Furthermore, it is important to bear in mind that essential oils can be found within virtually any part of a plant, whether flowers, seeds, leaves, roots, bark, or wood. There is thus great scope for methodical screening of all plants in order to discover new and improved sources of essential oils; scientists need to search for variability of supply not only among plant species, but within each species as well. In turn, this offers much promise for genetic improvement. Certain crops, such as those that produce citronella and palmarosa, usually feature only small amounts of oil in their tissues, whereas a few strains have been found to contain three or even five times as much. Similarly, there is opportunity for increasing the quality of the oils produced, in accord with their end-product applications—a further incentive for the crop geneticist to draw on as much genetic variability as he can find. And of course the plant breeder and genetic engineer will need to bear in mind the age-old aims of the cash-crop grower, to seek enhanced resistance against diseases and pests, and against quirky weather.

EDIBLE OILS

Let us now move on to look at those oils that are edible and are used not for human consumption but for industrial application.

We are familiar enough with coconut oil and palm oil. Both are used not only in margarine, ice cream, mayonnaise, bakery products, and other foods, but also in candles, detergents, glycerine, tinplate, and lubricants for jet engines and other precision machinery—well over one hundred distinctive products. In light of experience to date, it would greatly reward us to investigate the rest of the 28,000 species of palms for their economic potential. Because of its diverse applications and its economic importance, the palm family ranks among the top three of all plant families, the other two being the Gramineae (grasses)—including corn, wheat, and rice—and the Leguminosae—beans, peas, and many other foods, also many medicines. Yet the palm family is surely the most threatened of all major plant categories.[7]

For an illustration of the multiple end-products that we can enjoy from a single species, let us consider the Babassu palm that grows wild in some 140,000 square kilometers of Amazonia—a mere 3 percent of the region, but an area almost the size of Florida. A tree that grows 20 meters tall, the Babassu bears fruit with a higher proportion of oil than the coconut palm; and being very similar in makeup to the oil of the coconut palm, the Babassu's oil can be used for margarine, shortening, general edibles, fatty acids, toilet soaps, and detergents. Moreover, the Babassu's oil can serve industrial purposes in the manufacture of plastics; and during World War II, its oil was converted to a liquid fuel that worked well in both internal and external combustion engines. The prolific oil output of the Babassu can sometimes translate into a yield as high as 40 kilograms of oil per tree per year, or the equivalent of 1 barrel of oil from every 4 trees—a stand of 500 trees on one hectare could produce 125 barrels of oil per year. After the oil has been extracted from the kernel, the remaining seed-cake, containing 27 percent protein, makes an excellent animal feed, so the same stand of 500 trees could produce 5.7 tons of edible protein together with 250 barrels of ethanol as a by-product. As a final bonus, the leaves yield high-value waxes.

As mentioned, the Babassu is known only in the wild. Nobody has yet managed to domesticate the palm, so there has been no genetic breeding or agronomic improvement. But we can visualize what could lie ahead when we consider the accomplishments of palm geneticists in Malaysia, where tissue culture has led to cloning of palm oil trees. They have produced a plantation made up of identical specimens, all with short trunks for easier harvesting, plus 20–30 percent increased yield of oil. Fortunately the Babassu, like many other palm species, reveals great genetic diversity among its populations, with some strains starting to produce fruit earlier, and then producing much larger yields, than other strains. Regrettably, the Babassu is losing much of its habitat in its Amazonia environments, which almost certainly means that the species is suffering rapid genetic attrition.

EXUDATES

Exudates, such as gums, resins, and latexes, supply many consumer goods. They appear in our daily lives in pharmaceuticals, paints and varnishes, foods, water colors and inks, paper sizing, printing materials for textiles and calicos, soaps and other toiletries, and waterproofing for fabrics. They also turn up as prime ingredients in those perquisites of the good life for many a twentieth-century citizen, the golf ball and bubble gum. In the natural state, many of these materials emerge from trees by simply oozing from the bark—hence their name, "exudates." Latexes, for example, flow as an opaque milklike liquid that eventually hardens into a dull-gray rubbery mass.

Gums make first-rate dispersants, stabilizing agents, emulsifiers, suspending agents, and gel-forming agents. On occasion, they are further used as coagulants, binders, lubricants, and film formers. So important are natural gums, with their water-soluble properties, that industrial applications are growing at 10 percent a year. Resins, when converted to refined forms of turpentine, are used extensively in paints, varnishes, and pharmaceuticals, whereupon they command a price of more than $1,000 per ton. Although our most important sources of plant resins to date are coniferous trees of the temperate zones, there are at least twice as many trees in the tropics that produce resins, most of them in upland forests.

Two categories of exudates are known as copals and dammars. Found in many parts of tropical forests, they comprise large groups of resins that are characterized by their hardness and high melting points, making them very suitable for paints and varnishes. The main difference between the two categories lies with the fact that only dammars, derived notably from dipterocarps, are readily soluble in turpentine and coal-tar hydrocarbons, which makes them specially useful in the manufacture of spirit varnishes. Moreover, dammars are lightly colored, so they are widely used in white-paper varnishes and white enamels, also to enhance the luster of celluloses. In addition, dammars serve such disparate purposes as caulking materials for boats and binding constituents in ointments.

As for latexes, every reader of this book will be acquainted with chicle gum derived from the white resinous latex of a Central American tree. Although chewing gum is now manufactured from many other gums, generally synthetics, chicle still serves as a first-rate base for some of the world's commonest and most popular confections. Another latex—a coagulated one known as balata—comes from a tall tree of Guyana and Surinam; its latex product is mainly used for golf ball covers, to which it lends a toughness, durability, and elasticity unmatched by any synthetic substitutes. Another leading ingredient in golf ball covers is gutta percha, a polyisoprene available from many forest plants. With research and development, gutta could win broadscale applications as both a thermoplastic and a thermosetting resin; already it is produced at prices competitive with petroleum-derived polymers.

The most famous of the latex products is rubber. Derived from a Brazilian tree with the scientific name of *Hevea brasiliensis,* natural rubber serves our needs in literally hundreds of ways each day. Disregarded if not scorned only a little over a century ago, natural rubber soon became prominent after a few quick investigations by industrial chemists—which prompts one to ask how many other seemingly useless plants could similarly start to support billion-dollar industries if they were to be screened for their potential economic applications.

The extensive rubber tree plantations of Malaysia and several other nations of Southeast Asia have enabled rubber to become the fourth-largest agricultural export from the Third World, now amounting to 3.5 million tons per year, worth about $3.5 billion. According to projections by the World Bank, annual global demand for natural rubber could grow to 6 million tons per year by 1990 at the latest.[8] In point of fact, the rubber plantations of Southeast Asia are not nearly keeping pace with the needs of global industry. Given the growing demand of people in all parts of the world for natural rubber, we shall almost certainly encounter severe shortfalls by the end of the 1980s.

Nor can the citizen consumer look for relief to synthetic rubber. During the era of cheap petroleum, synthetic substitutes for natural rubber captured 70 percent of the rubber market. But petroleum-derived synthetics are becoming increasingly expensive. Still more to the point, many of our needs cannot be met by synthetic rubber. Certain products are highly dependent on natural rubber, with its extreme elasticity and heat resistance. The tire industry, for example, absorbs more than two-thirds of natural rubber available. Lorry and bus tires, also car radial tires, require up to 40 percent natural rubber if they are to withstand large potholes in the roads, and to dissipate the heat generated in screeching halts. Aircraft and bulldozer tires consist almost entirely of natural rubber. The tires of the space shuttle Challenger are at least 95 percent natural rubber.

In short, present consumption patterns are likely to exert ever-greater pressure on supplies of natural rubber. Moreover, a number of new applications of the material, exploiting its unusual qualities, will increase the pressure still further. We can even expect that the price of natural rubber will double during the 1980s, after which it could skyrocket as market demand steadily outgrows the expected production of rubber plantations in Southeast Asia.

The leading consumer of natural rubber is the United States, accounting for almost one-quarter of the world's total, at a cost of around $1 billion per year. In view of tightening supplies from Southeast Asia, the U.S. Congress has recently authorized an intensive effort to find alternative sources of natural rubber, thus triggering a search among the 2,000 plant species that yield rubber of one sort or another. Although a leading candidate is an arid-zone plant, the guayule shrub of Arizona and Mexico, we know that many tropical forest plants yield materials closely akin to our present rubber, and could eventually be developed to take over a share of the market.

In any case, the original species of rubber tree in Amazonia, of which a mere twenty-six seedlings eventually reached maturity in Malaysia (an extraordinarily narrow genetic base on which to build the present plantations with their millions of trees), retains many related strains and ecotypes in the wild. The samples originally collected in 1876 certainly did not include the best variations, either for quantity or for quality of rubber output. Nine additional species of the same *Hevea* genus offer other valuable characteristics such as disease resistance. Moreover, the present proliferation of elastomers and plastics indicates that industrial chemists frequently need several sorts of latex, which, while of little use in themselves, prove valuable as fillers to mix with synthetic products in order to alter their physical properties.[9]

It is resistance to disease that could hold the key to the future. In the Amazonian habitats where rubber originated, there is a potent fungal parasite known as South American leaf blight. It is precisely this disease that has prevented the establishment of commercial rubber plantations in the native environment of Amazonia. Were the disease ever to make its way, through an accident (or "accident"), to Southeast Asia, the blight could cripple the natural rubber industry within just a few seasons. Fortunately, a number of wild species of the *Hevea* genus offer resistance to the blight—just as others offer resistance to other diseases, and still others offer tolerance to cold, or to too much or too little moisture. But the survival of these germplasm resources is now threatened by a series of development projects sponsored by the Brazilian government in Amazonia. Of 15 such projects, 11 are located in the original forest habitats of the genus *Hevea,* and at least 8 of them overlap with *Hevea brasiliensis.* [10]

MINOR FOREST PRODUCTS: ECONOMIC VALUE

So much, then, for "minor" forest products, i.e., products other than timber, that can be harvested from tropical forests— the great bulk of them found only among the biological richness and the ecological complexity of primary forests. As we have noted, these minor products are exceedingly diverse. In Peninsular Malaysia, for example, we know that the predominant form of forest, lowland primary forest, contains at least 1,283 non-timber plant species of identified use to humans, constituting roughly 16 percent of all native species in the Peninsula.[11] If we extrapolate from this data base, albeit a meager one, to other forests of similar sort, we may estimate that about 1 species in 6 could serve non-timber purposes of one kind or another. This means, in theory at least, that a minimum of 15,000 plant species in the tropical forest biome could well offer potential to meet our material needs. Furthermore, the reckoning with regard to Peninsular Malaysia does not take account of many additional food plants that are garnered from the forest in small quantities; nor does it take account of all medicines of whatever sort; and it includes only a few items from several other major categories. So the true proportion could be much greater than 1 plant species in 6. Some experts, drawing on decades of field experience, believe the ultimate figure could be twice as large.

As for the economic value of these products, we have noted that rattan exports from Indonesia are worth at least $5 million per year. In addition, Indonesia's exports of citronella oil were worth $2.2 million in 1980; its exports of sandalwood oil, $1.7 million; of patchouli oil, $11.6 million; and of all essential oils, $21.1 million.[12] Together with exudates and sundry other products, totaling 80,000 tons in all, Indonesia's minor forest products earned the country $32 million in 1980.[13] Moreover, these figures reflect only a crude calculation of total revenues. Many other products, categorized by Indonesia's statistical surveys under various commodity headings, would push the total a good deal higher.

For a more systematic and comprehensive assessment, let us look at the case of India.[14] In 1977 the total net revenue accruing to India's forestry sector, from all sources including commercial timber, amounted to $336 million. Of this total, minor forest products accounted for $134 million, or 40 percent (and their share of forestry exports, 63 percent). Since an estimated three-fifths of all minor forest products are used by local people, that is, are consumed on the spot, they do not enter the cash economy and, hence, are not incorporated into national accounting figures. This suggests that a realistic figure for the value of minor forest products would be more than $200 million. Among leading categories in 1977 were medicinals, drugs, and pharmaceuticals, worth $38.4 million; lac and lac products, $19.8 million; gum, resins and balsams, $14.6 million; bamboos, $6.8 million; and essential oils, $5.9 million.

Equally important, the rate of growth in revenues from India's minor forest products during the period 1970–77 amounted to 15.6 percent per year, way ahead of that for commercial timber. In addition, minor forest products were generating much employment, more than 70 percent of the 2.3 million man-years in the forestry sector overall. The true figure for employment, including those man-years not counted by official surveys, could have been as high as four million.

FOREST INDUSTRIAL COMPLEXES

Many of the categories of items listed above are best grown in plantations—which themselves can form part of a forest industrialization strategy (see below). But many more of the items listed do not lend themselves to domestication of any sort. They simply cannot be cultivated, or persuaded to grow at all, outside their native habitats in primary forests. They must be collected in the wild, drawn from tracts of undisturbed forest by discriminating exploiters. In this sense, a primary forest can be made to yield a harvest of considerable economic value, with next to no disruption of its ecosystem. The materials would be assembled at a central collecting point, for processing through small-scale factories that produce primary phytochemicals for supply to sophisticated refineries; in turn, the larger-scale refineries would provide materials to integrative centers for manufacture of finished chemical products. We would thus have the forest industrial complexes that we visualized at the start of this chapter. In a few moments, we shall see how this phenomenon might work out in practice.

But first, what about those products that can best be grown in plantations? How do they fit into the industrialization strategy? The crop species will be heavily dependent on steady supplies of fresh germplasm from wild gene reservoirs—from those same tracts of undisturbed forest that supply the arcane cornucopias of raw materials that are available only from the natural forest. Due to factors of their biology, plantation crops for industry, like those for agriculture, sustain their productivity only as long as they receive regular infusions of germplasm from their wild relatives. Virtually all plantation crops need to be constantly safeguarded through "topping up" of their genetic makeup, not only to enhance their general vigor, but to maintain their resistance to diseases and pests, to make harvesting easier, and so forth. Breeding improvements along these lines depend on the amount of genetic variability available from outside sources, which places a premium on plenty of stocks of diversified germplasm in primary forests. Thus plantation forestry, as a means to produce botanochemicals for supply to the processing factories, fits right into the same industrialization strategy that depends, for its essential rationale, upon expanses of virtually undisturbed forest.

As for the industrial facilities themselves, what sort of installations shall we envisage? Fortunately, and by contrast with the case for petrochemical factories, we need not think in terms of outsize processing plants. Facilities for botanochemical products can be both small and dispersed. By way of illustration, let us consider a supplier forest that overall might cover 10,000 sq. km., 8000 sq. km. of which would be natural forest, and 2000 sq. km. (best located in a central zone), plantations, or man-made forests. Set at the heart of the cluster of plantations would be the processing factory, no more than 25 km. away from the furthest point of the plantations. Hence the great bulk of the entire zone would comprise more-or-less virgin forest to provide the multitudes of materials (non-timber products, phytochemicals, genetic resources, and the like) that cannot readily be produced in plantations.

What can we anticipate in terms of revenues? Regrettably, we have no experience in tropical forests on which to draw. The only indications we can derive are from planning calculations for industrial-crop establishments in the United States. These reckonings are preliminary and approximate to a degree, and they are presented here with the sole purpose of enabling us to "get a handle" on the economics involved. In many respects, the range of crops intended for the United States parallels the crops we envisage for a tropical forest zone: whole-plant oils, polyphenols, glucose, xylose, and a lengthy list of related materials (of course, from temperate zone plants). These United States industrial-crop plantations would also supply, through integrated production systems, a number of energy products, such as liquid fuels, which would help with the economics. (A tropical forest facility could do the same, as recounted in the next chapter.) In light of some exploratory arithmetic, these oil-and-hydrocarbon-producing crop plantations in the United States could supply raw materials amounting to some 2.5 million tons of feedstocks, for processing into products with a value of some $250 million per year.[15]

Thus a forest industrial complex need not be so large that it demands vast injections of capital. In marked contrast to a petrochemical plant, the phytochemical counterpart can be relatively small in size; and it can be launched with a surprisingly limited amount of start-up investment. As for labor, the processing of plant-derived materials does not lie beyond the capacities of semi-skilled and unskilled workers—precisely the kinds of labor that are all too readily available in countries of the humid tropics.

A further advantage is that the manufacture of phytochemicals releases far less pollutant heat into the environment than is the case with petrochemicals. Nor do they generate nearly so many toxic emissions. The by-products of fermentation processes, such as characterize phytochemical factories, can be converted into fertilizer and other useful materials—in contrast to the wastes of petrochemical industries, which tend to be persistent toxins that present threats to the environment in general and to human health in particular.

We have seen some of the innovations that would offer more

efficient and systematic use of tropical forests. An approach along these lines would mark an advance beyond present practices, which have been described by Dr. K.F.S. King, former Director of the Forestry Department under the U.N. Food and Agriculture Organization, as "primitive, costly and wasteful." When we consider the spectrum of valuable products available from tropical forests, compared with the very few that currently dominate our commercial use of forests, we can even say that, to date, these forests are over-exploited and under-utilized. Let us look forward, then, to the day when we shall see extractive and processing facilities in tropical forests, making methodical and broad-ranging use of raw materials drawn from a surrounding expanse of forest that remains virtually intact.

13

Green Gasoline and Other Energy Forms

THE MOST INNOVATIVE USE OF tropical forest plants could eventually become the most pervasive of all. What if we could persuade forest plants to generate vegetation of sorts that can be converted into "green gasoline" and other forms of energy?

This prospect is not so outlandish as it may sound. After all, it was photosynthesis that originally gave rise to startpoint the materials for geologic formation of oil, coal, and natural gas. What a giant step forward it would be for us if we could cut out the one-third of a billion years that have transformed ancient green plants into fossil fuel and instead harvest the stored solar energy of present-day plants. Photosynthesizing plants offer by far the simplest way for humankind to collect solar energy, and then to process the vegetative material into "bio-crude."

There seems little doubt about the potential. Each year photosynthesis is reckoned to store energy equivalent to 80 billion tons of coal (roughly ten times the world's consumption of fossil fuels), while energy now stored in the total biomass of all plants is considered to be roughly equal to the total of proven reserves of all fossil fuels. Of course, almost all of the solar energy reaching the Earth is wasted, insofar as plants absorb only a tiny fraction of the sun's energy. But what if we could devise plant types that are super-efficient at gathering up the sun's free gifts?

We already consume a good deal of biomass energy. This form of energy provides about one-seventh of the world's energy each year, or roughly four times the amount supplied by hydroelectricity and nuclear power combined. Although most of this biomass energy is in the form of fuelwood, it provides the equivalent of about 20 million barrels of oil per day—more than Americans consume. Even though the principal form of this bio-energy is fuelwood, the proportion of alternative forms is likely to grow rapidly. The Volkswagen Corporation estimates that by the end of the century, more than one car in three will be powered with fuel from a "petroleum plantation."

Were we to achieve the breakthrough of replacing petroleum with "phytoleum," we would achieve a quantum advance in our approach to fuel supplies. To date, we have been inclined to scrabble around for oil, coal, and gas—an approach that resembles the hunter-gatherer's lifestyle before agriculture. Perhaps we are now ready to "domesticate" those wild plants that can generate bio-fuels in commercial quantities on energy farms.

TECHNOLOGIES AVAILABLE

There are several technical options to convert biomass into fuels.[1] When once enough biomass is assembled at one spot, it can be processed anaerobically, that is, in oxygen-free containers, via one of three principal conversion strategies. The first is biomethanation, which produces a fuel gas composed of methane and carbon dioxide, otherwise known as biogas. The second is fermentation, to produce liquid fuel in the form of an alcohol called ethanol. A third is pyrolysis, or thermal decomposition of carbonaceous materials at high temperatures, to produce hydrocarbons in the form of gases, solids, and liquids (not unlike crude petroleum).

Through these three techniques, we obtain energy-rich products of types that are suitable for a range of purposes. Biogas is appropriate for small-scale operations, notably at village level in developing nations. Ethanol can be mixed with conventional gasoline, at a 10 to 20 percent level for the ethanol, to produce gasohol. Pyrolysis releases an oil akin to crude petroleum, with 75 percent of crude's calorific value, as well as a mixture of gases and a flaky char. Moreover, the fuels are of good quality. The sulfur content of plant matter is generally below 0.1 percent, compared with an average of about 2.5 percent for coal. The ash content of terrestrial plants is typically between 2 and 5 percent, compared with an average of about 14 percent for coal. Generally speaking, plant biomass is as easily burned or gasified as coal, though sometimes not so easily handled. In contrast to fossil fuels, biomass-derived energy is a non-polluting form of energy; and it would be associated with none of the hazards associated with nuclear energy.

Of course the energy value of biomass is lower than that of oil. Burning one ton of dry wood yields about 4 million kilocalories, while one ton of oil yields at least two and a half times as much. (Alternatively stated, one ton of wood generates energy substitute for two barrels of petroleum.) Nonetheless, biomass-derived fuel is a type of energy that makes use of Earth's single truly renewable resource, sunlight. A petroleum plantation need never run dry like an oil well.

Among suitable sources of biomass, the front rankers at present are starch crops, such as corn and cassava, or sugar crops, such as cane and beet. These crops underpin the gasohol programs of the United States, Brazil, and several other nations. Other herbaceous (non-woody) plants could prove suitable too: for example, Napier grass (elephant grass), Bermuda grass and Sudan grass, also certain swamp reeds, which when grown in the year-round warmth of the tropics, generate more biomass per hectare than super-swift growing crops such as sugarcane (for more details of these grasses among other suitable sources of biomass, see below).

Given our experience to date, we can assume that we have not yet identified the most promising species of plants for bio-energy. We are sticking, thus far, with the conventional crops with which we know we are safe. What if we were to launch an across-the-board survey of all plant species on Earth to see which ones would prove the best bets? What if we were to search in particular those plant communities that, by virtue of their evolutionary background, possess the most species with capacity to produce massive amounts of biomass: tropical forests? Surely here we shall find many possibilities for bio-fuels of the future. In fact, in light of our knowledge thus far, albeit based on the very few plants that have been methodically scrutinized by scientists, it seems a statistical certainty that we shall find many first-rate species in tropical forests. It seems equally a statistical certainty that if present extinction rates persist, we shall lose several thousand, perhaps as many as 10,000 of tropical forests' 90,000 plant species by the end of the century, and probably twice as many again by the middle of the next century.

WOODY PLANTS

Hitherto, such scant scientific investigations as have been mounted for exploration of potential materials in tropical forests have tended to focus on woody plants, on the grounds that trees lock away virtually all of the biomass produced in tropical forests. Generally speaking, one ton of pyrolized wood can be expected to yield at least 12 liters of methanol (methalated spirit), 35 liters of wood oil and light tar, 330 kilograms of charcoal and coal-like residues (to be used in lime and cement kilns, in the steel industry, also in domestic households), and 140 cubic meters of gas (used mainly to energize the production processes), plus a number of by-products of industrial value, notably 25 liters of acetic acid (used as a basis for acetone, and for manufacture of textiles), almost 12 liters of creosote oil (used for timber preservation), more than 7 liters of esters (used for methyl acetate and ethyl formate), and 33 kilograms of pitch (used for waterproofing, road surfacing, and as a caulking material).[2]

Of these many products, the first, methanol, is the one most likely to serve our needs at the gas station. Sometimes known as wood alcohol, methanol can, like alcohol of whatever kind, be used to drive an ordinary car when mixed with gasoline in a ratio of 1:9, while only a few adjustments to the car's engine enable it to function on neat alcohol. Compared with gasoline, methanol is a clean fuel, emitting fewer reactive hydrocarbons and nitrous oxides and reducing the prospect of smog over Manila and Nairobi. The leader in the gasohol field, Brazil, uses mostly ethanol as an additive for gasoline, but it plans to introduce methanol as well into automobile gas tanks. Many a car in Rio de Janeiro is partially fueled with alcohol, and the Volkswagen assembly plant in Brazil installs engines that are adapted to use nothing but alcohol. An alcoholic VW, with exceptionally high octane fuel in its tank, surges along the road with never a knock.

A key question arises with regard to methanol. Which tree species can supply the greatest stocks of biomass? Several trees grow fast enough to yield wood in suitably large amounts: for example, several eucalyptus and pine species, now growing in numerous plantations in both the tropics and temperate zones. But perhaps the most promising species available to us so far is one that has gained prominence in just the last few years: the giant ipilipil, known scientifically as *Leucaena leucocephala*.[3] This evergreen legume, a native of Central America, has spread to become one of the most versatile plants on Earth. It can grow to 4 meters high in six months, almost 10 metres in two years, and more than 15 meters in six years. A one-hectare plantation can easily produce 35 cubic meters of wood per year for pyrolysis into liquid fuels (or for direct use in electricity generators and industrial boilers). A rate of 35 cubic meters matches most other fast-growing species, while experimental plantings show that the ipilipil can yield twice and even three times as much under suitable conditions. In order to generate an estimated energy equivalent of one million barrels of oil a year, a fuel-thirsty nation need think in terms of no more than a 12,000-hectare plantation, or an area about twice as large as New York's Manhattan Island.

At the same time, an ipilipil plantation can serve our needs in other ways. A one-hectare stand yields huge amounts of palatable forage for livestock. The dry leaves contain 27 to 34 percent protein, with a balance of amino acids much as in alfalfa, and with one of the highest vitamin A contents ever reported in plants. When cattle are fed on a mixture of grass and leaves from young ipilipil trees, they put on weight as fast as cattle from the best pastures anywhere. Alternatively, the foliage can be used for green manure—the leaves from one hectare suppling hundreds of kilograms of nitrogen, potash, and phosphorus fertilizer. Furthermore, the seed pods can be eaten raw, dried and ground into flour, made into coffee substitutes, or converted into dyes; while the gum makes a fine commercial thickener for foods such as ice cream and mayonnaise. A many-talented tree.

To see how the ipilipil's potential works out in practice, let us briefly look at a couple of case studies. In the Philippines, the National Electrification Administration has launched a fuel-from-wood program that, by 1984, is expected to save at least $30 million per year in oil import bills. To date, almost all the 100-plus electric cooperatives in the Philippines' countryside have been using oil-fired generators, each consuming annually more than 6,000 barrels of oil. Each cooperative is growing 100 hectares of ipilipil trees, which should, it is hoped, meet their energy needs —with implications for the Philippines' adverse balance of payments, and with further implications for the overseas banks, notably American banks, that have made huge loans to prop up the Philippines' economy.

In Hawaii, experimental ipilipil plantations are being harvested after only four years, when the trees reach about 12 cm. in diameter. This size allows the trees to be harvested by a giant mowing machine, looking like a cross between a grain combine and an army tank, which rumbles along at less than 1 km. per hour, chopping off the trees at a rate of one every six seconds, at a height about 15 cm. above ground level, thus leaving the stump to regenerate by coppicing. Each cut tree passes through a chipper (the foliage is mechanically sucked out of the chips and dried for use as animal feed, worth $150 per ton); the harvested material, amounting to about 250 tons of wood per hectare, is then fed into boilers. At a rough conversion rate of 1.8 kg. of wood chips to produce one kwh. of electricity, the harvest should generate an acceptable output of well over 8 million kwh. At a fossil-petroleum price of $30 per barrel, the operation, with an annual cost of about $330,000, is expected to be economically competitive: eight million kwh. represent the equivalent of 18,000 barrels of oil or diesel fuel, worth $450,000.

What will count in the future for the ipilipil—as for grain crops and wild plants that offer potential for domestication in our bio-energy scenario—is the extent to which plant breeders improve the genetic content. Clearly a premium rests on selecting the genetic strains from wild trees that are best adapted to petroleum plantations. We might suppose the plant breeder will search almost entirely for genetic variability that offers the fastest growth rates. But there is more to it than that. If petroleum plantations are to produce super-sized harvests of ipilipil wood, season after season, without massive infusions of fertilizer, pesticides, and the like, the geneticist must also look for germplasm with resistance to disease, capacity to thrive without artificial fertilizer, and tolerance to environmental stresses such as too much or too little moisture.

To date, most ipilipil plantations are scarcely more than wild trees planted in rows. If we can track down enough genetic diversity among ipilipil trees in natural forests, we shall develop tree varieties whose productivity will soar way beyond today's levels, remarkable as these already are. Were variations of ipilipil to be developed with capacity to produce 200 cubic meters of wood per hectare per year, they could generate 1,000 barrels of oil per hectare. Dreamlike as such yields might sound to us right now, they are not to be dismissed as beyond the realm of possibility. Certain tree varieties, like certain tree species, are much more efficient than others in photosynthesizing carbon dioxide. In 1960, loblolly pine, growing in the sound conditions of South Carolina, used to yield about three dry tons per hectare per year. Through selection of superior genetic strains, in conjunction with better forest management, the harvest has been increased to 11 tons, while yields of 20 tons are well this side of the horizon, and ultimate yields of 30 tons are not ruled out of prospect. The eucalyptus in Brazil, growing on a seven-year cycle, formerly produced "only" 23 tons per hectare, but through genetic selection the yield has been improved to 33 tons on average, and 40 tons in unusually favorable conditions—while the best clones, developed through genetic engineering, produce well over 60 tons.

HERBACEOUS PLANTS

As with woody plants, so with herbaceous plants. There are plenty of potential bio-energy candidates, and there is plenty of scope to improve their productivity.

We have already noted, at the beginning of this chapter, the role of conventional crops such as sugarcane and cassava. Not so well known are, for example, Napier grass, which, when grown year-round in the warmth and moisture of the humid tropics, can generate an exceptional 88 tons (dry weight) of biomass per hectare per year, well ahead of sugarcane, which rarely surpasses 75 tons. Among the plethora of tropical forest plants, especially grasses, which ones, we might wonder, are waiting to be developed for bio-fuels?

A category with special promise is the water plants, notably swamp-dwelling plants. We already know that a number of tropical reeds can yield around 60 tons (dry weight) of biomass per hectare per year. (By contrast, a tropical forest generally generates only 35 to 60 tons of biomass per hectare per year, and a deciduous temperate forest of North America, only 15 tons.) In the flood forests of Amazonia, which sometimes extend at least 50 kilometers from the main river channel, there are large numbers of such swampland plants, many of them still to be documented, let alone investigated. The slope of the Amazon Basin is, for several thousand kilometers, less than the slope that enables water to drain out of our bathtubs; hence the description of Amazonia as an ultra-thin lake covered with plants. Moreover, were swamp-dwelling plants to prove suitable for petroleum plantations, they would obviously not compete for land with conventional agricultural crops.

Nor need there be a threat to established agriculture from plants that flourish in open bodies of water such as lakes and rivers. The water hyacinth is notorious in many parts of the world as an aquatic weed with capacity to proliferate at prodigious rates, until it covers entire sheets of water. A mere 10 plants can multiply to 600,000 in just eight months, forming a carpet a quarter of a hectare in size and thick enough for a person to walk across.[4] In the southern United States, hyacinths produce 70 to 100 tons of dry-weight biomass per hectare per year, and in the tropics as much as 120 tons, while hyacinths grown on sewage, a medium that is rich in nutrients, exceed 150 tons. Through anaerobic fermentation, hyacinth biomass yields methane at a rate of at least 30,000 cubic meters per hectare per year.

The hyacinth possesses other virtues. In common with numerous other water-weeds, it yields, as a by-product of fermentation, materials that make fine fertilizer and animal feed. In addition, the hyacinth extracts many compounds, both organic and inorganic, from water, thereby purifying it—a characteristic that could enable the plant to serve as a clean-up agent when grown on its favorite habitat, sewage lagoons.

HYDROCARBON TREES

Still another strategy for energy farming involves those plant species that do not emphasize carbohydrates like sugar in their tissues, but rather produce hydrocarbons that are akin to oil.[5] Whereas carbohydrates are plentiful in plants, they are difficult to convert to hydrocarbons because of their high oxygen content. By contrast, those plants that produce hydrocarbons and hydrocarbonlike materials can be processed into liquid fuels far more readily. But whereas carbohydrates must be converted to fuel (via microbial or thermochemical processes such as those described above) before they can be utilized for energy, hydrocarbons often prove suitable for direct use as fuel.

Plant hydrocarbons take several forms. We have long used two of them, turpentine and rubber. Pine trees, for example, yield oleoresins, which have been tapped in the United States for many years as a source of turpentine. The same applies to two tropical pine species in the Philippines, plus several other pines in other parts of the humid tropics. The Swedish auto corporation Saab has developed a turpentine-powered car, running it off oleoresins from pines that constitute the major conifer species in Scandinavia.

Rubber, the other well-known plant hydrocarbon, derives from a member of the tree family Euphorbiaceae. Several other members of the family produce significant amounts of milklike sap, or latex, that is actually an emulsion of 30 percent hydrocarbons in water. These hydrocarbons are similar to those produced by the rubber tree, though of much lower molecular weight (in fact, of a weight order that is favored by oil engineers), and with a size distribution that resembles that of hydrocarbons in petroleum. Euphorbia hydrocarbons are even superior to those of crude oil, in that they are practically free of sulfur and other contaminants.

All in all, at least 30,000 species of plants produce latex. But the family Euphorbiaceae, with some 5000 species, could prove specially suitable for growing gasoline. These plants range in size from small herbs that hug the ground so closely they can easily be overlooked, to trees more than 30 meters tall. One of the main ranges of the Euphorbias is the tropical forest biome, where dozens of species could prove to be leading candidates, for example *Eucalyptus trigona,* a native of Brazil and Puerto Rico.

Similarly promising is a member of the legume family, known scientifically as *Copaiba langsdorfii,* a tree of central Amazonia that grows almost 30 meters high and with a bole 1 meter across. According to preliminary inquiries, the tree's trunk can be tapped for its hydrocarbon fluid, in much the same manner that New Englanders and eastern Canadians tap sugar maples for syrup or Malaysians tap rubber trees for latex.[6] A single-cap bore of tapping can yield between 10 and 20 liters of hydrocarbon fluid within two hours. The sap's makeup turns out to be so close to that of diesel fuel that it can be put directly into the tank of a diesel truck. According to experimental plantings, albeit of a very preliminary and exploratory kind, one hectare of 250 mature trees might produce 60 barrels of fuel per year. Fortunately, the tree can be grown in areas too swampy or boggy for conventional agriculture. The Brazilian government has established experimental plantations, and we shall soon see whether the species lives up to its promise.

Other members of the *Copaiba* genus are worth evaluation for their hydrocarbon content, notably several species that flourish in Amazonia, in the West Indies, and in West and Central Africa. Thus there is plenty of variety, both among and within species, for geneticists to work on.

While *Copaiba langsdorfii* yields may seem low at present, geneticists believe they could double production through seed selection alone in just a single year; agronomists believe they could achieve similar increases in yield within a few years. In the case of certain plants, up to 10 percent of biomass weight is made up of hydrocarbons, a proportion that offers plenty of scope for increase through selective breeding and gene manipulations.

After all, the rubber tree in 1945 was producing a mere 225 kilograms of rubber per hectare, and scientists pushed the amount upward ten times within the space of 20 years (they have even bred a few trees that yield 40 times as much), all through selective breeding alone, before the advent of genetic engineering.

Finally, let us consider a plant with an established track record. During World War II, the Japanese used the "petroleum nut tree" of the Philippines for fuel for their tanks: the tree yields a highly volatile oil. Today, the oil's best use may not be as a fuel for combustion engines, but as a fuel for household needs.[7] A mere half-dozen trees planted in a rural backyard can produce 300 liters of oil per year, serving a variety of fuel purposes (cooking, lighting, etc.). The nuts grow twice a year, 1 kilogram producing about 70 grams of oil when pressed. The oil contains hydrocarbons of a type rarely found in nature, and it is sometimes so flammable that when the nuts are freshly picked, they can be lighted with a match. Fortunately, the tree seems to thrive in secondary or disturbed forests, which means there would be plenty of opportunity to establish bio-fuel plantations without encroaching on a single square meter of primary forest.

Not only Brazil and Philippines, but many other tropical forest nations surely possess tree species, and many other plants as well, that could generate bio-energy as a means to get them off the "fossil-fuel hook." Between 1970 and 1990, those nations of the tropical forest zone that do not produce their own fossil fuel are projected to increase their imports of petroleum at least four times over, and in terms of cost as measured by constant 1980 dollars, their bill is projected to expand more than 20 times over. Not surprisingly, several of these nations are investigating the outlook for developing bio-crude and plant-derived fuels from whatever source. In the tropical forests of South America alone, there are believed to be more than 1,000 possible plant candidates, of which 95 percent occur in Amazonia, 10 percent in the coastal forests of eastern Brazil, and 10 percent in the Pacific Choco region of Colombia.[8] Brazil alone possesses at least 700 of these plant species: an excellent opportunity for this nation to grow its own energy in infinitely renewable form. Who knows, within just a few years Brazil may eliminate the single most crippling item in its imports bill—and eventually the nation may even turn itself into an exporter of plant-derived fuels. Those who doubt this prospect should remember the dismissive remarks made around one hundred years ago of that apparently useless tree of Amazonia, the rubber tree.

14

Environmental Services

IN THAILAND, a man can be put against a wall and summarily dispatched if he is caught poaching timber. Reason: Thailand is learning, through bitter experience, that when its forested landscapes lose their tree cover, the environmental repercussions can be worse than serious for millions of citizens who suffer through the illegal logging of a few. As water catchments become denuded in Thailand's hilly zones, rivers no longer produce regular amounts of irrigation water year-round for the downstream farmers who sustain the nation's principal export commodity, rice.

In the Philippines, President Marcos, after the flooding in January 1981 that caused 331,000 people to be evacuated and inflicted damage to crops, livestock, and buildings worth $30 million, has declared that deforestation is a national emergency. According to the Philippines' National Environmental Protection Council and the Council for Agriculture and Resources Research, deforestation has led to such a decline in river flows that urban communities, such as those of Metro Manila, lack sufficient amounts of good-quality water for household and sanitary purposes.[1] A persistent threat of pandemic diseases thus persists through several of the Philippines' conurbations.

In Kenya, where forest cover has declined by two-thirds during just the past twenty-five years, soil erosion and siltation of hydropower facilities are causing severe setbacks to the country's hopes of escaping the energy crunch. In Amazonia, there are signs that deforestation in headwaters catchment zones disrupts river flows thousands of kilometers downstream, with costly effects on floodplain farmers. In the Caribbean, where many islands have lost virtually all their forests, cyclone damage is greater than it would be if there were tree cover to buffer the effects of freak weather.

These are just a few illustrations of environmental consequences when tropical forests are mis-used and over-used, or even used up. In all sectors of the biome, we find instances of the unexpected prices people pay—often high prices, all too often paid by those who can least afford to pay them, that is, peasantry —in the aftermath of deforestation. As forests disappear, so do their watershed functions and other environmental services. These services are rarely recognized ahead of time, even though they are significant. They are rarely accorded a realistic evaluation when development planners do their cost-benefit arithmetic, since it is tough to put a dollar figure on the values that a forest represents through its mere existence—even though it is only too easy, however, to put a dollar tag on the damages that result when the forest cover and its environmental services are eliminated.

In this chapter we shall look at various types of environmental services—and at one of the most perplexing of all forestry challenges, an economic assessment of these services.

REGULATION OF FLOODS AND DROUGHTS

Although tropical forests cover a mere one-sixteenth of Earth's land surface, they receive almost half of Earth's rainfall on land. The forests exert a sponge effect, soaking up moisture before releasing it at regular rates. While the forest cover remains intact, rivers not only run clear and clean, they flow throughout the year. When the forest is cleared, rivers start to turn muddy, then swollen or shrunken. In several major river basins of the biome, notably those of the Ganges, Brahmaputra, and Mekong, rainy-season supplies of water tend to be released in floods, followed by months-long droughts.

How does the sponge effect actually operate? The multi-storied structure of the forest, together with its vast amounts of foliage, helps break the impact of a tropical downpour. Much of the water trickles down branches and tree trunks, or drips off leaves in a fine spray, so that when the rainfall reaches the ground, it percolates steadily into the soil or runs off into streams and rivers at a gradual rate. An undisturbed dipterocarp forest in Southeast Asia intercepts at least 35 percent of rainfall, whereas a logged forest intercepts less than 20 percent, and a tree plantation, such as rubber or oilpalm, only about 12 percent.[2] In the Tai forest of southwestern Ivory Coast, in West Africa, the rivers flowing from a primary forest release twice as much water halfway through the dry season, and between three and five times as much at the end of the dry season, as do rivers from a coffee-plantation zone; thus the forest rivers greatly reduce any drought effect for valleyland farmers who depend on uninterrupted supplies of irrigation water for their multiple crops each year.[3] In Peninsular Malaysia, where the government wonders whether it has allowed too much forest to be eliminated during the past two decades, a drought in late 1977 and early 1978 reduced the rice harvest by well over one-quarter, a loss estimated at almost $95 million; moreover, and because of insufficient water flow through hydro-power facilities, the drought led to routine rationing of electricity in several urban areas.[4]

The region where deforestation is most advanced is Southern-Southeast Asia, and it suffers floods that increase in range and intensity year by year.[5] As an indication of how high flood costs here could climb, let us look at recent experience in the Ganges Plain. The river basin, covering 956,000 square kiometers, contains 70 percent of India's flood-prone lands, together with an even larger proportion of Bangladesh's. Since it is the main grain-growing area of both countries, the Plain now supports more than 500 million people, a total that is projected to double in less than 30 years—all of which makes the Ganges Plain the greatest single ecological hazard on Earth in terms of numbers of people involved.[6] Together with the Brahmaputra, the Ganges river system receives over half of India's total rainfall. During the past thirty years, forest cover in the upper catchment territories (not only in India but in Nepal as well) has been reduced by at least 40 percent, with the result that monsoonal flooding followed by drought conditions now cause marked damage. River courses flowing out of forestlands on the border between India and Nepal become 1.5 kilometers wide during the rainy season, before dwindling to a mere 40 meters at the height of the dry season. Flood damages used to total around $120 million per year until 1970, when they started to soar. During the 1978 monsoon season 65,712 villages were inundated within just a few weeks, 40,000 cattle washed away, and extensive croplands ruined; total damage to property was estimated at more than $2 billion, apart from the incalculable costs of 2,000 human lives lost through drowning. Since that exceptional episode, flood costs have subsided to a mere $1 billion per year.[7] India now undertakes massive flood-prevention measures, reaching $100 million or so each year. But flood control is tackled almost entirely through major engineering works such as embankments, dykes, and large dams, with hardly one-hundredth as much spent on forest conservation.

In other parts of Southern and Southeast Asia, a similar situation obtains: for example, in Bangladesh,[8] in Burma,[9] and in Sarawak.[10] Of course this part of the tropical forest biome, often more heavily populated than other parts, tends to be more susceptible to floods that damage property. Whereas total costs of flooding can be roughly estimated at $12 billion per year throughout the Third World, we can surmise that all but a small fraction of the costs are incurred in humid sectors of tropical Asia.[11]

Even a small amount of excess water can cause problems, notably for agriculture. In one-third of the rice-growing lands of Southern-Southeast Asia, it is common for flood waters to run at least half a meter higher at some stage of the growing season, more than can be tolerated by the Green Revolution varieties of rice with their short stems.[12]

Still other sectors of the biome appear to be following a similar track. Even in Amazonia, where deforestation is not a fraction so advanced, there are early signs of trouble.[13] During the 1970s, the population of Amazonian Peru, for example, doubled, causing the elimination of some 51,000 square kilometers of forest. Most human settlements in Amazonia are located along rivers, and their lifestyles are intimately related to the annual flood cycles. These farming communities occupy the relatively flood-free sedimentary terraces known as "restingas," which are formed by alluvial deposits bordering the rivers. They pursue their agriculture in critical equilibrium with the timing and extent of floods: crops are often lost when the yearly flood arrives earlier or lasts longer than usual. Many people in Peruvian Amazonia, inhabiting areas downstream from the deforested territories, find that the floods are larger than they used to be, causing severe damage to their crops.

Furthermore, the upland sectors of Peruvian Amazonia, with their generally rich soils, continue to attract millions of subsistence cultivators at ever-more rapid rates, arriving from crowded parts of the country on the other side of the Andes. A similar pattern applies in Amazonian sectors of Colombia and Ecuador. As a result, we can expect increased flooding to overtake many human communities—and not only in upper Amazonia, but in localities thousands of kilometers downstream in Brazilian Amazonia, where floodplain cultivators now encounter growing disruption for their agriculture.

LANDSLIPS, ROCKFALLS, AND
EARTHQUAKE DAMAGE

Deforestation aggravates disruptive phenomena such as landslips and rockfalls. In several parts of the Himalayan foothills, patches of land cover are sheering away at a rate eight times greater than was recorded a mere forty years ago, that is, before broadscale deforestation began. Worse still, deforested lands are less able to withstand the devastation of earthquakes. In late 1974, a moderate earthquake, registering only 5.5 on the Richter scale, struck the mountains of Indus Kohistan, in northern Pakistan. Although the geologic upheaval was not exceptional, it caused devastation over an area of some 750 square kilometers, killing thousands of people and injuring several times as many, razing whole villages, and severely disrupting the entire economy of the region. Under the impact of debris sliding down deforested slopes, sections of the Karakoram Highway were swept away at hundreds of points along a 64-kilometer stretch of road. Extensive sectors of agricultural terracing, carefully constructed to halt soil erosion, were swept away, much of the material ending up in the reservoir of the Tarbela Dam, 120 kilometers down the Indus River.[14]

REGULATION OF WATER FLOWS
FOR IRRIGATION

In several parts of the humid tropics, the greatest limitation on our hopes for more food in the immediate future stems from problems of water supplies for irrigation.[15] This is especially the case in much of Southern-Southeast Asia and Central America, as well as in parts of tropical Africa. Many tropical forests, notably in Asia, are located above rich alluvial valleys, several of which support some of the highest-density agricultural communities on Earth, farming primarily through means of irrigation. Two-thirds of Southern-Southeast Asia farmers, and two-fifths of all developing-world farmers, live in such valleys, which leaves them heavily dependent on the activities of the one-tenth who live upstream in the same rivers' catchment zones. Not surprisingly, most of Earth's civilizations have become established in such fertile river basins, and we can expect that the bulk of humankind will continue to make its livelihood in these key localities.

Irrigated croplands now constitute about 15 percent of all arable lands, while producing about 30 percent of our food. If we are to grow twice as much food during the last twenty years of this century, as we must if all developing-world people are to go to bed each night with a full stomach, we must reckon that well over half of that increase must come from irrigated croplands. Irrigation agriculture uses seven times more water than the next two water-demanding activities combined: industry and domestic needs. Several irrigated crops are extremely thirsty: for example, wheat, which requires 1,500 liters of water to produce 1 kilogram of grain, and rice, which requires three times as much (and rice accounts for roughly two-thirds of all food grown in the developing world). To supply a one-hectare field with 20 milimeters of water requires 200 cubic meters of water (a greater volume than is contained in a large living room); in terms of conventional domestic needs, this is enough to supply a developing-world family of six persons for almost four years. All in all, an irrigated field is likely to need at least 900 milimeters of water for each harvest, or 9,000 cubic meters per harvested hectare.

Tropical humid Asia, not counting China, accounts for more than three-fifths of irrigated land in the developing world. The total now amounts to 800,000 sq. km., a little larger than Texas and Oklahoma; and this area needs to be expanded to 1,400,000 sq. km. by the year 2000, or an increase equivalent to another Texas. In order to grow twice as much rice in the region by the year 2000 (of all irrigated croplands, 85 percent are given over to rice), the Trilateral Commission believes that we may have to invest as much as $54 billion by the year 2000;[16] moreover, if we are to meet all irrigation needs throughout the Third World, we may need to think, according to the U.S. Department of Agriculture, of an investment of $230 billion (both figures in 1975 dollars).[17] Yet in several of the better irrigated areas of Asia, notably in parts of the Indian sub-continent, Thailand, Indonesia, and the Philippines, the Green Revolution is losing momentum as farmers find they can no longer rely on sufficient flows of irrigation water for their multiple crops of bumper-harvest rice. In the Ganges river system, for example, dry-season water flows have declined by almost one-fifth during the five-year period 1973–78. Yet despite the exceptional emphasis that must be directed toward water supplies, neither of the two agricultural proposals referred to above, with their multi-billion-dollar budgets, allocates a dime to safeguarding of forest cover in water-catchment zones.

A similar story can be told for much of Central America.[18] Costa Rica hopes to expand its irrigated croplands during the last twenty-five years of this century by 180 percent, El Salvador by 230 percent, Nicaragua by 300 percent, and Panama by 340 percent. Yet in all these countries, deforestation is proceeding apace, with one-third of the region's forest cover having been eliminated during the past three decades (see chapter 7). Not surprisingly, these countries' water needs are already catching up with stable runoff from watersheds, in several localities almost as critically as is the case in virtually the entire Indian sub-continent.

We encounter a parallel situation in Africa along a broad belt bordering the tropical forest expanse of the equatorial zone and the strip along the Gulf of Guinea. For much of its extent, this belt is under great and growing pressure from human populations, which progressively degrade and destroy the forest cover. Several countries are projected to encounter water shortages for irrigation agriculture well before the end of the century.

CONTROL OF SOIL EROSION

In those parts of the tropical forest biome that receive over 3 meters of rainfall a year, a half-hour thunderstorm can produce

25 milimeters of rain, up to forty times more water than an average shower in the northeastern United States. The impact of these tropical downpours causes more soil erosion in deforested areas than almost anywhere else on Earth.

The most notorious instance is probably Nepal, where the Department of Soil and Water Conservation estimates that between 30 and 75 tons of soil are washed away annually from each hectare of deforested land. This means that the country altogether loses as much as 240 million cubic meters of soil per year —a highly precious export that Nepal unwittingly dispatches to India.[19] The economic costs are substantial, causing a loss to agriculture that in the long run could amount to billions of dollars each year.

Almost as extreme is the deforestation-caused erosion in Ethiopia. Less than thirty years ago, forest covered about 10 percent of Ethiopia's territory, an amount that has now fallen to 4 percent. As a result, huge amounts of topsoil flow from the Ethiopian highlands each year, some of it being carried down the Blue Nile until it silts up the Roseires Dam many hundreds of kilometers westward over the border in Sudan.

To consider the economic repercussions of erosion, let us look at the case of Indonesia. Eroded territories, formerly forested, now exceed 400,000 square kilometers, or about one-fifth of national territory. The agricultural costs of this erosion are sizable, especially in the island of Java, which, with an area of 132,470 square kilometers (roughly the same as New York State, or Greece), and with 95 million people, possesses only 15 percent of its original forest cover. Java's croplands lose at least 770 million tons of topsoil each year, worth a rice output of more than 1.5 million tons, or equivalent, at an average consumption rate of 100–130 kilograms per person per year, to the food needs of 11.5–15 million people per year.[20] A similar decline in cropland productivity is overtaking those parts of Indonesia's outer islands, notably Sumatra, Borneo, and Sulawesi—principal locations for Indonesia's transmigration program, under which 12–15 million

Carefully terraced hillsides, such as these in Nepal, remain productive for centuries, generating at least two rice crops per year. But the caring maintenance of generations can be destroyed in just a day if landslips caused by deforestation allow terraces to be carried away. *(Professor Jack Ives, University of Colorado)*

When a forest is cleared, and its canopy no longer protects the ground from the sun's heat and from thunderstorms, the soil quickly becomes impacted and eroded, as illustrated here in new settlements in northern Sulawesi, Indonesia. *(Michèle Dépraz, World Wildlife Fund)*

subsistence farmers are being resettled from Java and other heavily overcrowded islands.

CONTROL OF SEDIMENTATION

When topsoil washes away, rivers become burdened with suspended sediment that can turn them into turgid streams of mud. The Ganges, for instance, carries an average annual load of sediment amounting to 1,544 tons per square kilometer of drainage basin, or fourteen times higher than that of the Mississippi. The Brahmaputra carries almost as much, 1,429 tons; while the Irrawadi carries well over 900, the Indus over 500, and the Mekong almost 500 tons.[21] Siltation in the Ganges system is so pronounced that a number of river beds are rising at a rate of one-sixth of a meter per year, grossly aggravating floods. Newly formed shoals have rendered several sectors of the main river unnavigable, while certain industrial installations in the downstream plain suspend activities for several months of the year due to lack of water. The ports of both Calcutta and Dacca are silting up.

Perhaps the most remarkable consequence of all is the gigantic "shallow" that is building up in the Bay of Bengal. Caused by the accumulative buildup of sediment during recent years, this shallow, covering some 50,000 square kilometers, will, when it breaks the surface of the sea, be called either New Moore Island or South Talpatty, depending on which country, India or Bangladesh, gains the right to name the island. Nepal, which has contributed most to the new phenomenon, is not being consulted by India and Bangladesh, which are busy enough with their dispute over sovereignty of the emergent territory.

A related problem threatens the Panama Canal. About half of the Canal is made up of two man-made lakes, Gatun and Alajuela. The two lakes account for about one-tenth of the Canal's total watershed of 3262 square kilometers. In the early 1950s, when 85 percent of the watershed was covered with its original rainforest, there were only a few thousand people living there, mostly pursuing an agricultural way of life in an area that offered sufficient space to obviate any permanent damage to the forest. By 1970 the watershed's populace had grown to 60,000, a good number of them living in urban areas, but with growing throngs practicing forestland farming of degradative sort; and by 1980 the total had grown to 90,000 people, with the result that a full 1,000 square kilometers of former forest cover had been burned for crop growing and for cattle raising. If the deforestation trends continue, virtually all the original forest will disappear by the year 2000. In turn, this deforestation will cause an increase in annual sedimentation in the Canal from 9.5 million cubic meters in 1980 to 16 million by the year 2000—and Lake Alajuela, which has already lost 5 percent of its storage capacity through sedimentation, stands to lose 40 percent by the year 2000.

These problems are already affecting the Canal's operations. In early 1977 a serious drought compounded the sedimentation problem, causing water levels to decline too far for full operation of the Canal and diverting some cargo ships around Cape Horn. Were the remaining forests to be protected, and were denuded areas to be reforested, the annual sedimentation could be reduced from 9.5 million cubic meters to well under 8 million within five years—and Lake Alajuela's storage capacity, which is projected to decline to a mere 20 percent by the year 2040, could remain as high as 70 percent. All the more welcome, then, is the recent AID initiative to safeguard the watershed through a $10 million contribution to a $17 million project.[22]

Furthermore, deforestation-derived siltation proves a serious problem for water impoundments in several parts of the humid tropics.[23] In the Philippines, the Ambuklao Dam is silting up so fast that its useful life is being reduced from its planned 56 years to 32 years, due to deforestation in the Agno River watershed; if there is no change in the present situation, the reduced operational capacity of the dam will, as early as 1985, incur costs of $25 million. In Pakistan, the Mangla Dam, completed in 1967, receives so much silt from the Jhelum River watershed that its operational life is being reduced from more than 100 years to less than 50 years.[24] Also in Pakistan, the world's largest dam, the Tarbela Dam, is losing its storage capacity of 12 billion cubic meters at a rate that will leave the dam useless within just 40 years.[25] In Ecuador, the 100-million-cubic-meter Poza Honda Reservoir, constructed in 1971, is losing its capacity at a rate that will leave the installation useless within only another 25 years; a conservation program to reforest that half of the 175-square-kilometer watershed where forest cover has been eliminated would cost only $1.8 million, extending the reservoir's life to its planned 50-year life span and producing benefits of at least $30 million.[26]

This adverse experience with reservoirs is all the more regrettable in that many parts of the humid tropics offer promising sites for hydropower. Nepal, with its mountainous terrain, has a theoretical potential of 83,000 megawatts, about the same as the installed hydropower capacity of the United States, Canada, and Mexico. But because of deforestation-caused sedimentation, Nepal has so far brought on-stream less than 50 megawatts.[27] For the same reason, among others, Indonesia has exploited only 2 percent of its potential 31,500 megawatts; Thailand, 4 percent of its 22,344 megawatts; and Philippines, 9 percent of its 7,903 megawatts.[28] In addition to these countries, several others of the tropical forest biome possess unusual potential: in Asia, the countries listed, plus India, Burma, and Papua New Guinea, could enjoy more than 300,000 megawatts of hydropower; in Africa, Cameroon, Gabon, Zaire, and Madagascar possess potential for 236,500 megawatts; and in Latin America, Peru, Ecuador, Colombia, Venezuela, Guyana, and Brazil possess potential for 192,-000 megawatts. These regional totals add up to 728,500 megawatts, which amount to almost 60 percent of the potential of the entire developing world, which in turn represents almost two-thirds of the global potential of just under 2 million megawatts.

Another "hidden" environmental cost of sedimentation is the

impact on offshore fisheries. Around the Philippines' 7,000 islands, with their 18,000 kilometers of coastline (only slightly less than the United States'), there are 44,000 square kilometers of coral reefs, which supply about one-tenth of the country's fisheries catch in commercial terms and a further one-tenth in subsistence terms. Several parts of these fisheries are declining, due to deforestation far inland.[29] Along other sectors of the Philippines' coastal zones, mangrove ecosystems are been suffocated through silt carried down from watershed catchments. The country now exports $100 million of oysters, mussels, clams, and cockels each year—and again, these valuable fisheries are being impoverished through the ecological backlash effects of deforestation.[30]

LINKAGES WITH PUBLIC HEALTH

Everybody everywhere needs water every day for drinking, cooking, washing, and sanitation. A minimum amount is 20 liters per day. As the standard of living increases, so does the demand for water, until a better-off person in an urban community of a developing nation may consume 200 liters per day. According to the World Health Organization and the World Bank, there are now 100 million more developing-world people without access to clean water or adequate sanitation than in 1970.[31] Were these two services to become more widely available, they would help to cut down on the toll of water-related diseases. Typhoid, cholera, amoebic infections, bacillary dysentery, and diarrhea, among related diseases, cause an estimated mortality of 25 million people each year, and morbidity relating to as much as 80 percent of all sicknesses.[32]

Insofar as large numbers of these afflicted people live in the humid tropics, tropical forests, by assuring dependable supplies of good-quality water for domestic use, make a substantial contribution to campaigns for better health. Interestingly enough, the worst-off communities are in Southern and Southeast Asia, where two-thirds of all people do not enjoy even minimal standards of water supplies and sanitation. To meet these needs for the entire Third World would cost some $300 billion during the course of one decade. Were a mere 1 percent of this sum, $3 billion per year, to be allocated to safeguarding watershed systems, it would represent as sound an investment as in water piping, stand taps, sewers, and the other conventional equipment that development agencies now envisage for the Water and Sanitation Decade.

Yet in a tropical forest territory with some of the highest rainfall on Earth, Peninsular Malaysia, water is rationed for part of the year in Kuala Lumpur and several other urban areas.[33] Water demand in the Peninsula is projected to double before the end of the century; and as water supplies decline in face of rising demand, so costs increase for the Malaysian consumer. The price of water from a catchment with undisturbed forest increases twofold when the forest becomes subject to controlled logging, and fourfold when the forest becomes subject to uncontrolled logging.

In similar fashion, public-health programs in Bangkok, Manila, Lagos, Abidjan, and several other conurbations of the humid tropics are being set back through deforestation-caused declines in quantity and quality of water supplies. In Bangkok, for instance, with a population of five million, water demand has now topped 1 billion liters per day. The city is projected to double in size by the year 2000, and its thirst to increase at least four times. At present, about one-third of the water supply comes from the giant Chao Phraya River, which rises 1,200 kilometers away in the northern mountains of the country. The rest of the city's water comes from local wells. Both these sources of water are declining.[34] The river's basin covers 177,550 square kilometers, or more than one-third of the entire country, and supports almost half of the population—many of whom are forestland farmers who destroy the tree cover year by year. As a result of the disrupted river flow, that is, a regime of too much water followed by too little, the river is less and less capable of supplying the needs of Bangkok. So the citizens turn increasingly to their 11,000 groundwater wells, with the result that, at present rates of extraction, the underground stocks are expected to give out by 1990. Worse still, the massive pumping from subsurface reservoirs is causing the city to sink. Bangkok is literally declining, at a rate fourteen times faster than Venice. Because most of the city lies a mere 1 meter above sea level, the situation gives cause for more than concern. According to the Bangkok Metropolitan Waterworks Authority and the Asian Institute of Technology located in Bangkok, the present situation, if allowed to proceed, will leave the city way below sea level by the year 2000. As a preventive measure, city planners are trying to bring in water from other parts of the country, at a cost of many millions of dollars each year. One wonders what could have been achieved if just a few million dollars had been spent on tackling deforestation before it got out of hand.

Metropolitan Manila, capital of the Philippines, totals eight million people. Local watersheds on the island of Luzon retain only one-fifth of their forest cover, with the result that water levels in major reservoirs repeatedly drop to a point where they trigger water shortages, raising the threat of several pandemic diseases at once.[35]

At projected rates of population growth and urbanization, the region of Southern-Southeast Asia will, by the year 2000, contain three times as many people living in towns and cities as it did in 1975. Also by the year 2000, at present rates of deforestation, there will be only a small fraction as much tree cover in watersheds as in 1975.

CONTROL OF CYCLONE DAMAGE

The humid tropics are the site of most cyclones that occur in the world. The adverse impact of these phenomena is aggravated by deforestation, in that the "buffering effect" of forest cover is reduced, thus fostering environmental damage generally. The costs to life and property can be much greater than those caused by other natural disasters. During the period 1966–75, for example, the worst earthquake in the world took a toll of 54,000 lives, but the cyclone that struck Bangladesh in late 1970 accounted for at least 200,000 lives, possibly twice as many. During the course of the many climatic disasters that have overtaken humankind, including floods and droughts, cyclones are thought to have accounted for four-fifths of all lives lost.[36]

Known as typhoons in the northern part of the Pacific, as hurricanes in the Caribbean, and as cyclones in the rest of the humid tropics, these revolving storms generally form in zones between 8 degrees and 25 degrees on either side of the Equator. They originate above oceans with warm surface layers, from which they derive their enormous stocks of energy. After "birth," the storms move in a westerly direction at a speed of at least 15 knots per hour, often twice and sometimes three times as fast. Eventually they encounter a land mass, over which they expend their energy speedily. This means that the territories worst hit are islands and coastal zones along the eastern edge of continents, with an impact area extending as much as 300 kilometers inland.

Cyclones are feared for three destructive attributes: high-speed winds, torrential rain, and storm surges. A huge whirl or vortex of air rising up to 15,000 meters, and with a diameter of between 100 and 800 kilometers, the cyclone generates wind speeds well above 100 kilometers per hour, sometimes as much as 200 kilometers, occasionally approaching 300 kilometers. Rainfall reaches astonishing levels; during a 24-hour period in mid-July 1911, Baguio City in the Philippines received 1,170 millimeters of rain (roughly one and a half times as much as New York City receives in one year). As for storm surges caused by sea water piling up in front of the wind and being propelled shorewards, these phenomena tower several meters high and inflict vast damage when they reach coastlines. It was a storm surge that primarily caused the mass drowning of huge numbers of coast-dwelling people in Bangladesh in 1970.

Each year some 80 to 100 cyclones form over tropical oceans.

Of these, an average of 19 hit the Philippines, around one dozen strike the Bay of Bengal, and another dozen create havoc in the Caribbean, while the rest are scattered throughout the cyclone zone. Altogether, they are believed to kill at least 20,000 people and cause damage worth more than $8 billion each year. According to the U.N. Economic and Social Commission for Asia and the Pacific (ESCAP),[37] the region of Southern-Southeast Asia now suffers injury of around $4 billion per year, or six times more than two decades ago—the increase being partly due to the expansion of human settlements that thus offer more numerous targets, and partly due to the clearing of forests that no longer mitigate the impact of cyclones. Of course these are rudimentary reckonings, guesstimates at best. But insofar as they err, they are surely on the low side; some observers believe that the actual economic costs of all kinds are twice as large. According to the World Meteorological Organization, the Philippines has been suffering annual costs of around $80 million per year.[38] In Bangladesh, where 20 million out of the populace of 100 million live in 20,000 square kilometers of coastal zone, the costs are estimated to average more than $325 million per year. In much of the ESCAP region, we can expect costs to soar, insofar as the majority of human communities live close to coasts (with numbers growing faster than populations overall), and insofar as littoral forests are being cleared more extensively than most other forests.

15

Climatic Linkages

We have seen that tropical forests act as "buffering effects" to stabilize many environmental processes, notably water flows. Significant as these environmental safeguards are, none of them compares with the contribution of tropical forests to probably the greatest environmental process of them all: climate. At both regional and international levels, we are learning that tropical forests make considerable contributions to climate. When forest cover is maintained, climate tends to persist with the predictable patterns established through hundreds and thousands, if not millions, of years. We know what we can expect from climate: generally speaking, the same as before, with perhaps some marginal trends that take many millennia to express themselves. Whenever climatic changes are in the making, tropical forests serve to modulate their shifting course, smoothing out the "wrinkles." But when tropical forests are eliminated, we can find that climatic changes swiftly ensue, some of them so abrupt and sizable that we have next to no time to adapt to their arrival.

We also know that we can anticipate further climatic changes in the future as a result of deforestation. What we do not know is the nature and scale of these changes. We have all too little knowledge to help us make accurate predictions about these changes: how swiftly they will overtake us, how profound their impact will be, and so forth. Fortunately we already have some clues to help us peer into a deforested future and discern some likely outcomes. In this chapter, we will examine one set of regional repercussions, in Amazonia, and two sets of repercussions that will affect the global community. It is this factor of climatic linkages that, perhaps more than any other, could cause deforestation to generate wide-ranging consequences for lands far outside the tropics, on a scale that would trigger deepseated changes in lifestyles for many developed-world citizens.

AMAZONIA'S HYDROLOGICAL CYCLES

Amazonia, the wettest region of any size anywhere, contains two-thirds of all fresh water on Earth (in the sense of "free" fresh water above ground; much fresh water is "locked up" in polar ice caps and in subterranean reservoirs). This water is constantly falling from the sky, percolating through the vegetation, being absorbed by plants, running away into streams and rivers, and evaporating (through the sun's warmth) or being evapotranspired (by plants) back into the atmosphere. Despite the huge size of the Amazon River system, which discharges one-fifth of all river water that makes its way into the Earth's seas and oceans, more than half of the region's moisture remains within its own Amazonian ecosystem. As fast as much of the moisture falls to the ground, it is returned, through the respiratory activities of trees and other plants, into the skies, whereupon it gathers for a fresh series of thunderstorms. Day in, century out, the water cycles round and round, remaining within the bounds of Amazonia. True, some rainfall derives from circulation patterns outside the region, primarily from the Atlantic, but at least half originates within the region. In western Amazonia, a full 88 percent of water reaching the ground is falling for at least a second time, and probably a third or fourth time, from the atmosphere.[1]

The implications are profound. Were a substantial amount of forest to be removed, the climate would become drier than at present. Each time a sizable sector of forest is cut down, the remainder is less capable of evapotranspiring as much moisture as was circulating through the ecosystem before—and all this makes for a steadily desiccating ecosystem. At what stage could the forest start to be transformed into a different kind of forest by virtue of the drying-out phenomenon? Has the process already begun? If so, how far has it gone? Could it still be reversed? To date, we scarcely know how to formulate the correct questions, let alone to supply the right answers.

Equally to the point, we are now learning that the climate of extensive territories in Brazil outside Amazonia depends, in part at least, upon the same hydrological cycles that are so critical to the persistence of the rainforest as we presently know it in Amazonia. In the vast cerrado woodlands to the south, for example, extending as far as Brasilia, there could well be less rainfall than now if Amazonia were to be fundamentally transformed. Much the same could even apply to Brazilian territories still farther south and west, including the principal agricultural sectors of the country. Fortunately, the Brazilian government is starting to take note of these climatic linkages, in their full scope (so far as they can be discerned at this early stage), and Brazilians are becoming less inclined to look on the Amazonian forest ecosystem as a "valuable asset going to waste."

ALBEDO EFFECT

The hydrological cycling of moisture in Amazonia is a phenomenon the repercussions of which would be confined to a relatively small part of the planet. What a different picture we encounter when we examine the albedo effect—a phenomenon related to the proportion of sunlight that the Earth's surface reflects back into space. Let us suppose, for the sake of argument, that the sun radiates 100 units of energy per minute in the direction of the Earth. When this energy reaches the outer limits of the atmosphere, the atmosphere may well absorb 20 units, allowing only 80 units to strike the Earth's surface. If, due to the vegetation pattern that occupies that part of the Earth's surface, some 40 units of energy are reflected upward and out into space, then the albedo is reckoned at 50 percent. In point of fact, the amount varies a great deal. Fresh snow, with its high reflectivity, reveals an albedo of 85–90 percent; and deserts, with little vegetation to absorb and retain the sunlight, generally reveal 25–30 percent. Croplands, with their moderate amounts of vegetation, span a range of 12–20 percent albedo; while evergreen tropical forests, with their dense masses of vegetation, only 7–15 percent. Because of its impact on convection patterns, wind current, and hence rainfall regimes, the albedo effect constitutes a basic factor in controlling our climate.

Although we have no conclusive scientific evidence as yet, on the grounds that we do not understand the precise workings at issue, many scientists believe that widespread clearing of tropical forests would cause an increase in the "shininess" of the Earth's surface, with the result that the deforested lands would reflect greater solar heat.[2] Much of the tropical forest zone, being a broad band on either side of the equator, corresponds with a zone where there is a pronounced uprising of air, a process that introduces energy into the large-scale circulation movements of the general atmosphere. If this process becomes disrupted through deforestation, the repercussions could well extend, through altered patterns of air circulation, to the temperate zones—especially those of North America and Eurasia (not so much in the Southern Hemisphere, because of the modulating effect of the ocean expanse).

So far as climatologists can tell, a not unlikely outcome would be a decrease in rainfall in the equatorial zone itself, an increase in rainfall for territories between 5 and 25 degrees North and South, and a decrease for lands between 40 and 85 degrees in the North. In short, greater rainfall for the southern half of the Sahara, India, and much of the arid sector of Mexico, but reduced rainfall for the northern half of the United States and Canada, and for most of Europe and the Soviet Union. These climatic dislocations could prove traumatic for food-growing territories of the northern temperate zones, notably the grainlands of North America, Europe, and the Soviet Union.

CARBON DIOXIDE BUILDUP

We often think of carbon dioxide as an insignificant gas, of main relevance to our daily lives when it supplies the "fizz" to soft drinks. But carbon dioxide has a much greater effect than that. While it constitutes only 0.03 percent of gases in the Earth's atmosphere (by contrast with nitrogen, 78 percent, and oxygen, 20 percent), it absorbs radiant energy from the sun at infrared wavelengths: and by retaining some of this energy at the Earth's surface, it creates a greenhouse effect, that is, it traps heat and warms the Earth's atmosphere. This greenhouse effect has been mostly moderate, stable, and acceptable. But in recent times, the amount has been increasing, and at a rate that is itself increasing. If this buildup of carbon dioxide in the global atmosphere continues, the result will probably be warmer temperatures for the planetary ecosystem as early as the second quarter of the next century, together with basic shifts in rainfall patterns for extensive sectors of humankind's habitats. It could further lead, in the longer term, to some melting of polar ice, with a marked rise in sea levels. Overall the economic consequences, plus the "political fallout," of this phenomenon could prove far-reaching indeed.

The issue concerns us here because tropical forests contain a vast amount of carbon: some 340 billion tons, equal to about half of all the carbon in the atmosphere. When tropical forests are removed, they often release carbon into the skies. Of course not all deforestation engenders this effect. When trees are cut for commercial timber, the carbon in them is merely sequestered in another location, where it remains locked away in the form of housebuilding materials, furniture, parquet floors, and the like. But when tropical forests are felled and burned, as is generally the case during deforestation (due to the broadscale activities of cattle raisers and forest farmers), huge stocks of carbon are released into the atmosphere. As we have seen, the long-range repercussions of carbon dioxide buildup could trigger an environmental perturbation on a scale greater than anything else we can foresee, short of nuclear war. Since the impact could be so profound, let us examine the prospect in some detail.

The Earth's carbon cycle involves a stock of roughly 40,000 billion tons, the great bulk of it in the deep oceans. The planetary atmosphere contains almost 700 billion tons, which now works out to over 335 parts per million—a concentration that is increasing at a rate of about 1.5 parts per year. Finally, the Earth's biotas: living things contain about 800 billion tons, while other sectors of the biosphere contain at least 1,000 billion tons (possibly much more) in the form of soil humus and peat. To give an idea of the way the biosphere's carbon stock is divided up among categories of living things, plants are believed to contain about 800 billion tons—of which 80–90 percent is in forests, almost half of it in tropical forests. Humankind, by contrast, with over 4.6 billion individuals averaging less than 50 kilograms in weight and a carbon content of about 4 kilograms each, contains well under 20 million tons.

At roughly midpoint of the last century, when the Industrial Revolution was gathering momentum, the atmospheric concentration of carbon dioxide was about 290 parts per million, probably rather less—as low, some scientists now believe, as 265 parts per million. As industry stepped up the burning of fossil fuels, the concentration increased. By 1960 it had risen to 310 parts per million, and the past 24 years have seen an increase of almost a further 30 parts per million. At the current rate of increase, which is itself increasing, the pre-1850 "natural" amount could well double by the year 2030 or soon thereafter.

The principal culprits fingered right now are the factory and the automobile. Through these two devices, we are burning fossil fuels at an enormous rate of almost 5 billion tons per year. We are also increasing our burning rate at between 3 and 4 percent per year, and if we continue with this level of growth (albeit there has been a slight slackening since the start-up of OPEC), we could eventually cause the atmospheric concentration of carbon dioxide to rise to 600 parts per million. If we slow the growth rate of fossil-fuel combustion to only 2 percent per year, the doubling time will be extended by no more than 15–20 years. If we were to throttle back on our increasing use of fossil fuel, and maintain today's level into the indefinite future, the doubling would be delayed awhile longer, though not much longer.

In the overall picture, what is the role of burning of tropical forests? Many observers believe it is a significant source of carbon dioxide.[3] When a tract of forest is burned, not only is carbon from living plants released, but further carbon becomes available when the soil is exposed to the tropical sun, and soil humus with its organic carbon becomes oxydized. Yet despite the vast stocks of carbon that are constantly released from tropical forests, certain scientists consider that the forests, with their immense capacity to soak up carbon dioxide through photosynthesis, could actually be serving as a sink, rather than a source, for carbon dioxide.[4] Some forest cover that is disrupted by small-scale farmers grows back again, to make up a forest formation with fewer species but with greater capacity to produce large amounts of plant biomass in short order. To this extent, a successional restoration of forest cover will counterbalance some of the depletion activities. On the whole, however, there is much evidence to the effect that forest farmers generally serve to reduce the amount of forest biomass; their net impact amounts to an increase in carbon dioxide in the global atmosphere.

How much carbon may we suppose is released from tropical

forests? We are far from reaching any conclusive answers. But for purposes of this analysis, let us consider that a best-judgment estimate, proposed here strictly as a working appraisal, could consider that the net amount of tropical forests eliminated through burning each year amounts to around 1 percent of the biome. When we consider the amounts of biomass involved, together with the countervailing factors at work, this probably translates into an amount of at least 2 and possibly 4 billion tons of carbon—or between two-fifths and four-fifths as much carbon as is released through combustion of fossil fuels. Furthermore, we can anticipate that as forest farmers become much more numerous in the next few decades, they will obviously step up their rate of burning. It is far from improbable that tropical forest burning could eventually match fossil-fuel combustion as a source of carbon dioxide in the global atmosphere—for a few decades anyway.

So much for the global carbon budget. Let us move on to another area of analysis, where the situation is beset with still more uncertainties—and where the outcome, in terms of upheavals for humankind's welfare, could be significant to an extreme degree. We have already noted that, given the greenhouse effect, a doubling of atmospheric carbon dioxide beyond the baseline figure (265 parts per million as of the middle of the last century) could lead to an increase in global temperatures of a mean 5 degrees Fahrenheit, causing temperatures to rise toward a warmth not known on the planet since the Mesozoic, the age of the dinosaurs. Specifically, there would be an increase of only 2 degrees F. or little more at the equator, around 5 degrees F. in temperate latitudes, and as much as 8, possibly up to 12 degrees F. at the poles.

Such temperature changes would almost certainly be accompanied by shifts in rainfall patterns. There could be an overall increase in rainfall of about 7 percent; but—and here is a key factor—there would be a good deal of differentiated distribution. At present the main rainfall zones are as follows: first an equatorial zone of heavy rainfall; then a sharp switch to the anticyclone zone with its low rainfall (these belts, on either side of the equator, contain almost all the world's hot deserts); next a zone of prevailing westerly winds and temperate moist climate; and finally the polar regions with their markedly reduced precipitation. Were a "greenhouse effect" to come into play, these climatic zones would tend to shift toward the poles, more so in the Northern Hemisphere (with its larger land mass) than in the Southern Hemisphere. Certain zones that are now dry, such as the Sahel along the southern fringe of the Sahara, could receive the benefit of equatorial rain, but the Sahara itself could extend farther north, with adverse repercussions for conventional agriculture in parts of the Mediterranean basin—as in the extensive food-growing territories of the Soviet Union, where drought is already a pervasive threat. Still farther north, a warmer climate could extend the presently short growing season for grain crops in both North America and Eurasia (each increase in average temperature of 1 degree F. can lengthen the growing season by about one week). But whereas there are good soils available for the Russians in Siberia, and we may eventually see wheat fields in much of Scandinavia, even Iceland, North America's grain belt cannot readily migrate northward, since it will encounter thin and infertile soils. A temperature increase of just 2 degrees F. could reduce the U.S. corn crop by at least 11 percent, when linked with greater evaporation and a decline in August rainfall; while the same climatic dislocations could depress the wheat-growing industry by at least half a billion dollars (1980 prices) per year.[5]

Let us examine how these climatic scenarios might develop by looking at precipitation patterns during a period of so-called optimum climatic conditions 6,000 years ago (Fig. 2). At that stage, there appear to have been wetter conditions in much of Europe, the Middle East, China, India, and parts of Australia, whereas (and as is predicted by the carbon dioxide model) much of North America appears to have been drier. Were the rainfall map in the middle of the next century to resemble that of 6,000 years ago, many countries could benefit, while many would suffer. India

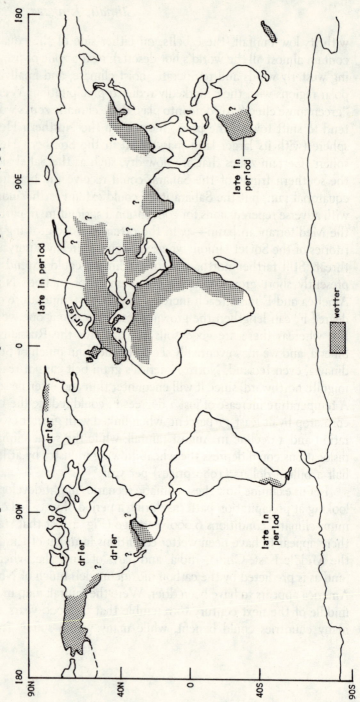

FIGURE 2. Schematic Map of the distribution of rainfall during the Altithermal period of 4,000 to 8,000 years ago, when the world was generally several degrees warmer than it is now. If the global climate grows still warmer in the future, precipitation patterns may again shift, perhaps dramatically. Blank areas do not necessarily indicate regions in which rainfall has remained unchanged. Meteorological information about the distant past is still far from complete, and many gaps remain.

Source: W. W. Kellogg, 1977, *Effects of Human Activities on Global Climate*. World Meteorological Organization, Geneva, Switzerland.

might find itself with more rain, while neighboring Pakistan and Bangladesh could find themselves with less: the political implications could be so extensive that it is difficult to foresee the power upheavals that would result. Similarly, Mexico might turn into a bread basket, thus adding much to that nation's politico-economic clout. Several Islamic countries of northern and northeastern Africa, as of the Middle East, could find themselves with a food weapon to restore their international muscle after their oil wells run dry. Perhaps most significant of all, the Soviet Union and China might find themselves with such enhanced capacity to grow food that they could hope to dictate terms to the rest of the world on a whole range of issues.

These scenarios remain, however, little more than speculation at present, because we simply do not have enough information about how climatic changes are likely to evolve, let alone what the agricultural backlash might be. A carbon dioxide buildup might have many unanticipated positive impacts. For example, it could exert a fertilizing effect on crops by stimulating photosynthesis, thereby enabling certain areas with reduced rainfall to nevertheless maintain their agricultural productivity. Certain agronomists calculate that if total plant biomass were to increase by only about .75 percent, that would take up all the carbon from fossil fuels, 5 billion tons, added to the atmosphere each year; and probably an increase in plant biomass only half as much again could well account for all carbon emitted from tropical forests.[6]

While much remains unknown, we do know that during the past one hundred years, global temperatures have increased by about 1 degree F., half of the increase occurring in the recent short span from the mid-1960s to 1980. These temperature increases are consistent with the calculated greenhouse effect that results from increases of carbon dioxide.

Significant as are changes in precipitation patterns for agriculture, there could emerge a set of far more serious environmental repercussions if global temperatures were to rise to a level where they triggered a melting of the polar ice packs. A mean increase of much above 9 degrees F. would probably cause the ice of the Arctic Ocean to disappear completely during the summer, though parts of it may reappear in winter. An ice-free Arctic Ocean is a phenomenon that has not occurred at any stage during the past one million years. Similarly, the Greenland ice cap could start to melt, though the process would be slow and protracted. By contrast, if the West Antarctic ice sheet, lying on bedrock below sea level, broke away from its "moorings" and slid into the sea, it could break up rapidly. By the second quarter of the next century, this huge Antarctic ice pack could be disintegrating, and it could well melt away within as little as one hundred years. Such changes in the polar ice packs would cause sea levels to rise by between 5 and 7 meters before the end of the next century. In the longer-term future, that is, during a period of several centuries, both polar ice packs could conceivably be eliminated, which would raise sea levels by 50 to 70 meters.

Suppose sea levels were increased by 5 to 7 meters. Many low-lying land areas, such as Florida and the Netherlands, would lose much of their territory beneath the waves. In fact, at least 40 percent of Florida's population would be affected, and more than 11 million people in the United States as a whole.[7] Presumably the flooded-out people could be settled elsewhere, albeit at massive cost, and supposing there were plenty of time for planning. Similar problems would arise for coastal cities and other concentrations of human communities in other parts of the world, involving two-fifths of all humankind: more than 30 percent of all people live within a 50-kilometer zone adjoining seas and oceans. The American economy could presumably afford to move New York inland, and the same for Los Angeles, New Orleans, and other coastal conurbations. A similar process would have to be implemented in other countries, that is, for London, Glasgow, Tokyo, Osaka, Montreal, Stockholm, Copenhagen, and so on. But people in Calcutta, Shanghai, Jakarta, Cairo, Lagos, and Rio de Janeiro would have less room to maneuver, since their hinterlands are already crowded to the bursting point, and the national economies in question would have far fewer funds to use for mass migrations.

One obvious response to these scenarios is a curbing of our appetite for fossil-fuel energy. This is in fact occurring in North America and Western Europe, which now experience growth rates of somewhere between zero and 2 percent per year. But Eastern Europe and the Soviet Union (which could prove to be beneficiaries from a carbon dioxide buildup) are expanding their fossil-fuel consumption at 4 percent per year. As for the developing nations—which might interpret a call by the developed nations for a cutback on use of fossil fuels as a further ploy of the developed world's "conspiracy" to block their progress toward industrialization—they are stepping up their appetites by 5 percent per year. Thus North America, now accounting for almost 30 percent of all commercial energy used, might by the year 2025 be using only 10 percent fossil fuel, while the developing nations could increase their share from 13 percent to 40–50 percent.

With regard to tropical forests, we can sense a powerfully expanded premium on confronting the challenge of both slash-and-burn farmers and (though to much smaller extent, because of their very much smaller numbers and limited impact) cattle raisers in tropical forestlands. Of course there are deepseated humanitarian reasons for assisting the plight of the many millions of forest farmers; the issue of carbon dioxide merely strengthens our incentive to reach a speedy resolution, notably through strategies such as traditional home garden systems of agriculture, and through agroforestry (see chapter 8).

The day may come when we find that a major approach to the carbon dioxide problem lies with, quite simply, planting trees. Man-made forests can soak up immense amounts of excess carbon dioxide from the atmosphere. As we have seen in chapters 5 and 6, present tree plantations amount to well below one-fifth of what reflects immediate needs for timber, let alone future efforts to safeguard climatic stability. Ultimately, however, it may be advantageous for us, in terms of climatic stabilization, to establish extensive plantations in the humid tropics, where growth of biomass is fast and continuous. But how many trees, roughly speaking, would we need to plant?

We know that the present increase in atmospheric carbon dioxide each year amounts to about 2.5 billion tons. It is realistic to project that, if the burning of fossil fuels and of tropical forests continues to expand at recent rates, the quantity could increase to four times as much by the year 2000 or shortly thereafter. For present purposes, let us work with a figure of five billion tons per year. Were we to embark on a gigantic "soak-up" campaign, we could use, for the sake of illustration, a tree such as the leucaena, or giant ipilipil, which grows faster than most other species used in many plantations of the humid tropics. Wood, of course, is not a straight accumulation of carbon. It is a complex mixture of carbon and hydrogen, together with some oxygen and nitrogen. One hectare of healthy leucaenas can consume around 30 tons of carbon (dry weight) each year, or a good deal more than an unusually fast-growing temperate-zone tree, the American sycamore, which achieves only about 7.5 tons. So leucaenas can transfer carbon from the atmosphere and into tree tissue at a net rate of around 3,000 tons of carbon per square kilometer per year until they reach maturity. This means that in order to soak up enough excess carbon dioxide, we would need to think in terms of almost 1.8 million square kilometers, or an area a good deal larger than Alaska, and roughly equivalent to half of Brazilian Amazonia.[8] Of course these figures are presented only for ease of ready reckoning, and they should be viewed as no more than very preliminary estimates. All plantations in the humid tropics now comprise about 100,000 square kilometers.

Fortunately, the leucaena is a nitrogen-fixing legume, so we would not have to worry about nitrogen fertilizer. But it sometimes appears to require quantities of a fertilizer that is in short supply: phosphate. The chief producers of phosphate rock today are the United States and Morocco, and at current rates of consumption by agriculture (which are likely to expand), their reserves appear likely to last for only another two decades. Worse still, the land costs could total several billion dollars per year (even after we mobilize the half a million square kilometers of deforested lands that have become too impoverished for convential agriculture), while establishment and maintenance of the plantations would cost many more billions of dollars. For the sake of rough-and-ready reckoning, we would probably need to think in the order of at least $20 billion per year—some experts consider the figure could be several times higher, even twenty times higher. As large as these sums sound, however, they would constitute only a fraction of the costs that would arise in the wake of a melting of the polar ice caps, tentatively estimated at a minimum of $1 trillion.[9]

IV. WHAT WE CAN DO

16

Role of Tropical Forest Countries

IF WE WERE to note all the options for action described in this book, we would come up with a lengthy list. Notably, more careful logging, in select localities; greater processing of timber by producer countries; many more plantations, both for commercial lumber and fuelwood; better use of forestry waste; greatly improved management of established pasturelands; systematized screening of genetic resources for their potential applications in agriculture, medicine, industry, and genetic engineering; a better deal, in a dozen ways, for the forest farmer; and detailed documentation of the many environmental services that forests perform through their mere existence. The list would be, as I say, lengthy—and that would be for starters, since this book has not tried to cover the waterfront. In other words, there are many ways to help the situation right now. Instead of going into them further, however, we shall look at some fresh aspects of the problem, starting with an intriguing question: Why does forestry enjoy such a poor image in tropical forest countries?

THE CINDERELLA SYNDROME

When I used to visit the Ministry of Agriculture in Nairobi, I would approach an impressive-looking building agleam with chrome and glass. Much the same with the Ministry of Commerce, or of Transport, and lots of others, all of them stately structures, with fountains in their forecourts and flagpoles lining the approach-way. But when I visited the Department of Forestry, I would have to leave the main downtown area, and wander off across a sort of no man's land located between a public park and a church, at the far side of which, half hidden among a bunch of bushes, I would come across a collection of wooden huts. The huts were arranged around a dusty compound, which would sometimes be used, at whatever hour of the day, as a football field. Whereas the staff in the other Ministries would generally appear alert and "involved," those at Forestry would seem demoralized, with little sense of wishing to be otherwise.

This, I used to reflect, epitomizes the difference between forestry as an economic sector in Kenya, and the other sectors. It has ever been thus. When I first arrived in Kenya, back in 1958, Forestry was a yet more lowly department, hardly allowed a presence within the confines of Nairobi. Now it has been vouchsafed a place with the rest of Kenya officialdom, but with very much of a sideline status. Meanwhile Kenya's remnant forests have declined by at least half in the past two decades, and may all but disappear by the end of the century. Given the scant recognition accorded to the guardians of Kenya's forests, the plight is not surprising.

In country after country of the humid tropics, I have encountered a similar scene. Not only is the actual Forestry office inclined to be a hole-in-the-wall affair, overshadowed—literally as well as figuratively—by the larger bureaucracies. All too often,

Forestry ranks virtually at the end of the governmental pecking order. Generally it is a mere division, and treated as a sub-division, of the Ministry of Agriculture, where officials are engaged in "true priorities" such as growing food—sometimes by means of chopping down forests to clear new lands for crops. Most of these countries, not being industrialized, are based on strongly established agricultural traditions, so when the government finds its people are running short of food, it takes the traditional route of fixing the problem by mobilizing whatever "spare lands" are available, notably forestlands.

In Indonesia, for instance, the government has been spending well over half a billion dollars per year to import 2 million tons of rice to feed its citizenry, especially in Java, where two-thirds of the populace resides. In order to make way for the fast-growing population, Java's forest cover has been shrinking year by year, a trend that has degraded watershed mechanisms that should supply paddy-rice farmers with dependable amounts of irrigation water for three bumper crops of rice each year. Whereas short-term "food fixes" keep Java's citizenry adequately nourished, and thus disinclined to protest or riot, the country's foreign-exchange balances take a beating (the cost of the rice imports eliminates a large share of the country's earnings from timber exports), and the long-run capacity of Java to grow more food is steadily undermined. Forests are seen as part of the immediate problem rather than as a factor in an eventual solution. The Indonesian government now derives about $450 million per year in royalties, levies, and taxes on its timber exports. Yet it devotes only about $100 million to the forestry sector, the funds being spent mainly on replanting trees in devastated areas, rather than on forestalling further devastation: curative rather than preventive medicine.

Let us note, fortunately, that this situation in Indonesia is now changing. A spanking new building has been established for Forestry, located next to the Office of the President. Moreover, Forestry has been upgraded for a full Ministry on its own, with a soundly increased budget that allows foresters to stand tall, as well as to do a better job. Not everything in the garden is yet rosy, not by a long way, and especially in some of the further reaches of Indonesia, where timber cutting leaves more to be desired than almost anywhere else in the biome. But compared with the dismal setup ten years ago, Indonesian forestry has come a long way. Let us hope that in another ten years we shall be still more astounded at the further progress it has made. Much of the recent advance is due to the efforts of a man who was appointed to an entirely new government post in the late 1970s: Dr. Emil Salim, who is Minister for Environment and Development Control. In turn, Salim is strongly supported in his consultations with his Cabinet colleagues by a widespread environmental movement that has sprung up among Indonesia's citizens, with no fewer than 400 separate organizations, able to make their voices heard in top government ears by banding together in a coalition called the Indonesian Environmental Forum.

Yet this encouraging turnaround in Indonesia is very much the exception. Almost invariably, forests have rarely been perceived as resources that can permanently contribute to economic development in multiple ways. On the contrary, they are often looked on as obstacles in the path of expanding civilization. As a result, Forestry Departments have been inclined, by tradition, to be "Cinderella agencies," dominated (if not derided) by more powerful and entrenched agencies, such as Ministries of Agriculture and others. In response to this unfortunate situation, Forestry Department officials have tried to generate political clout for themselves by playing the development game according to the old-established rules, which means that their sector must produce as much cash as possible, fast. This has fostered an attitude of once-and-for-all exploitation of forests for timber, in disregard of broader benefits for the long-term future.

The track record is dismal. In Thailand, the officially recognized contribution of forestry to the national economy is about 2.5 percent.[1] In fact, however, this figure is way below the true figure, since it ignores illegal logging (considered to be at least twice as much as legal logging), the bulk of fuelwood collecting (which does not generally constitute a marketplace transaction, hence is not recorded by economists), the supply of dozens of so-called minor forest products that are used at subsistence level, plus the many environmental services that help to keep Thais well fed, healthy, and supplied with electricity—and even help to keep Bangkok afloat (see chapter 14). According to development economists who take account of what actually happens as well as what is officially recorded, a realistic assessment of forestry's contribution to Thailand's Gross National Product is more likely to be 7 percent. Yet the share of the national budget accorded to forestry is only 1.4 percent. By contrast, agriculture in Thailand is reckoned to contribute just over 20 percent of GNP, while enjoying about 24 percent of the national budget.

A similar tale can be told with respect to the Philippines,[2] India,[3] and a lengthy list of other countries. Even in Peninsular Malaysia, which tries to accept forestry as a full participant in its development strategies, the sector has not been incorporated into national economic planning with the same creative sophistication that is applied to agriculture, manufacturing, and mining. Otherwise we would not find that the Peninsula's forests have been so over-cut that the situation has been described by the Deputy Prime Minister as a crisis—the Peninsula will soon become a net importer of hardwoods. Many forests are cleared in order to make way for agriculture, which is regarded as a "more economically productive" sector. So powerful is the influence of agriculture bureaucrats, that some 8,000 sq. km. of forest, equivalent to about one-tenth of the Peninsula's remaining forests, were cleared in the mid-1970s, yet still awaited their farming settlements several years later.[4]

Nor is this story tale of neglect confined to tropical forest nations and their governments. International agencies suffer from the same myopia. The Asian Development Bank has recently boosted its lending for forestry, but it still assigns only one dollar for every twenty it allocates to agriculture and rural development.

Even within the Food and Agriculture Organization of the United Nations, the latest shareout of the budgetary pie reveals that Forestry receives only 3 percent (about one-ninth as much as Agriculture), down from several times as much in 1965. As for the Commonwealth, its ministers have never once, during the last two decades of their annual meetings, placed tropical forestry as a topic on their agenda, whereas agriculture leads time after time.

Amid this bleak picture there is one outstandingly bright spot. The World Bank used to treat tropical forestry the way the other big agencies did. During the period 1968–77, it loaned only $13 million to the sector. Then came a reappraisal of policy, after which there has been a tenfold increase in funding. To date, the Bank has loaned more than $1 billion for tropical forestry, with about half of all projects being given over to protection forestry (watershed management and the like) rather than production forestry (mainly timber harvesting). A similar story can be told for the U.S. Agency for International Development.

MIS-EVALUATION OF FORESTS'
PRODUCTS

The problem with forestry is not failure to produce the goods. Rather, the problem lies with our incapacity to measure all the goods generated. It is easy enough to reckon the amount of hardwood timber cut, and to put a price tag on its export value. The same applies to forestland beef and other products registered through the marketplace. The difficulties arise when we start to consider goods not amendable to straightforward appraisal— those that cannot easily be quantified and measured in monetary terms. Were these errors of omission to be set right, and were we able to evaluate all outputs of tropical forests in dollar terms, we would find that forestry should rank as anything but a poor competitor in government councils.

This applies especially to environmental services. As we have seen in chapter 14, forests confer on society an entire array of benefits, through, for example, safeguarding of watersheds and tempering of climates. Yet these many services tend to be taken for granted, due primarily to the fact that the pay-off does not arrive in the form of recognizable and quick cash via the market-place. The many people who benefit from watershed functions cannot use their dollar votes to express their appreciation; and the same for those who benefit from other environmental amenities supplied by forests in all manner of shapes and forms. In Peninsular Malaysia, cost-benefit analysts calculate that the financial return from the fastest-growing natural forest is only one-third of what would be available were the same forest to be converted into rubber or oil-palm plantations. But the calculus omits a key component of the economic arithmetic: the environmental services of the natural forest, which are way ahead of those supplied by plantations. When I visited Malaysia's capital, Kuala Lumpur, in 1980, I was surprised to read in my hotel shower that I was requested to turn off the faucet while soaping myself: in one of the wettest localities on Earth, there can be shortages of water! Yet I, in common with thousands of other hotel guests and several million urban residents, knew of no organized way to make our preferences known in the concise and direct fashion that can be registered by the rubber or oil-palm customer.

Hence, there is a constant reenforcing of the Cinderella syndrome. For many reasons, politicians, economists, development planners, and others are little inclined to keep a close eye on the resources, phytochemicals, environmental functions, and the like, which may not supply their benefits so super-speedily as timber. After all, politicians and their brethren are accustomed to dealing with recognizable concerns such as timber revenues—and with elections. Meanwhile, many people suffer through deforestation, people who number in their millions—far more numerous than those who abuse the forest resources. But because the former suffer only in slow and not readily noticeable fashion, we do not see them assembled in rioting throngs outside Forestry Department offices.

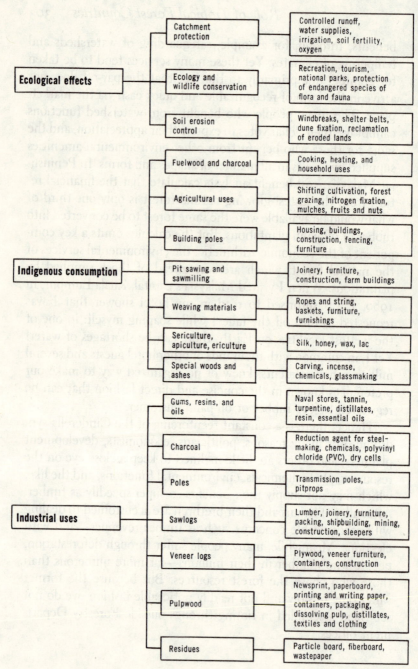

Ecological effects	Catchment protection	Controlled runoff, water supplies, irrigation, soil fertility, oxygen
	Ecology and wildlife conservation	Recreation, tourism, national parks, protection of endangered species of flora and fauna
	Soil erosion control	Windbreaks, shelter belts, dune fixation, reclamation of eroded lands
Indigenous consumption	Fuelwood and charcoal	Cooking, heating, and household uses
	Agricultural uses	Shifting cultivation, forest grazing, nitrogen fixation, mulches, fruits and nuts
	Building poles	Housing, buildings, construction, fencing, furniture
	Pit sawing and sawmilling	Joinery, furniture, construction, farm buildings
	Weaving materials	Ropes and string, baskets, furniture, furnishings
	Sericulture, apiculture, ericulture	Silk, honey, wax, lac
	Special woods and ashes	Carving, incense, chemicals, glassmaking
Industrial uses	Gums, resins, and oils	Naval stores, tannin, turpentine, distillates, resin, essential oils
	Charcoal	Reduction agent for steel-making, chemicals, polyvinyl chloride (PVC), dry cells
	Poles	Transmission poles, pitprops
	Sawlogs	Lumber, joinery, furniture, packing, shipbuilding, mining, construction, sleepers
	Veneer logs	Plywood, veneer furniture, containers, construction
	Pulpwood	Newsprint, paperboard, printing and writing paper, containers, packaging, dissolving pulp, distillates, textiles and clothing
	Residues	Particle board, fiberboard, wastepaper

Source: World Bank, 1978, *Forestry: Sector Policy Paper*, World Bank, Washington D. C.

In sum, the true overall value of tropical forests tends to be ignored, or discounted, or otherwise given short shrift. To appreciate the full range of goods and services derived from tropical forests, we can consult a diagram prepared by the World Bank (Fig. 3). Only a tiny minority of these products find themselves traded in the marketplace, hence accorded a value that is recognized by those people who calculate the "real values" of tropical forests. The marketplace, which serves our daily lives in millions of creative ways, does us a severe disservice when it tries to handle a stock of resources with the exceptional attributes of tropical forests.

THE FORESTER'S FINGER IN MANY PIES

Moreover forestry, as a multiple-product activity, is a "mushy" affair. Agriculture is straightforward: it simply produces food. Medicine keeps people healthy. Most economic sectors have their own definable bailiwicks. A Minister of Fisheries knows what his job is; he confines his attention to it, and that is that.

A forester, by contrast—at least, a modern forester—must concern himself with a variety of activities, some of which entail collaboration with other officials and technocrats. As we know, government officials feel secure when they operate within their own bureaucratic empires, without "interference" from outsiders. The switched-on forester, almost by definition, needs to be an interferer. He must point out to the Minister of Agriculture (often an ancient foe) that the Green Revolution will lose steam unless the farmers can look to watershed forests for their irrigation water released in regular supplies throughout the year. He must point out to the Minister of Health that the best preventive-

---

FIGURE 3.   Broad-ranging as this diagram is, it nonetheless presents well under half of the goods and services that already make a solid contribution to people's lives. If we were to re-draw the diagram in 20 years' time, and again in 100 years' time, how much more expanded would the various categories be?

medicine campaigns are likely to lose punch unless the citizenry can look for dependable supplies of clean water for drinking, cooking, and sanitation. He must inform the Minister of Energy that without tree cover to hold catchment-zone soil in place, hydroelectric projects may become short-lived through sedimentation. He must consult with the Minister of Fisheries and tell him that, say, the extraordinarily productive shellfish breeding grounds of coastal mangrove forests may become suffocated under a blanket of silt unless more attention is paid to protection forests in headwater territories far inland.

The modern-day forester must give at least as much of his time and effort to what goes on outside his forests (sometimes hundreds of kilometers outside) as to what goes on inside them. But an aptitude of this sort runs against the grain of many a forester. Like most civil servants, he prefers to keep out of trouble, to let other departments go their way. By training and by background, he is more experienced in dealing with trees than with people—people of any sort, let alone the ambitious personalities that lead government departments. Far from being a gregarious character who behaves like a creative coordinator, the forester tends to be a loner, content when he is surrounded by huge trees which do not argue or answer back or do anything but support his professional interests.

So the principal priority of a campaign to get a better deal for tropical forests is to upgrade forestry's image. The problem has been not so much the failings of past foresters in the tropics, but the defects in the way that government systems handle a nation's affairs. Instead of recognizing forestry as a sector that is often fundamental to across-the-board interests of the national economy, forestry has tended to be treated as an illiterate creature of the boonies. Thus the number one recommendation, as I see it, is that foresters embark on activities which have little to do with their traditional endeavors—engage in a public relations exercise. They must learn to become trumpet blowers.

BROAD-SCALE ECONOMIC EVALUATION

To be an effective trumpet blower, you have to have a concise tune to play. What shall this consist of?

One way to re-write the music for modern forestry is to undertake economic analysis of a scale that includes the entire range of goods and services available from tropical forests. As noted, we have a concise grasp of the value of hardwood timber and forest-zone beef. But we have next to no idea of the ultimate value of the forests' many other outputs, even though it is often possible to attempt some minimum-value estimates (for sample figures, see chapters 10–15). These estimates, however preliminary and approximate, help adjust the asymmetry of present evaluation procedures, which highlight established products, such as timber and beef, while ignoring less conventional outputs, such as genetic resources and watershed functions.

A few efforts are underway to tackle the new analysis proposed, sometimes known by the technical title of expanded cost-benefit analysis.[5] But plausible as these attempts are, they tend to be limited not only in number but in scope: they are far from matching the nature and scale of the problem. The many fine-grain analyses of, for example, timber demand-and-supply patterns should be complemented by equally rigorous analyses of the costs that will arise if the forests continue to disappear. In addition, the analytic methodologies should be geared to social equity as well as economic efficiency. The benefits of, say, timber harvesting tend to accrue through only a limited sector of the community, whereas the costs of deforestation fall on the community as a whole—and they extend further into the future than do the benefits of unsustainable timber harvesting.

In particular, we need to give much more attention to the costs of continuing on our present track. Of the various economic analyses available, some of them quite comprehensive in scope, not one deals, in systematic or even substantive fashion, with the costs that result from broad-scale deforestation. Perhaps the finest scientific center in the tropical forest biome is the Forest Research Institute in Dehra Dun, India. While the Institute proudly claims that it has accomplished a great deal of first-class research on forest industries, and on the timber stocks needed to feed the industry, it has conducted scarcely a single study on the overall impact of deforestation in the country, even though flooding and the like can wreck the livelihoods of half a billion people in the Ganges Plain alone. With the notable exception of the World Bank, the international agencies have done little to help (though there are signs of changes ahead). During the late 1970s, FAO, AID, and other large donors apart from the World Bank were spending almost $1 billion per year on tropical forestry. Of this sum, however, 54 percent was being spent on industrial activities, that is, forest-removal activities, less than 10 percent on forest conservation projects, and less than 3 percent on research, with hardly a couple of million spent on research into the billions of dollars' worth of damages sustained throughout the biome each year as a result of deforestation.[6]

By way of illustration, let us look at the cost that forest farmers unwittingly inflict. In Thailand, they burn commercial timber worth $170 million per year.[7] In Sarawak, a territory only one-fifth the size of Thailand, every log of high-value timber exported is matched by an equivalent that is needlessly burned, at a net cost of $120 million per year.[8] Yet neither Thailand nor Sarawak appears to wield these figures as a measure of the urgency of the forest farmer's problem in relation to other forestry priorities—or even other development activities outside the forestry sector.

Without basic calculations of these sorts, no country can claim that it is even trying to get a grasp on forestry's true place in the overall development picture. It has taken us ten years to move away from talk of forestry in the traditional sense of mere stocks of timber, to "development of forestry" as a sustainable enterprise. The time is over-due for us to make another quantum advance in forestry thinking, and that is to look at forestry's role in overall development. When we achieve that massive leap in our thinking, we may well find that development experts will start to accord forestry the Prince Charming treatment.

PARKS AND OTHER PROTECTED AREAS

Let us conclude this chapter with a look at a salient way for tropical forest countries to safeguard exceptional ecosystems in their forests: by according them absolute protection in the form of parks and other protected areas. How well is this strategy working out?

In many ways, far better than had once been anticipated. At the time of the Second World Parks Congress, held in Yellowstone in 1972, I heard speaker after speaker bemoan the fact that tropical forests, 10 million sq. km. of them, contained only about 120,000 sq. km. of parks and reserves. Many types of forest, in each of the three main regions, received no protection whatsoever. By contrast, preliminary estimates suggested that at least 10 percent, and perhaps as much as 20 percent, of tropical forests need to be preserved, in selected localities covering distinct ecosystems, in order to ensure preservation of sample biotic communities with their endemic components. Moreover, since many species in this highly differentiated biome are characterized by localized distribution, a few large parks and reserves will not suffice to protect the range of biological diversity. An extensive and strategically sited network is needed.

But a vast series of small areas will not suffice either. Park planners must take note of a crucial criterion for forest ecosystems, the minimum critical area. In Malaysia's lowland rainforests, for example, wild fruit trees occur at low densities. A sample survey of 676 hectares reveals that only a few species have more than 25 individual trees per 100 hectares, while many have fewer than one dozen trees per hectare. So if, as seems likely, each species requires 10,000 trees to provide an adequate gene pool, we shall have to think in terms of sizable areas to provide for these species, or we shall have to consider some "interventionist" types of management to maintain their genetic diversity (by, for example, shifting around germplasm artificially).[9] In the same region, hornbills that possibly require minimal populations of 5,000 individuals to ensure adequate gene pools will need between 2,000 and 10,000 sq. km., while monkeys with similar-size populations will need anywhere from 250 to 3,000 sq. km.[10] If a tiger population can survive with as few as 400 individuals—probably the smallest number we can contemplate if we are to avoid the emergence, through inbreeding, of "degenerate" tigers in a few dozen generations' time—the area required to support them may need to be very large indeed.

To help us comprehend the scope of this problem, let us look at Amazonia. According to best scientific calculations, we can hypothesize that were we to save 1 percent of Amazonia's forest, that might correspond, very roughly, to saving 25 percent of the region's species; and if we save 10 percent of the forest, that could correspond to 50 percent of the species. So to make a sound job of safeguarding pretty well the entire spectrum of species (except for those with highly localized distribution), we can suppose that a parks network covering 20 percent of the forest, or roughly 1 million square kilometers, might do the job. Of course these calculations are crude in the extreme, and they are presented as no more than an informed first guess about the size of the challenge that faces conservationists in Amazonia.

How are things working out? Fairly well—for a start. There has been a splurge of new parks during the last few years in Venezuela, Peru, Ecuador, Bolivia, and especially Brazil. The amount designated to date totals 110,000 sq. km.—an area about as large as the state of Mississippi, or Austria. Still more encouraging, Brazil entertains ambitious hopes to expand its network year by year, according to a methodical plan that will take account of all eight distinctive plant zones in Brazil's sector of Amazonia, each with its characteristic faunal communities.[11] All these localities contain exceptional concentrations of species, with large numbers of endemics, making them worthy targets for priority conservation. In point of fact, Brazil has already established several outsize parks in Amazonia, at least one of them measuring 10,000 sq. km.

Fine as this accomplishment may be, the parks do not coincide well with the critical localities defined by species concentrations. Worse still, at least three and perhaps four of the areas targeted for forest conservation fall entirely within those parts of Amazonia that have been earmarked by the government for intensive development through agriculture and other forms of exploitation, while major parts of other high-priority conservation areas overlap with development foci. If any one of these remarkable ecosystems were to be eliminated, a large number of species would go with it. Let us note too that some of Brazil's longest-established parks are proving to be little better than "paper parks." The Xingu Park has been reduced by one-quarter of its size by virtue of a highway project, and another park, an area rich with endemics, is grazed by large numbers of cattle, is burned every few years, and contains a system of trucking roads.

Let us also note some encouraging news from Indonesia. In this country, which accounts for roughly one-tenth of the biome, parks and reserves now account for 112,000 sq. km., and the eventual goal is about 250,000 sq. km.—out of a total forest expanse covering no more than 1 million square kilometers at most! While not all the protected areas are parks (many are reserves, and some are conservation units), they amount to 6 percent of national territory, to be compared with a world average of only about 2 percent.

All in all, protected areas in tropical forests now amount to some 400,000 sq. km. The countries named, together with others such as Malaysia, Thailand, Zaire, Gabon, Congo, Cameroon, and Costa Rica, are establishing an impressive record in this field. Others have scarcely made a start. The backsliders sometimes protest that it is a costly business to establish a park in a tropical forest: not only in terms of setting up park buildings, establishing a ranger force, and so forth, but mainly in "opportunity costs," or the potential timber revenues that are sacrificed when the area is declared off limits to development. Suppose one hectare of dipterocarp forest contains 25 major marketable trees, and each tree could sell on the international market for $500 (125 cubic meters of timber per hectare, by no means an exceptional amount); this means that a park of 100 sq. km. would effectively deprive the nation of $125 million for the value of the timber. In some places, the number of trees taken could be only half as large; but at the height of the timber boom in the mid-1970s, one ton of timber was fetching almost $200. So, using a "swings and roundabouts" calculation, the figures stand up—and $125 million is no trifling amount. Hence, the attempt to link parks with economic advancement: if parks can help to sustain development, then their case is made. A splendid example of such a linkage has been established in northern Sulawesi, where the World Bank has funded a large rice-growing project; as a component of the project, the Bank has required that a sizable patch of forest overlooking the rice-growing valley lands shall be preserved as the Dumosa-Bone National Park, in order to safeguard the watershed services that supply the critical irrigation water.

Parks need not only be places where one gazes eye to soulful eye with an orangutan. Parks offer many spinoff benefits, whatever their rationale at the time of establishment. An obvious advantage of parks is that they safeguard genetic resources, which may one day—perhaps one early day—offer a boost to modern agriculture or medicine or industry. Yet the benefits of genetic resources are likely to accrue, for the foreseeable future at any rate, to those nations with the technological knowhow to exploit genetic variability, that is, the advanced nations. These in particular are the nations that might consider a gesture toward sharing the financial burden of parks. At the World Parks Congress in late 1982, the Minister for Environment and Development Control in Indonesia, Dr. Emil Salim, put it thus: "When oil runs out, plant-derived chemicals will replace petrochemicals. The benefits will be huge. But these benefits will be obtained mainly in the developed world. Who should pay the costs of conservation in the tropics, and how? The basic premise should be that whoever is getting benefits right now in the way of plant-derived chemicals should pay a premium to ensure continued future benefits."

The question of who shares the costs of tropical forest conservation is so important that we shall defer it to the next chapter, on what outsiders can do to help the cause.

17

Support from Outsiders

THE PRIME RESPONSIBILITY for tropical forests belongs with the countries immediately concerned. It is their rightful role, and all the rest of the world put together can do no more than support such efforts as tropical forest countries are prepared to undertake. But now that we see that a good number of tropical forest countries are picking up interest in the cause, outside countries can do a great deal more than merely shout encouragement. Herewith a review of some avenues of support.

ENHANCED STATUS FOR FORESTRY

It is not only tropical forest governments that have traditionally looked on forestry as a pretty unglamorous activity. Many aid agencies, whether of individual rich nations or multilateral organizations such as the United Nations, have done little better. Great Britain, with a huge fund of expertise in the form of foresters who worked in the tropics during the days of Empire, allows much of this talent to go unused: the government allocates less than 2 percent of its foreign aid budget to forestry. So how about some specific measures to upgrade the appeal of tropical forestry? The effort need not be costly, and the results could be highly positive.

As emphasized in the previous chapter, the first step in this process is to recognize the multi-purpose character of tropical forestry. Were outsider governments to wave a flag in behalf of this aspect of the forestry cause, they could probably achieve more than through supplying fancy technology. It is the spirit of the enterprise that counts. To reiterate the guts of the message, forestry is not just about sawn lumber and pine seedlings. It is about a host of other products and activities, as we have read at many points throughout this book. I remember the scene at the Eighth World Forestry Congress, organized by the Food and Agriculture Organization and held in Jakarta, Indonesia, in late 1978. The symbol for the Congress, placed front and center on the podium, was a tree stump with an axe sunk in it. When the Ninth Congress opens its doors in early 1985, what symbol shall we see then? Can we hope for something that demonstrates the broader-focus approach of the modern forester? Can we look, above all, for something that highlights the main attribute of the modern forester, the outreach spirit that reveals he is at least as preoccupied with people as with trees?

Way back in the Jakarta days of 1978, forestry schools in tropical nations were still teaching their students that they should adopt an unequivocal stance with regard to local communities. They should protect their forests against interlopers, invoking the full force of the law if need be. How different should be the attitude today, when the with-it forester should actually seek out surrounding villagers, and explain how the forest serves the needs of the community in multiple ways. Obviously, the forester of today needs to know not only about what makes trees tick, but about what makes people tick. Yet how many forestry schools teach their students the arts of persuasion, of selling their cause among local people? Until recent years, the only contact a forester would have with a local villager would be when he was arresting the man for nefarious activity within the forest. Now the forester must extend a hand of understanding to the local villager.

All this represents a basic departure for forestry schools of the tropics. Many of these schools receive help from outside in the way of funding and technical support. How many international agencies, one wonders, recognize that, important as it may be to include courses on wood chemistry and silviculture, it is just as important to include courses on public communication and forestry interpretation generally. Without these critical skills, all the expertise in tree genetics and sawmill mechanics will no longer serve the real interests of the forestry cause.

As a parallel initiative, outsiders should promote the concept that forestry is not an activity on its own. Rather, it integrates with other sectors of the development field, several other sectors. This is a difficult challenge, to operate in the no man's land zone between government departments. The dilemma has been well expressed by the Forestry Adviser at the World Bank, John S. Spears, who points out that technocrats prefer to adopt their own sectoral viewpoints; whether they be agronomists, ecologists, soil scientists, sociologists, botanists, zoologists, meteorologists, land-use planners, politicians, whoever, they all incline toward parochialism. "The volume of specialist literature which has grown up around the subject of tropical forestry would fill St. Paul's Cathedral, whereas the number of papers which tackle this issue in a 'holistic' way could probably fit into my briefcase."

FORESTRY AND DEVELOPMENT: AN ENLARGED APPROACH

These dilemmas notwithstanding, let us look at some of the more expansive approaches that could be adopted toward tropical forestry, and that could be strongly promoted by "specialist outsiders." As we noted at many points in Part II of this book, tropical forests tend to be exploited for only a few products, with disregard and with adverse repercussions for their many other outputs, whether material goods or environmental services. At the same time, and as we noted throughout Part III, there are many other products available from tropical forests, such as phyto-chemicals, genetic resources and other raw materials, watershed

Wood processing—for, say, veneer slices as pictured here (in the Philippines)— increases the value of raw lumber at least twice, sometimes several times over. This enables tropical-forest countries to earn much more from every tree cut and thus reduces the incentive to over-harvest forest stocks. Today, however, the main con-sumer countries (developed-world nations) prefer to carry out the processing them-selves and thus to capture the related large profit margins. They impose heavy tariffs and other trade restrictions on processed imports from tropical-forest countries. *(Duncan Poore, World Wildlife Fund)*

functions, and so forth. Thus while tropical forests are over-exploited, they remain under-used. Clearly a stock of natural resources of their exceptional diversity and potential value should play a full part in the development campaigns of the countries concerned. Hence the challenge: How can tropical forests best be put to use?—and let us note that this challenge focuses on *use*, rather than preservation, with emphasis on the full range of uses available. After all, despite the many goods and services they offer, tropical forests rank among the least developed of all natural resource stocks on Earth—"developed," that is, in the proper, broad sense of the word, definable in this instance as subjected to sustainable use for whatever outputs, including both material products and environmental service, that will best serve the long-term interests of human communities concerned. In turn again of course, this means that the basic question we should ask is not how to safeguard all tropical forests, but which sectors of tropical forests shall we use in which ways, to meet whose needs of what kinds, and at what cost to whose opportunities for a better life through alternative forms of utilization?

Within an analytic framework of this scope, extraction of commercial timber can sometimes constitute acceptable exploitation, provided it treats the entire forest as a renewable resource stock. In occasional circumstances, wood harvesting can even entail transformation of a sector of natural forest into a man-made forest of fast-growing trees, enabling ten times as much wood to be grown as in a patch of natural forest, and thereby relieving excessive exploitation pressures on remaining tracts of natural forest. In similar style, development can occasionally include replacement of natural forest with plantations of oil palm, rubber, tea, and other crops, which, like the "tree farms" mentioned above, can sometimes maintain environmental services at acceptable levels. It is always advisable, of course, to establish plantations in areas other than primary forest, supposing the land remains available—and it is particularly appropriate to establish plantations on forestlands that have been degraded by commercial ranching or subsistence cultivation.

In a few other select instances, development can even require that sectors of natural forest be replaced by permanent croplands. Parts of the Amazon River floodplains, for example, with their rich alluvial deposits, covering over 65,000 sq. km. (roughly the size of West Virginia or Ireland), could, without costing a penny in fertilizer, feature an agriculture as rich as any known in the great river civilizations of the past, such as those of Egypt, Mesopotamia, and China. True, we would need to safeguard the abundant fisheries stocks in the river systems, dependent as some of them are on fruit food from riverbank trees.[1] But through judicious planning, it should be possible to enable both the flood-plains and the water systems to become much more productive, in commercial senses, than they now are.

In particular, let us note that development of tropical forests, as postulated here, includes use of forest ecosystems for less established, though equally valid, purposes. Within the framework of an expanded approach, with its emphasis on sustainable overall outputs of tropical forests, the maintenance of watershed services can be seen as a form of development that ranks alongside timber harvesting. A national park is as legitimate a "use" as a paper-pulp plantation. Genetic reservoirs count together with forestland agriculture. In certain localities, "use" can entail outright preservation of forest ecosystems for scientific research, especially applied-biology research as it relates to new modes for reaping Earth's natural bounty in tropical forests as the most productive ecosystems on Earth.

MORE RESEARCH, BETTER RESEARCH

Yet in order to enable tropical forests to play their full part in development, we must learn much more about how they work. At present we exploit tropical forests in much the way that Stone Age man exploited his natural environments—as a hunter and gatherer of such wild products as can be found. If we are to refine this crude approach, we need to undertake massive amounts of basic research. Right now we know less about the functionings of tropical forest ecosystems than we do about any other biome on Earth. In view of this ignorance, we need to adopt a cautious approach to our use of tropical forests. As in other circumstances of uncertainty, it will be better to find that we have been vaguely right than certainly wrong.

What sorts of research are most pressing? Several items spring to mind immediately. In plantation forestry, we need to know

Southeast Asia's export of rattans is now worth about $60 million per year, while the value of the end-product trade approaches $4 billion. If we were to undertake systematic surveys of all fibrous plants that, being low in volume while high in value, were able to be sustainably harvested without damage to forest ecosystems, we would surely find dozens of suitable plants. *(Duncan Poore, World Wildlife Fund)*

much more about suitable tree species, instead of relying on the handful of types that we have used so far. We need to learn how to make much better use of disturbed forests—their timber, for example—rather than assuming that secondary forests are somehow second-rate forests. We need to gain a far better grasp of how forests cycle their nutrients, of how farming systems can mimic the natural complexity and stability of the natural forest, and how we can improve our capacity to monitor depletion patterns of forests around the tropics. When we have a better understanding of any of these topics, we can do a much better job of putting tropical forests to wise use—and all these research topics were recognized as urgent way back in the mid-1970s.

There is, of course, some tropical forestry research being done in ninety research centers of one sort and another around the world—though, ironically, over half of them are outside the tropics. Not nearly enough use is being made of these research centers. So the most promising approach seems to lie with a beefing up of existing research centers, which would then be encouraged to share experiences and swap ideas. Were all the additional cost to be undertaken by rich nations of the temperate zones, in light of their increasingly apparent stake in the future of tropical forests, the bill need be no more than $10 million per year, or double the present support for research from whatever source.[2]

We might note, moreover, that the amount of money now contributed by all sources to agricultural research in the Third World is around $1 billion per year. This massive effort could be strengthened, and we could look for a better return on each of those dollars invested, were this research to be linked to forestry research, especially in the humid tropics. For example, irrigation water is a prime component of Green Revolution agriculture, and hydrological systems are often no better than the forest cover in their catchment zones. The tropics receive twice as much solar energy as do the temperate zones, yet they grow only half as much food: if we are to learn to harness the power of the sun to better effect, we shall need to pursue our inquiries with those parts of the tropics that do the best job of exploiting the sun's energy, the forests. As we have seen in chapter 10, tropical forests contain an abundance of plants that could offer us new foods, plus plenty of materials for genetic improvement of existing foods. Herein lies the scope for a Gene Revolution that could prove far more revolutionary and productive than the Green Revolution. But to exploit the plant productivity and genetic variability of tropical forests in this manner would require a systematized research effort for investigating tropical forests, on a scale far surpassing anything of the past. We need answers to basic questions such as the numbers and sorts of species in different categories, the ways in which they all live together, the ways by which they make most efficient use of moisture and soil nutrients as well as sunlight, and so forth. These questions, fundamental as they are, remain almost entirely uninvestigated. Of the several million species that inhabit tropical forests, we believe we have identified only one in six. Indeed we have next to no idea of the scale and value of the stocks of natural resources that lie waiting for us out there—except that experience to date tells us that these stocks must be rich to a degree. Yet total funding for research in tropical biology of these basic kinds is no more than $30 million per year worldwide, around half of it supplied by the United States (the New York Botanical Garden, a private organization, supplies almost $1 million, and the Missouri Botanical Garden well over $1 million). According to a recent appraisal by the National Research Council, we could achieve answers to many of these questions through a program of intensive research in just four ecosystems located at strategic points in the tropical forest biome.[3] A five-year effort could be accomplished for way under $40 million, and, through supplying us with insights into basic biology, would give a powerful boost to modern agriculture and forestry.

SUPPORT FOR PROTECTED AREAS

Some people assert, naturally enough, that we should concentrate on conservation rather than research. If we could only get hundreds of key ecosystems locked away as parks and reserves, they agree, we could conduct our research in them at later leisure.

With all its obvious merits, this strategy is far more expensive than a research program. As we have seen in the previous chapter, we shall need to safeguard an awful lot of areas in tropical forests if we are to do a sufficient job of conservation—and the effort does not come cheap. Nonetheless, a number of tropical forest nations are making a solid pitch at the challenge, and it provides a first-rate opportunity for the developed nations to support the efforts of tropical forest nations. An imaginative proposal along these lines has been formulated by Dr. Ira Rubinoff, the Director of the Smithsonian Tropical Research Institute in the Panama Canal zone. Rubinoff argues that tropical forests constitute "critical areas that are unique to humankind," and thus merit a collective effort on the part of the community of nations to safeguard these resources for everybody's indefinite use. Drawing on his lengthy experience in tropical forests, Rubinoff proposes a network of 1,000 parks and reserves of an average 1,000 square kilometers each, amounting to 1 million square kilometers, or a full one-tenth of the biome. Extravagantly large as this expanse may sound, it is, according to ecological investigations, a minimum that we need to safeguard if we are to assist most (not even all) species in tropical forests. In principle, the program would be financed by contributions from all nations. In practice, however, the burden would be mainly borne by those nations that can afford it (those that will draw major advantage from the program in the foreseeable future, through, for example, their genetic engineering industries), viz., the advanced and affluent nations.

A conservation initiative on this sizable scale would be costly: a cool $3 billion, Rubinoff estimates, spread over ten years. The payments would serve to support such systems of parks and reserves as tropical forest countries wish to establish, meaning that participation in the program would be entirely voluntary, and would not affect these nations' sovereignty sensitivities. The amount paid would be based on a country's areas under protection, in return for the country's undertaking to act as custodian, plus an agreement that maintenance activities could be monitored by panels of independent scientists.

A price tag of $3 billion for a protected areas network of 1 million square kilometers might seem high: ostensibly, $3,000 per square kilometer. But not all the funds would be spent on protected areas. In fact, far from it. Rubinoff has wisely pointed out that if the protected areas are to be not merely established and maintained in the usual manner, but to be safeguarded against population pressures in the years ahead, we shall need to do a much better job on agriculture throughout the humid tropics. By the year 2000 we can anticipate that there will be two billion peasants living in this zone. If they can be enabled to practice a productive and stable form of agriculture, especially one that does not entail an endless sequence of slash-and-burn cultivation within forests, then, and only then, the protected areas within forests will remain protected. Some of the funds, in fact a sizeable share of the funds, should thus be utilized to improve agriculture, through enhanced crops and the like. The funds represent a trifling part of the amount required each year to do a proper job on rural development, estimated by the World Bank at $6 billion per year.[4] Still, the measure would demonstrate that Rubinoff's proposal is not a gratuitous gesture that derives from a spirit of "save wild things from people." It is based, from start to finish, on looking out for the interests of both wild things and people, as equally legitimate occupants of the humid tropics.

How would the bill of $3 billion be shared out? Rubinoff suggests that we look to those nations with a per-capita GNP greater than $1,500 per year, and that we impose a progressive tax, beginning with $0.5 for GNP of $1,500, rising to $5 for GNP of $10,000 and more. In practice, the two dozen nations of the Organization for Economic Cooperation and Development, sometimes known as the rich nations club, would thus contribute $241 million; the seven nations of the centrally planned economies group, better known as Communist nations, would contribute $50 million; and the seven richest nations of OPEC would contribute $9 million. All in all, $300 million per year.

FURTHER FINANCIAL SUPPORT

Of course little is accomplished in the long run if we safeguard one-tenth of existing forests in the form of parks and reserves, and leave the remaining nine-tenths to be exploited out of existence. What kind of financial support would we have to think about in order to do a better job overall?

First of all, we shall need to invest much more in tree plantations, both for commercial timber and fuelwood. According to the Food and Agriculture Organization, the total sum required for all such plantations throughout the Third World is a full $5 billion per year. Commercial timber plantations will mostly be required in the humid tropics, while fuelwood plantations will be more appropriate for drier zones. But since the humid tropics contain many more people than do the drier tropics, the bulk of the investment will be required for tropical forest countries. Let us suppose, then, that the sum for the humid tropics amounts to $3 billion per year. Of this amount, we can reasonably hope that a fair share will be supplied by private capital via timber corporations, bio-energy enterprises, and the like. In addition, a solid proportion will be provided by Third World governments themselves. So let us suppose, for the sake of an educated estimate, that a reasonable proportion of the burden to be undertaken by developed nations is $500 million per year.

Second, tropical forest nations aim to go into the business of hardwood processing far more thoroughly in the future. As we have seen in chapter 5, the profit on, say, 1 cubic meter of hardwood in the form of finished furniture is between five and twenty times greater than that on raw logs (the range in profit margin reflects the luxury level of the end-product). The more processing that tropical forest countries undertake themselves, the greater the revenues they derive from each hectare of logged forest; and thus the greater the incentive for them to log the forests in a controlled way, as resources of exceptional value that deserve to be managed for sustainable harvests into the indefinite future. Timber industries can be expensive, however. Capital equipment, such as sawmills, plywood factories, and finished-product plants, gobble up investment in huge amounts. According to the Food and Agriculture Organization, tropical forest countries need as much as $3 billion per year for at least ten years if they are to complete the strategic advance from timber producers to timber processors. Again, we can reasonably expect that the largest proportion of the investment will come from the private sector, and much of the rest will be borne by Third World governments themselves. A reasonable estimate of development support needed from advanced nations is perhaps $100 million per year.

Third, a catch-all category including research, training, watershed management, and the like might require an estimated $100 million per year. So the grand total for the outside world to contribute in support of the global heritage in tropical forests is, very roughly speaking, $1 billion per year for the next ten years, declining thereafter when the main start-out investments are in place.[5] More than peanuts. Yet far less than peanuts when we compare it with some of our other regular outlays. If we were to divide up most of the $1 billion among the rich nations (leaving the rest to be picked up by the Soviet bloc and the more affluent OPEC nations), it works out at less than two-thirds of a dollar per person, or less than $3 per family. Each American family spends almost $100 per year on flowers and potted plants, a fine way to bring color to the home or office. Obviously an extra few dollars each year on tropical forests would not be begrudged, especially in light of the prospect that in this zone alone we may lose many thousands of plant species within the foreseeable future, many of them ranking among the most brilliant blooms on Earth.

The problem is not "We can't afford it." The problem is rather that people generally do not know how much is at risk. If they did know, their question would then become "How can we not afford it?" If the flower lovers of America were to be made aware of how much is threatened in tropical forests, and how relatively little is needed to save it, they might show themselves ready to support the cause. Plainly, Americans are not miserly people. They contribute an average of $180 each to charities yearly, placing them among the most generous people anywhere. The missing factor lies with the fact that most Americans are less than informed about tropical forests.

We can look at the situation another way. A sum of $1 billion is equivalent to what the world spends on armaments in half a day. Many military conflicts erupt over natural resources. How much less secure will the world be in a few decades time if some of its richest stocks of raw materials have disappeared?

Probably the best way to raise funds for developed nations to contribute to the tropical forest cause is through taxes on tropical forest products. For instance, a 2 percent value-added tax on hardwood imports would generate $80 million per year from developed-world citizens. This would make hardwood products marginally more expensive to the consumer, and so it would be slightly inflationary. Yet this is no more than reasonable. At present, the consumer acquires his hardwood products at a price that does not take full account of all costs that go into their production, notably environmental costs in the lands of origin. These additional costs, which presently remain outside the scope of transactions between producer and consumer, should be added in. The European Economic Commission is considering an import levy on tropical hardwoods right now, the proceeds to be spent on reforestation projects in West Africa and other traditional sources of hardwoods for the EEC; the United Nations Conference on Trade and Development is working on a similar proposal within the scope of its proposed International Tropical Timber Agreement between producer and consumer nations.

A parallel approach could apply to imports of so-called cheap beef from Central America into North America, together with beef from other forestlands in Latin America destined for Western Europe, Japan, and the rest of the developed world. The funds could be used to help ranchers in Central America and Amazonia do a better job of producing beef from existing pasturelands, thus eliminating the incentive to clear yet more forest to establish yet more pasturelands. A 5 percent value-added tax would generate $13 million per year. We should remember, of course, that the beef is currently imported precisely because it is so "cheap" (environmental costs are ignored), and a tax would shave off a little of the price advantage. But the measure would remain rational. In any case, our main aim should not be the eventual cut-off of beef exports from Latin America's forest zones. It should be the improved management by cattle raisers, so that they can continue to serve the export trade and assist their fellow citizens' nutrition as well. The forests that have been converted into pasturelands cannot be changed back into forests, not at least for a very long time. So let's concentrate on doing a far better job with areas where the forest has already been burned away.

The tax approach could be extended to many other products. For instance, medicines, drugs, and pharmaceuticals that derive, in one way or another, from materials in tropical forests. The trade in end-products can now be roughly estimated at $20 billion per year throughout the developed world. So a 1 percent tax would raise $200 million.

All in all, we can surely conclude that there is no problem in principle for developed nations to raise the funds necessary to assist tropical forests. A sum of $1 billion per year lies well within these nations' means. It also lies well within their responsibilities, if not duties. Furthermore, it will represent a first-rate investment —one of the best gestures that our grandchildren will perceive we could make to their future.

OUR WILLINGNESS TO PAY

Were support funding on this scale to be made available by developed nations, it would need to be subject to periodic appraisal. Suppose citizens in the donor countries began to object to the financial strain: the program would have to be scaled down, or even abandoned. Citizens in developed nations would then act in conscious recognition of what prospects face tropical forests. The entire exercise would enable them to gain a clearer grasp of their responsibility in the situation—an insight that would stand in marked distinction with the present position, where people in affluent nations have little chance to appraise their responsibility in anything like clear terms. In essence, the process would serve as a measure of the public's willingness to pay for the amount of tropical forest conversion that it sometimes proclaims it would like to see. Eventually it could develop into a framework that reflects costs and benefits as perceived by participant parties: a pay-for-what-you-want approach would permit a graduated response. The community at large could express its financial support for those parts of the tropical forest biome that it believes merit the gesture in comparison with other goods and services—thereby relieving the "either/or" type of situation where conflicting issues are seen only in terms of absolutes.

So whatever our reactions to the details of these proposals for a helping hand from developed nations, we should bear in mind that we are being presented with a choice that is spelled out in explicit terms. If we decide that the funds are too great, or that the matter must wait until next year, we are effectively deciding against an improved prospect for tropical forests. We shall no longer be able to plead that we have not been told. We are being asked to face a set of specific proposals. If we turn away from them, we cannot kid ourselves that we do not know what it is we are rejecting. We shall be turning thumbs down on a way for us all to join hands in an endeavor to save one of the glories of our common natural heritage. And we shall be turning thumbs down deliberately and effectively, and with knowledge aforethought.

"Our common natural heritage" is a phrase that supplies the rationale for collective action in support of collective values. It transcends concepts of developing worlds and developed worlds, and other divisions that have characterized the last two chapters. It merits detailed consideration in the following chapter.

Let us conclude the present chapter with a shopping list of items that the developed-world citizen may wish to consider if he or she wants to take up cudgels in support of tropical forests. The list is directed only at the developed-world reader, on the grounds that this book will be circulated almost entirely in the developed world.

- Support conservation organizations that specialize in tropical forests: for example, World Wildlife Fund-US, 1601 Connecticut Avenue NW, Washington DC 20005, and World Wildlife Fund International, Avenue du Mont Blanc, 1196 Gland, Switzerland; the Sierra Club, 530 Bush St., San Francisco, California 94108; Friends of the Earth, 1045 Sansome Street, San Francisco, California 94111; and the Nature Conservancy, Suite 800, 1800 N. Kent Street, Arlington, Virginia 22209.

- Maintain pressure on your government, through, for example, letters to Congress, to give ever-greater support to tropical forestry through AID, the World Bank, U.N. agencies, and other international development bodies.

- Press your government to require environmental safeguards for quasi-governmental agencies that sometimes dabble in tropical forestry, for example, the Export-Import Bank and the Overseas Private Investors Corporation in Washington DC.

- Do what you can to persuade giant corporations to do a better job in tropical forests. Timber concerns can be induced, through stick-and-carrot techniques, to be much more careful in their harvesting patterns; and to phase out logging of natural forests in favor of plantation forestry (tree farms). Certain hamburger corporations use no beef from tropical forest zones; why should the rest?

- Do your homework on tropical forests. Learn all you can about these celebrations of nature—what makes them tick, how we can keep them ticking. The references at the end of this book supply stacks of suggested reading, and any conservation organization will be delighted to send you dozens of additional ideas.

■ Ponder your own involvement in the fate of tropical forests. Apart from your use of hardwoods and your choice of hamburgers, consider other possible connections between your material lifestyle and tropical forests. For example, one of the best ways to safeguard tropical forests—indeed, to protect any natural environment in the tropics—is to get people off the land and into the cities. This depends of course on their being able to enjoy a tolerable level of living in urban environments; and this depends in turn on their prospect for employment. Third Worlders can manufacture a host of internationally sought goods, such as footwear and other leather items, clothing and textiles, electronic sub-assemblies, and so forth. But if enough manufacturing facilities are to be established to absorb even a share of the populations now crowding many rural zones in the Third World, the manufacturers need access to the world's main markets, and that means those in the rich nations. Far from these markets opening up to receive more and more Third World goods, they are actually shrinking, as governments impose import levies, tax barriers, and other trade restrictions. Each vote in Congress in favor of protectionism is in effect a vote against wild nature in the tropical Third World. Of course the shoemakers of Maine and the shirt manufacturers of Georgia have their legitimate interests, which must be protected just as much as tropical forests. But if a good-quality shirt can be produced by a Filippino or by a Brazilian at less cost than the equivalent can be made in Georgia, then there is gain not only for the survival outlook of tropical forests, but for the pocketbook of shirt purchasers throughout the United States (and a splendid stab against inflation). Surely the workers of Georgia, ranking among the most educated people ever to appear on Earth, should not busy themselves with activities that can be performed just as capably by semi-literate Third Worlders? They should move on to more sophisticated forms of employment—which can often be achieved only through a helping hand, in the form of re-training schemes, from the federal government.

■ At the end of each day, consider how far you have helped or hindered the cause of tropical forests. Constantly assess your responsibility. (For instance, when you buy, say, a tube of toothpaste: is that paper bag really necessary? Even if only indirectly, it helps to foster excessive exploitation pressures on tropical forests.) In turn, of course, this tends to raise the question that was pointed out to me during the course of a recent teaching stint at the University of California, when a student remarked that all these questions really boil down to "Who am I?" This question, first raised by the Ancient Greeks as the only question that matters, is taking on a new significance today.

18

The Global Heritage

As THIS BOOK has made plain time and again, tropical forests form part of the global heritage. We all find our daily lives enhanced by virtue of their existence. We shall all lose if they disappear. Just as much as Venice and the Taj Mahal, the Grand Canyon and the Victoria Falls, the *Iliad* and the *Bhagavad-Gita*, they form part of everyone's birthright. They "belong" to us all.

Of course they no more belong to everyone in a litteral sense than Venice belongs to the Indonesians or the *Bhagavad-Gita* to the Eskimos. But all of us can sense satisfaction that we share, each and every one of us, in a planetary patrimony that includes tropical forests. The Serengeti Plains have been declared part of the World's Natural Heritage, and Tanzania takes pride in the selection. The temples of Borobudur in Java have been declared part of the World's Cultural Heritage, and Indonesia expresses pleasure that all other nations share her joy in these marvels of human creativity. The spirit is entirely constructive and positive all round. Neither Tanzania nor Indonesia perceive any trace of outsiders' encroachment on their sovereignty; and the world has not the least desire to exert any vestige of control over these unique spectacles. The world demonstrates a collective delight in these exemplars of humankind's heritage, and Tanzania and Indonesia respond in a like spirit.

We can assert, then, that tropical forests are of major interest to more than just the few nations where, by geographical accident, they are located. True, they remain natural resources over which these few nations exercise national jurisdiction. The claims of sovereignty are to be respected. But as unique manifestations of nature, and as ecosystems with capacity to influence other ecosystems many horizons away, the sense of ultimate concern cannot be limited to those few nations that happen to exercise physical possession over the forests. While Brazil, Zaire, and Indonesia exercise legitimate property rights in their forests, the rest of the global community can express an equally legitimate interest in them. After all, the future of these tropical forests will impinge upon the future of the Illinois corn farmer, the Florida orange grower, the Swedish furniture maker, the British plastics manufacturer, the Caribbean banana grower, the Japanese gasoline grower, the Australian pharmaceuticals industrialist, the Soviet wheat growers, and the Arab leukemia sufferer. We are all in it together.

THE ISSUE OF SOVEREIGNTY

But there is no getting away from it: the biggest obstacle in the way of global efforts to safeguard the global heritage lies with sovereignty. Trees grow on particular portions of land that occur in the independent territory of this or that nation. As such, they are deemed to come under the exclusive jurisdiction of those nations.

So at least runs the theory—though we shall have more to say about the practice of sovereignty, and how it diverges from theory. Clearly the issue strikes to the heart of the problem of tropical forests. Many countries in question, especially those of Amazonia, and most particularly Brazil, are super-sensitive about sovereignty. And not without reason. Toward the end of World War II, the Allied leaders spoke about eventual "internationalization" of large territories, including Amazonia, that could one day become bread baskets for the world (little did they know that Amazonia should have been one of the last places to look toward for that purpose). More recently, there has been a Hudson Institute proposal presented to governments of North America and Western Europe, reaffirming the internationalization idea. Brazil has good cause, furthermore, to fear its immediate neighbors around Amazonia, seven of them with frontiers extending for almost 13,000 kilometers, or 80 percent of Brazil's total frontier extent. At one time or another there have been disputes between Brazil and most of these other countries—just as there have been between Ecuador and Peru, Venezuela and Guyana, and Colombia and Peru. So when Brazil launched its TransAmazon Highway project, with the aim of filling the "demographic void" of Amazonia (the region contains fewer people per square kilometer than the Sahara), the nation was trying to establish "live frontiers," driven by a slogan of "Let us integrate Amazonia so as not to surrender it."

By the same token, Colombia and Peru are seeking to assert sovereignty over their sectors of Amazonia by simply filling up the territories with people. Similarly the Mexican government is settling 26,000 people in a 1,800-sq.-km. bloc of the Chiapas forest in the southeastern sector of the country—the last remaining undisturbed forest in Mexico. By regrettable chance, the forest is located in an isolated area jutting into Guatemala; it is rich in oil; and it is populated at present with Indian farmers rather than with "true Mexicans."

Given the political pushing and shoving that takes place in Latin America, the safeguard measures are understandable, even if not acceptable. During the course of my repeated visits to Brazil, political leaders have observed to me that the Russians have expanded their network of the Trans-Siberian Railway in an attempt to establish jurisdiction over their vast remote territories in eastern Asia; while more recently, the Australian government has sponsored agricultural settlement in its northern tropical zone in order to "secure" the country's perimeter facing onto Asia with its teeming throngs. Brazil feels, moreover, that other nations of the world, including its most vociferous critics, demonstrate by their actions that they do not really care about any common natural heritage of the global community. Major powers contaminate large water bodies with chemical wastes, or with spillage from oil tankers and petrochemical refineries. A few of them even expose humankind to radioactive fallout from nuclear tests.

Thus there is a political motivation to the grandiose settlement schemes that are planned for certain tropical forests. Brazil may have lost a lot of money over its initial efforts to settle Amazonia. No matter, the country has made its point. Similarly, the Indonesian government may care little about the cost of its Transmigration Program that ferries excess people from Java to Borneo, Sumatra, and other little-populated outer islands. Although Indonesia could put the investment to better use by intensifying farming on Java instead of breaking new soil in other territories, it covertly wishes to "Javanese" the rest of the country with its diverse ethnic strains.

At the same time, a number of tropical forest countries are recognizing that, willy-nilly, they simply cannot maintain realistic control, let alone absolute control, over their forests. In the Indian section of the Ganges Delta, the Sunderbans Forest is dependent for its health, if not its survival, on the headwaters gathering zone in Nepal thousands of kilometers upstream. In Peru and other catchment areas of the Amazon River, deforestation is affecting the flood regimes, with all that means for low-lying forests as well as farmlands, in territories far downstream, not only in Peru itself, but in Brazil. Moreover, it is becoming apparent that if a large part of the Amazonia forest is chopped down, the remainder may, through diminished evapotranspiration, become a steadily drier forest, and thus a fundamentally different sort of forest from the rainforest that the countries of the region possess today.

It becomes increasingly difficult, then, for a government to assert that it exercises complete dominion over its forests, and that nobody else should be so impertinent as to comment on the matter. What goes on within the boundaries of a modern nation-state is influenced by what goes on in surrounding nation-states —and in some respects, by what goes on in nation-states around the back of the world. However much Brazil might claim that it will be beholden to its own inclinations alone if it chooses to put a match to its Amazonia forest, with all the climatic dislocations that such a step could impose on the rest of humankind, Brazil remains mindful that its leading crop, coffee, is dependent for its health and productivity on wild germplasm from the remnant forests of coffee's ancestral home, Ethiopia. By the same token, cocoa growers of West Africa must rely in part on cocoa genetic reservoirs in western Amazonia. In similar sense too, Indonesians and Bangladeshis, who import much rice from Thailand and Burma, can feel fair concern for what happens to the remaining upland forests of Thailand and Burma, with their critical control of irrigation water for paddy-rice farmers.

Despite their traditional powers as independent nations, tropical forest countries can scarcely consider themselves in a position to disregard the ultimate concerns of human beings far beyond their frontiers. They can hardly remain truly indifferent (whatever they may say) to the climatic impacts of tropical deforestation on farmers in the grain belt of North America, who could suffer from poorer growing conditions were tropical forests to disappear in smoke—plus the hundreds of millions of people in many lands who benefit from the surplus grain produced by those American farmers.

In other words, the concept and expression of sovereignty are changing. Only three decades after the number of independent nations around the globe has grown from about sixty to some two hundred, the practice of sovereign independence is, in certain respects, becoming something of a charade. However much this may sound an unrealistic interpretation of the current scene, it increasingly reflects the actual nature of nations' dealings with each other, by contrast with national leaders' rhetoric about independence. By virtue of trade patterns, monetary systems, and myriad other international relationships, nations are becoming ever-more intricately involved in each others' affairs. As long as governments and private entrepreneurs engage in transactions with organizations from beyond their national borders, foreigners are in a position to affect their national interests. It is an inescapable fact that every international transaction is subject to the influence of more than one sovereign state; and a hundred solemn declarations by governments to the opposite effect do nothing to alter the objective facts.

If this growing limitation on sovereignty applies clearly to developed nations, with their push for endless economic growth, it applies at least equally to developing nations, which, though they frequently wave a flag for their new-found political independence, seek rapid economic advancement for their impoverished peoples and thereby need to play a growing role in the international economy. Brazil, for example, is burdened with a foreign debt of some $90 billion, and in order to obtain relief from agencies such as the International Monetary Fund (IMF), it must accept far-reaching interventions in its economy, through regulation of its money supply, control of its inflation rate, crimps on imports, and many other measures that directly affect the daily life of every Brazilian citizen. Of all things—and what an incursion this is on Brazil's sovereignty!—the nation has had to bow to IMF wishes by foregoing a bid to stage the next World Football Cup tourney. In short, developing nations, like developed nations, are swapping certain images of independence for the benefits of interdependence.

Of course, hardly a single government wishes to admit as much. But tacitly most nations recognize that the age of independence in the "grand manner" has disappeared, and we are all involved in one another's fates. This is all the more pertinent to tropical forests, in that many of their ecosystems constitute international entities. The River Mekong basin comprises six countries, the Amazon eight, and the Zaire nine. Indeed it has been a spirit of "adaptive sovereignty" that has inspired the eight nations of Amazonia to formulate an Amazon Basin Pact that "reaffirms before the world that, as we have exclusive jurisdiction over this immense region, so also do we accept and fulfill the responsibilities of initiating its development without destroying its ecological balance and of using its immense potential riches in a rational way."

EVERYBODY'S HAND ON THE AXE

Ironically, Amazonia is a region that, more than all others of the tropical forest biome, exemplifies the role of foreign capital and technology in the use, and often the abuse, of its forests. Much of the logging is conducted by giant timber corporations from North America, Western Europe, and Japan, invited into Amazonia and allowed to operate pretty much as they please at the behest of those same governments that lose no chance to proclaim their absolute sovereignty. Moreover, these foreign enterprises harvest hardwood primarily in response to excessive consumerism on the part of the developed world. Much the same applies to outsize ranching concerns that raise beef in Latin America for overseas markets. Thus we see the pervasive hands of the remote entrepreneur and consumer in the exploitation of tropical forests. While Amazonia represents an extreme case of the foreigner's role, we can discern similar relationships at work in virtually all parts of the biome.

It has ever been thus. Whereas we often hear that tropical forests have experienced an orgy of mis-use and over-use only during the past couple of decades or so, this recent phase has been no more than an intensification of a process that has lasted for well over one century. It is illuminating to reflect on the extent to which, via trade and commerce, the forest cover of all three regions has been depleted at the hands of persons who live thousands of kilometers outside the tropics. Let us pause for a moment to examine this process.

In the lower Himalayas, widespread deforestation started as far back as the early 1800s.[1] The British sought the fine-grain and insect-resistant timbers of the area for a variety of construction purposes, stimulating a process that led to much over-cutting in the foothills. By the start of World War I, much of the western Himalayas had been subject to selective but extensive deforestation. Similarly in Burma, the delta zone of the River Irrawaddy was supplying hardwoods to the navies of several European nations as early as the fifteenth century. By the time Britain had depleted its oak forests around the end of the eighteenth century, Burmese teak came into still greater demand—as did the hardwoods of other tropical countries, such as Sierra Leone and elsewhere in West Africa, where many primary forests were felled by World War I.

Around the middle of the last century, moreover, the delta forests of the River Irrawaddy started to be cleared completely in order for lowland Burma to become a leading rice-exporting zone. Thousands of square kilometers of rich monsoon forest were eliminated in short order. In 1866 the main rice zone of Burma amounted to only 7,000 sq. km., but by 1950 it had expanded to at least 40,000 sq. km. Next door, in Thailand, a similar tale unfolded. In the year 1850, only about 9,000 sq. km. of Thailand was under rice, but thereafter the paddies steadily expanded to satisfy the export trade, topping 50,000 sq. km. by the end of World War II, and reaching almost 70,000 sq. km. today. In addition to these deforestation pressures caused by the rice trade, other cash-crop exports are today accounting for remaining forests in Thailand, notably cassava, used as livestock feed in Western Europe and Japan.

In the Philippines, there was greater deforestation before the year 1900 than between 1900 and 1950, due to the arrival of plantation agriculture in the form of export crops such as sugar, coffee, indigo, tobacco, and abaco. In the Caribbean, sugarcane plantations caused the elimination of virtually all sizable forests by World War II. (Since sugar consumes a great deal of fuelwood during processing, it accounts for at least as much forest in the grinding as in the growing.) Sugar, in conjunction with coffee, likewise took its toll in the Atlantic coastal forests of Brazil, which, being far removed from the Amazonian forests, once possessed their own distinctive and highly diversified floras and faunas, now virtually eliminated.

As a final illustration of these economic linkages across the seas, let us note that some of the last remaining forest habitats of the mountain gorilla, holding out in the zone where Rwanda meets Uganda and Zaire, have lost a full one-quarter of their extent since the European Common Market decided to emphasize pyrethrum-based pesticides rather than synthetic chemical pesticides. The pyrethrum plantations have been established almost entirely at the expense of natural forest. Thus, while there are many environmental merits to these more "natural" pesticides, the upshot has been a reduction in the gorilla's range to a final few hundred square kilometers.[2]

In addition to these direct economic linkages among the community of nations, there are various activities of a more diffuse type, more difficult to perceive and pin down. Yet as the global economy becomes increasingly integrated, international linkages can be expressed in some abstruse but nonetheless substantive ways.[3] Japan, within a strict context of its own domestic situation, does a fine job in safeguarding its forest cover—and thereby safeguards the nation's future timber supplies, together with its watershed services and other environmental amenities. Yet because this policy directs increased exploitation pressure at the forests of Southeast Asia, Japan's forestry strategy, however laudable within the narrow perspective of Japan itself, is less than helpful from a global standpoint. If 1,000 sq. km. of Japan's forests are subjected to gross disruption, the loss in species, genetic resources, and the like will be a mere fraction of a similar process overtaking 1,000 sq. km. of forest in Borneo. By way of comparison, we may recall that the elimination of the virgin forests of the eastern United States, covering at least 1 million sq. km. at the time of the settlers' arrival, resulted in the extinction of only three bird species, whereas the elimination of a similar expanse of Southeast Asia's forests would surely eliminate hundreds of bird species, together with thousands of plant species and tens of thousands of insects.

A similar consideration applies to Great Britain. A recent report proposes that the nation's upland forestry areas should be doubled, in anticipation of tightening timber supplies in global markets. Britain imports nine-tenths of its timber needs, a good part of them from tropical Africa—and British demand for timber is likely to double by the time that trees planted today become ready for harvest in fifty years. To date, only one-twelfth of Britain's land area is forested, by contrast with an average of about one-fifth for Western Europe. Yet the proposal for more forests in Britain has been resisted by, of all people, various conservation groups that want to enjoy open mountain landscapes in the manner to which they have grown accustomed; they fear for upland open-area wildlife, and see plantation forests as alien to the country's cultural traditions. Again, we find that a stance on national forestry that suits the wishes of many local people may not accord with the needs of the larger community.

Many more such linkages can be traced between developed nations and tropical forest nations. However unwitting may be the impact of the developed-world citizen, the repercussions are already pronounced, and they grow more numerous and more effective with each day that goes by. Thus we all play our part in degrading tropical forests: the common natural heritage becomes a collectively impoverished resource. The irony is compounded by a still more curious factor. The outsider's hand on the axe is often set to work with the conscious assent of tropical forest nations. Despite their assertions about absolute jurisdiction over their forests, governments in many parts of the biome appear only too anxious to have foreign enterprises of every stripe come and exploit their forests. Moreover, these governments extend a welcoming hand to the foreigner while knowing that the foreigner is as inclined as anyone to mis-use and over-use the forests. The foreigner possesses abundant capital, management skills, and fancy technology to hack down forests at a rate almost invariably beyond the capacity of local enterprises. Thus tropical forest nations are selling their native resources for a few one-time profits. Sometimes sovereignty does not seem to command much of a price.[3]

RESOURCES OF GLOBAL INTEREST

Even while it plays its part in despoiling tropical forests, however, the outside world perceives an interest in the future welfare of tropical forests. Increasingly, the global citizen plays a Jekyll-and-Hyde role in the saga of tropical forests. Supposing the global citizen wanted to contribute in a more constructive style to the future of tropical forests. What could he or she do that would not run up against a brick wall of sovereignty sensitivities?

An answer to this difficult question may lie with a more expansive approach to our strategies for managing global resources such as tropical forests, that is, resources that, while remaining the strict property of national jurisdictions, are a legitimate concern of humankind. With sufficient goodwill, there need be no intrinsic conflict between the precepts and practice of sovereignty. Slowly but surely, nations are coming to acknowledge a sort of public trust doctrine at an international level, or an awareness that provides an authoritative basis for basic principles of collective living in the global village. If nations are responsible not only to themselves but to other nations for what they do, we can apply this concept to the management of many high-value resources, not only exceptional ecosystems, genetic reservoirs, and other environmental resources that characterize tropical forests.

We can apply them equally to international waterways, the high seas, in fact to the Earth's air mantle, and to all that constitutes the single biosphere that sustains life for us all. If a sufficient concensus can be established along these lines, we may reach a stage where select categories of resources, such as tropical forests, can be designated as resources to be maintained by "countries in possession" as trust resources for the global community, now and forever. Of course the individual nations will retain the intrinsic substance of sovereignty (though perhaps less of the out-moded semblance) over their natural resources: these resources will remain the exclusive property of nations in question. But they will become resources in which other nations could express a reasonable interest—and even proffer gestures toward assisting with responsibility for the resources' welfare.

Of course, collective responsibility for the common heritage does not necessarily mean collective rights in particular resources. Rather we shall find that an individual nation will no longer be left to rely on its own isolated efforts to protect precious resources such as tropical forests. It will continue its own on-the-ground management of its forests, while the international community offers help in support of the common heritage.

In operational essence, the strategy will depend on incentives rather than coercion as a means to induce individual nations to align their individual rights with the interests of the broader community. Thus the strategy will not downplay sovereignty. On the contrary, it will emphasize it as a functional rather than an idealized concept. In fact, the management of exceptional ecosystems will depend in large measure on decentralized planning—and what better institution to accomplish this than the individual nation that exercises competence in its own bailiwick without overlooking its broader responsibilities among the community at large.

Sooner than we might anticipate, the strategy can help tropical forest countries to feel they thereby enhance their international image in terms of status and responsibility. Rather than feeling pressured in a manner that offends their sovereignty, they will find that their prestige and political authority within the global community will be reinforced.

Of course, far-reaching initiatives along these lines depend upon political will. In turn, this depends on the perception of common needs and opportunities. As this perception grows stronger, so nations will come to acknowledge reciprocal responsibilities. They will then become more inclined to take broader-ranging measures to meet new challenges of collective life within the global village. In short, perception is all.

Thus the first priority in addressing the issue of tropical forests is to establish the topic on political agendas of nations outside the tropical forest zone. An exercise in consciousness raising will entail an educational campaign through the mass media, as well as through forums for political leaders and policy makers. Here is a distinct role for citizen activist groups, such as the World Wildlife Fund, Friends of the Earth, the Sierra Club, and other conservation bodies that have been espousing the cause of tropical forests for years. The World Wildlife Fund, in conjunction with its sister scientific organization, the International Union for Conservation of Nature and Natural Resources, is conducting a two-year, $11 million Tropical Forest Campaign, which seeks, among other activities, to evaluate the urgent conservation needs of Thailand, to supply support for an entire raft of projects in Indonesia (ranging from threatened species to environmental education), to foster protection of endangered forests of Madagascar, to speed the management of species-rich montane forests in Tanzania, to help establish parks in Cameroon, to supply forest rangers' equipment in parts of Central America, to help develop a National Conservation Strategy for Malaysia, and to protect some of the most endangered, as well as some of the richest forests in the world, along the Atlantic coast of Brazil.

The publicity-generating power of citizen organizations is greater than we sometimes suppose. During the past two decades, the great changes in American life—civil rights, the women's movement, environmentalism, consumers protection, among others—have all originated at grass-roots level. Far from being a leader, the U.S. government has followed behind. So could it be in many other developed nations, with respect to tropical forests. The U.S. government is several years ahead of all other developed-world governments in beating a purposeful drum for tropical forests—and the conservation constituency in other developed countries could mark the pioneering efforts of their counterparts in the United States, who persuaded their government to give the issue a strong second look.

During the formative stages of a community-of-nations effort in behalf of tropical forests, it will behove "outsiders" to maintain a low political profile, broaching the matter in a way that reinforces concensus and cooperation while playing down politics and reducing divisiveness. They should demonstrate a spirit akin to that with which they handle other global resource issues, such as the high seas, the ozone layer, carbon dioxide buildup, and outer space—issues of collective concern where a premium lies on distribution of benefits and costs. Thus the most pertinent role for non-forest nations will lie with quiet diplomatic initiatives, plus financial contributions and technical backup for whatever international measures emerge, leaving tropical forest nations themselves to blow clarion calls for concerted action.

Given the North-South dimension to the issue, there will be much virtue in cooperative endeavor. At the same time, an increased emphasis by developed nations on the future of tropical forests might help induce that spirit of international cooperation necessary in addressing broader collective questions of natural resources and environmental values. A campaign in support of tropical forests could assist in articulating the common interests of nations—and it might even encourage governments to adopt a more collective approach to other collective challenges, leading to an enhanced world order. The more we agree to work together on behalf of tropical forests, the harder we shall find it to squabble over Antarctica, the ocean floor, carbon dioxide, and outer space.

19

A Final Word

IN EARLY 1979, the United Nations Environment Programme was asked to arrange a first international conference on tropical forests. I was recruited by UNEP to organize the event, and I learned a lot from the experience.

The initiative stemmed from several activities of the United States during the late 1970s. A number of American conservationist groups had taken a lead in alerting American political leaders to the accelerating decline of tropical forests. As this concern gathered momentum by mid-1978, the U.S. State Department convened a Strategy Conference on Tropical Deforestation. This meeting encouraged the U.S. government to seek ways to arrange an international gathering on the issue. Not being a tropical forest nation itself, the United States wished to retain a low profile on the matter. So it requested UNEP, as an international agency, to undertake the task. UNEP said fine, and we were off and running.

We decided not to go for a full-blown conference such as the mega-scale gatherings organized by the United Nations on food, water, deserts, and other resource issues. Due to the political sensibilities involved, we aimed for a low-key occasion, with a limited number of experts from select countries in all parts of the Earth. We wanted to emphasize that the community of nations stands ready to support the efforts of tropical forest countries as they seek to do a better job with their forests.

We planned to hold the conference in early 1980. In an Overview Document that I prepared by way of background, I stressed that we would not indulge in bemoaning the impending demise of tropical forests. Rather we would draw on the favorable experience of those tropical forest countries that were achieving some success in mobilizing their forest resources to promote their general development in all perpetuity—development viewed in a proper broad sense as "the utilization of resources for greatest sustainable advantage for all peoples concerned, primarily the present generation but including all generations into the future." With this approach, I intended to emphasize the many potential uses that tropical forests can serve, by contrast with a preservationist approach that would try to "build a fence" around them. Equally important, I hoped that this approach would help the conference to shift the tropical forest debate away from a negative clamor about their destruction, and toward a constructive appraisal of how they can best confer their manifold benefits on society at large.

I found that what happened before the conference and what emerged afterward were to prove as significant as what occurred at the conference itself. I and my colleagues at UNEP hoped that the participants would agree on an International Plan of Action, in recognition of the fact that many tropical forest nations do not possess all the funds and expertise they need to develop their forests, so they look for support from outside. Just as important, we wanted to acknowledge that what one country does with its forests may have ramifications for other countries, hence there is need for countries to work in close consultation with each other, to share experiences, to swap ideas. So our aim was to formulate a coordinated campaign, to come up with a comprehensive work program, and to portion out tasks among all parties interested.

The conference could not even agree, however, that an International Plan of Action was desirable. A handful of tropical forest nations hinted darkly that this could eventually lead to "interference" in their internal affairs and that it would be better not to start on a Plan at all. Through compromise, the conference finally decided to postpone the sticky question of a Plan until a second session to be organized two years later. But alas, the interim period saw such an outburst of bickering that the follow-up conference did not attract any representatives from Brazil, Venezuela, Colombia, Zaire, and several other leading nations, meaning that it reflected the views of countries making up less than half the biome. Regrettably, a good part of the bickering owed its inspiration to disputes among the sponsoring bodies of the first conference. UNEP naturally wanted to organize the affair in conjunction with other U.N. agencies, and the United Nations revealed that it can be a dis-united affair. Turf rivalries won the day, and the UNEP process that started with high hopes in 1979 went to its grave within three years.

SUCCESS STORIES

But after fifteen years of working with international bodies, I have found that it is the process of consultations that can sometimes turn out to be as important as the product. Much more emerged from the UNEP exercise to make it more worthwhile than the bald record reveals.

For instance, the daily sessions of the conference were punctuated by protests from Brazilians, to the effect that they had nothing to learn from foreigners about forestry; thus all dialogue was ruled out from the start. In the evenings, those same Brazilians would join me for a drink and ask what new techniques were being developed by forestry professionals elsewhere, what UNEP's experts thought of their approach to Amazonia, how I viewed their latest planning endeavors. During a mid-conference break, I went on a short safari with several officials from Malaysia; while watching a herd of giraffes watching us with assured stare, the officials reminded me that Britain had once been a sea of forest from one end to the other, and the British had cut down those forests in order to build a fleet with which to rule a full one-quarter of the globe, including Malaysia. So why did Britishers now suppose they could advise Malaysians on the virtues of retaining forests? Malaysians, they said, would not be so crass as to follow any misguided example of Britain's with regard to wholescale deforestation; they were formulating their own strategies. On return to the conference, they demonstrated, through an account of some of their activities in eastern Malaysia, that is, Sabah and Sarawak, why they are gaining a reputation for some enlightened practices in modern forestry.

In particular, it appears that Malaysia is making progress with logging of secondary forests. Hitherto this has constituted one of the main obstacles to the sustainable harvest of timber. The more Malaysia learns to extract harvest after harvest from secondary forests, the more it relieves exploitation pressure on primary forests. Malaysia is not alone in this pioneering work. Its efforts are being matched, in differing ways, by those of Burma, Gabon, and Guyana. None of these nations ranks among the leading ten nations as measured by forest expanse. Yet their innovative advances deserve to become better known throughout the biome, so that other nations can apply the new technologies and management strategies involved.

During the course of the conference, I discussed these success stories with forestry officials from a host of countries. I found that hardly one was much aware of advances in forestry know-how beyond his country's borders. After all, they protested with a single voice, they are preoccupied enough with problems within their own forest territories. Yet they also agreed with a single voice that a systematized diffusion of new information could help to improve forestry on all sides. As it turned out, the conference set off a process of information exchanges, both formal and informal, among dozens of countries—a process that flourishes with ever-growing interest from year to year, even though the conference-style consultations have come to an end.

Similar advances are being made on other forestry fronts. They are limited in scope and arouse little ballyhoo. In comparison with the scale of the problems to be tackled, they are small potatoes. But they serve to show that certain countries are not standing still. We heard how Thailand is trying to move away from a policy of unproductive persecution of its hill-forest farmers with their opium crops. Instead, Thai officials now sit down with local community leaders to explore prospects of growing other cash crops. Opium is being abandoned as less dependable than alternatives, and the farmers, no longer needing to keep out of sight of the law, are forsaking their former migratory way of life with its wasteful slash-and-burn practices. In country after country, agroforestry is becoming an established mode of restoring tree cover to denuded tracts. Through an imaginative experiment, the Philippines government, in conjunction with a loan from the World Bank, has enlisted the powerful support of private enterprise in fostering a commercialized variant of agroforestry. When we can mobilize the innovative skills and the financial muscle of the marketplace to promote tree planting rather than tree chopping, we achieve an advance to proclaim from the rooftops, if not from the treetops.

As for developed nations, the U.S. Agency for International Development has been moving ahead at a remarkable rate. In 1980, the Agency set up a new forestry program, leading to a 1,000-percent increase in funding during the period 1977–1982. In the late 1970s, the Agency had only five professional foresters at its headquarters in Washington DC, and fewer than twenty in the field. Now these totals have increased fourfold.

Many more success stories were related at the conference, both in the main assembly and over dinner tables. Taken all in all, these individual advances represent far more progress than was envisaged only half a dozen years ago, or was even dreamed of a dozen years ago. True, the problems are now much greater than they were only a short while back. But in a few countries at least, the situation is no longer being viewed as an overwhelming problem. Rather it is viewed as a creative challenge.

A KEY FACTOR FOR THE FUTURE

So much for the recent UNEP initiative. What of the future? I believe that a key factor will lie with the way we look at tropical forests, and at the larger world. I refer not only to our intellect, the way we "make sense" of our world. I refer to our feelings for our world, the way we "make sympathy" for it—or, rather, the way we engender empathy for it. I am not suggesting that we learn to love a tree a day. I am suggesting that we try to grasp what it is to be citizens of one world. I am also suggesting that we find ourselves pretty poorly equipped for the task, in terms of the human makeup supplied to us by our prehistoric past. In certain respects, we resemble a bunch of Neanderthals trying to pilot a spacecraft.

Let me explain it this way. Man became man when he became an established hunter-gatherer, a momentous event that occurred at least one million years ago. The profession of hunter-gatherer puts a premium on coordination with your fellows. You make out better when you operate as a member of a hunting band. You are more likely to survive, and thus to pass on your successful genes, when you throw in your lot with a dozen others, sinking your individual preferences in the greater good of the community. You also fare better when you learn to take thought beyond the here and now: you sharpen your spear tonight in readiness for tomorrow's foray beyond the hill.

During the course of man's formative stages, then, there presumably arose a force of natural selection in favor of those who could empathize with the world beyond the ends of their noses. In collaboration with others lay survival, just as it lay in conceptualization about tomorrow and lands beyond the horizon. This surely is one of the things we mean when we talk about man becoming human.

Regrettably, *Homo* has spent almost his entire time-span as a hunter-gatherer. In fact, man pursued that one lifestyle until just 10,000 years ago, when he started on agriculture. Thus we are stuck with a set of sensibilities that have suited us for at least 99 percent of our time on Earth, but may have left us sadly deficient for dealing with our present-day world. The remaining 1 percent of our time, a mere 400 generations, has not been enough for us to develop and refine the extra faculties we need to deal with a world far more complex than that of our primitive forebears. We now have to operate as a hunting band of almost five billion people. We have to think of an exceptionally long line of tomorrows. We have to visualize lands around the entire Earth. To achieve all this, we possess the equipment that evolved to help us with a job of kindred nature but of far smaller scale.

As I sit here at my Oxford desk, I ask myself: Do I really care about all those trees in Borneo that I described at the start of this book, those marvelous dipterocarps that fall to the chainsaw hour by hour? Of course in an intellectual sense I do care. The scientist in me is affronted at the wasting of nature's crescendo on Earth. The conservationist in me is equally affronted at the ravaging of a resource stock that could sustain many millions of people to the year 2000, and many billions to the year 3000. But as an individual, as a human being, how do I react? Do I identify with those remote forests and all they stand for?

I raise the question because I want to be sure of my personal response. How can I, a developed-world citizen of the temperate zone, talk about those forests with colleagues in the developing tropics unless I am rock-solid certain of my conviction? I question myself, I check my reactions time and again. Intellectual speculation is a luxury afforded to those who are fed. Are my feelings true, when I sense that I want those forests to survive?

Yes, I do want them to survive. If pushed to take a final stand,

I shall say that I want them to survive because they are a uniquely fine phenomenon on the face of the Earth, and there is no compelling reason to remove them. If I could leave Britain for a while and fly off to the dear old tropics, and if I could choose between watching lions in sunlit Serengeti and gazing at shady dipterocarps in Borneo, I would head for the forest. The first would be enjoyable, the second would be part of being alive.

None of this detracts from the argument that it makes sound economics to safeguard tropical forests. The materialist rationale loses not a scrap of force when we invoke a mystic response. True, this mystic dimension is much harder to describe. It cannot be reduced to neat columns of figures, as is often the case with economics. But for me, it counts just as much. The finest economic arguments do not stir me to take up arms in favor of tropical forests any more than does the premonition that when we lose tropical forests, we lose we know not what. If we are to advance beyond the perspective of the Neanderthal, we need an elemental sense that tropical forests represent more than is reflected through the economist's arithmetic. In this case at least, a gut conviction about the worth of tropical forests surely counts as much as do the most careful calculations. I can no more explain my feeling for tropical forests than I can dissect my reaction when confronted with the Grand Canyon or the Acropolis. Sure, we can make better "use" of the Grand Canyon by damming it for electricity, and of the Acropolis site by converting it into an office block—just as Central Park in New York is an "economic waste" of top-value real estate.

Incapable as I may be of expressing my deepest response to tropical forests, I specially like the remarks of a colleague, Professor Daniel H. Janzen, of the University of Pennsylvania, who gets close to target when he writes (in a paper for the Office of Technology Assessment, U.S. Congress, 1982):

If I told you that I was going to magically introduce a gene into the human population that would make all your grandchildren and all their descendants color blind, you would be less than pleased. The same is true of a gene that would eliminate all awareness of music. But what are color vision and music awareness? Traits that you would never know you had if the world was not colored and the air not filled with complex sonorous sounds. I maintain that you, dear reader, are an animal rich in mental and physical receptors for the complexity of nature; and by destroying that nature, you condemn your offspring to the sleep of never even knowing those receptors exist—and by destroying tropical nature, you destroy easily the majority of the signals those receptors are designed to receive. Of course, humans are good at generating mild complexity in their workings. But the level of complexity generated by humanity is to the complexity of a tropical forest as a mouse's squeak is to all of human music. The city-center dweller who feels no cultural and biological deprivation is simply unaware of what he is missing. If your response is, well, what he doesn't know won't hurt him, then I suppose that you won't mind if I eliminate color vision from your children at birth, and all their children after them. Humans have spent rather many millions of years inventing the ability to be very aware of what is around them. How ironic that just about the time they get themselves to where they can sit back and gawk rather than fear all of that, their quintessential human trait is removing the very thing that made them what they are. A human without senses is not even an animal.

If ever we allow tropical forests to disappear from Earth, we shall have to tell ourselves not only that we have lost something of value. We shall have to admit that in certain significant senses, we have not gained much since we left our caves. For all that we are now firmly established on our hind legs, our response to tropical forests serves as a measure of how far we are ready to stand tall as citizens of the Earth, as joint members of our planetary home.

THE ELEMENT OF HOPE

In whatever way we look at them, tropical forests are something special. As we have noted, various people are trying to safeguard what there is left. The efforts of these folks, in face of appalling prospects, serve to demonstrate that we can expand the single element that counts for more than the most sophisticated technology, the most refined planning, and the most stream-lined management, and that is *hope*. I am frequently reminded by questioners, during the course of public lectures and television interviews, of the forestry horror stories that are recounted more and more often in the press; and I am asked whether I do not agree that they presage a bleak future for tropical forests. I respond that there are still grounds for hope. True, it is often hope to be voiced through clenched teeth. Sometimes it is hard to see much light. But it would be harder still to heave up words to the effect that the cause is lost.

Let me end this book, with its message of hope, on a severely practical note. I am becoming convinced, on the basis of field evidence that I come across from time to time, that we can now meet virtually all our current needs from tropical forests by making more efficient use of cut-over forests, fallow forests, and deforested lands. We shall soon need not cut another tree in primary forests. Whatever our needs—commercial timber, paper pulp, fuelwood, rubber and other industrial crops, rice and other food crops (whether cash crops or subsistence crops), even beef—we now possess sufficient experience and technical know-how to produce all we want from forests that already bear the heavy imprint of man's hand. It will cost us, all of us: citizens of the affluent world may have to contribute up to 1 billion dollars per year. Do we want to pay the price? Right now, we enjoy something of a free ride from primary forests, in that the price we pay for their goods does not reflect the full costs of their production.

While we are asking ourselves whether we want to play fair with our tropical forest heritage, let us be assured that the untouched forests, the so-called virgin forests, can remain pristine pure with all their biological integrity, until such time as we learn how to put them to best eventual use, whatever that may be—and if that best eventual "use" is to leave many of them in virtually undisturbed form (for their climatic-regulation effects, if nothing else), we shall count ourselves fortunate that we have retained sufficient undisturbed expanses to meet our future purposes.

This is a far-reaching statement. It proposes that virtually all primary forests can remain primary forests. Despite the pressures of growing human numbers and growing human aspirations, we need convert hardly any more of these primary forests. In light of all I have learned from my travels and colleagues in tropical forest countries during the course of almost two decades, I am sure that the proposition stands up. Herein lies my greatest source of hope.

Will we in fact do it? Certainly not this year or next, probably not during the 1980s, possibly not before the year 2000—by which time the remaining primary forests could well be reduced by at least one-third. Let us not become too carried away with dreams and ideals, grounded as they may be in real-world experience. But let us continue our fight for tropical forests, with the spark of hope that we can yet win the day—and we shall be thanked for millennia to come. Let us be glad that we have been given the chance to undertake one of the most creative challenges ever to confront humankind.

Since completing the manuscript, I have come across various pieces of good news. While they are very far from constituting an answer to the overall problem, they represent a heartening response to the situation—much more than we would have deemed possible only five years ago. So they make a fitting finale to this book.

1. International Tropical Timber Organization

Under the aegis of the U.N. Conference on Trade and Development, the main nations involved in the tropical timber trade, i.e., the leading producer and consumer nations, have gotten together to appraise their common needs and interests. Both sides recognize that to continue as they have done to date will work only to the detriment of all parties. So representatives from Japan, the United States, Britain, and other major importers of tropical hardwoods met with representatives of the major producing nations, notably Indonesia, Malaysia, the Philippines, and so on, to ensure a sustainable supply of commercial timber in a manner that meets everyone's long-term requirements. A new agency, the International Tropical Timber Organization, will oversee a continuous negotiating process that will put the tropical hardwood trade on a better footing. These are early days, and we must see how the new agency works out. But the very fact that the effort has been made is evidence of a new spirit of cooperation between the two sides, in marked contrast to the spirit of confrontation that has marked their relationships thus far. It may be the first small brick, and potentially a solid one, on which to build a foundation for the future.

2. Tree Planting in India

In Gujerat State of India, a handful of primary schools started their own tree nurseries in 1980, in order to boost the planting of fuelwood lots around the countryside. Following on this fine initiative, seven hundred schools now produce well over 10 million seedlings per year, an average of 14,000 each. John Spears, the Forestry Adviser to the World Bank, points out that if 1 in 10 of India's 620,000 primary schools were each to raise 10,000 seedlings per year (feasible with a nursery measuring 20 meters by 20 meters), the effort would at least match the total number of seedlings produced by the government.

3. Commercial Timber Plantations

Brazil, a country sometimes criticized in this book, has undertaken a far-sighted program of commercial timber plantations in order to supply the needs of its burgeoning industries. Since beginning in the early 1960s, Brazil has established 40,000 sq. kms. of plantations, almost all in the southern part of the country in territories far from Amazonia, and at virtually no cost to natural forests. The plantations supply 60 percent of Brazil's needs, whereas the country's sector of Amazonia, an area 70 times larger, provides only 10 percent. By the mid-1990s, Brazil should be able to meet all its industrial demands from plantations.

Similarly, Ivory Coast has embarked on a plantation program amounting to 2,000 sq. kms., equivalent to less than one-tenth of the country's natural forest area. Early in the next century, these plantations will supply about 4 million cu. m. of hardwood timber per year for the export trade, as much as do all the country's natural forests today. Burma too is following this strategy, planting 400 sq. kms. of high-value teak trees, with capacity to produce more teak than do all the country's natural forests right now.

4. Linkage between Protected Areas and Agriculture

As chapter 14, Environmental Services, has indicated, agriculture in the humid tropics is often dependent for water supplies on forest cover in catchment zones in upstream hinterlands. This linkage is now being recognized by the World Bank, which has decided to support the World Wildlife Fund's activities in northern Sulawesi of Indonesia: the Bank is providing a $700,000 loan to help safeguard a 3,000-sq.-km. park and reserve that protects upland forests, a source of irrigation water for downstream farmers. In other localities, we find that development agencies are starting to perceive the connection between conservation and agriculture. In Java, the Chibodas Park supports the livelihood of several million lowland farmers and thereby attracts funding from agencies that have not usually accepted that protected areas can serve as a valid form of "development." Similar support is emerging for the Monteverde Cloud Forest Reserve in Costa Rica: the Reserve, which protects 320 bird species and 100 mammal species in its 2,000 hectares, safeguards water supplies for several thousand square kilometers of farmlands in the plains below. In parallel style, the Canaima Park in Venezuela's sector of Amazonia provides water for hydroelectric facilities that support extensive iron mines, while also serving as the principal source of water for the capital city, Caracas.

Epilogue to the 1992 Edition

Is THERE a renewed prospect for those forests? What hope of that, ten years later?

As I have indicated in the Prologue, part of me feels despondent about the outlook. After all, the forests are falling faster than ever. Nor are there many signs—hard, substantive signs in the workaday world—that our political leaders will do enough about it fast enough. But at the same time I feel heartened by the fact that public opinion is becoming roused about the issue, and that we shall shortly reach a stage when political "leaders" will have no choice but to follow on behind the John and Jane Doe Citizens in many lands who are demanding that something be done forthwith. Will that time come soon enough?

I suspect it won't be soon enough to save many more forests from the chainsaw and the matchbox. We are going to lose a lot of remaining forests, let's be realistic about it. We can't stop the forces of destruction just like that: they have worked up too much momentum to be halted cold turkey. But if we don't get a move on, those same forces will soon be working with a speed that will leave hardly any forests within just a few decades. No doubt about it.

There's also no doubt that we can still save vast amounts of forest. If we get our conservationist act together, we can still get on top of the problem before it gets on top of us. The resource in shortest supply is time. We must surely beat the problem into submission during the next decade at the latest; otherwise we might as well go to the beach until the last tree is chopped. In turn, this means we have less than three thousand days left—and we lose one percent of our maneuvering room every month.

Another key resource is available to us, and in infinite supply for those who want to exploit it. It is hope. Despite the gloom in my head, I feel a spark of hope remains alive elsewhere within me. Ten years ago I wrote there was little solid reason to hope, yet I knew I kept on hoping. Today, when many aspects have grown much worse, that spark continues to spark. I'm not sure why. Partly of course it is because many more people are now aware that remaining forests face terminal threat, and they have a good idea of what is at stake for us all. Partly too, I cling to the thought that humankind simply will not allow the pinnacle of nature to be turned into four-by-twos and chopsticks; nor do I buy the argument that we absolutely must clear the forests in order to raise more beef and grow more crops. We might as well accept that human welfare will be promoted by turning Hyde Park into a parking lot, or by dismantling the Eiffel Tower for railroad tracks, or by converting the Lincoln Monument into building blocks. So there must still be hope for tropical forests.

Yet that is different from saying there *is* still hope. Some days I climb out of bed and consider that yesterday another forest expanse equivalent to one hundred Central Parks was destroyed, and at least a similar expanse was grossly degraded. I am reminded of the person who defined an optimist and a pessimist, saying that an optimist is one who proclaims this is the best of all possible worlds, to which the pessimist responds that is probably true. Then I recall that when I first wrote this book there was only a single Rainforest Action Group in the United States, and today there are more than 170, and every ten days see the establishment of another one. So there is stacks of cause for us to be keeping on keeping on hoping.

Certainly there is a rising tide of public support for the forests. Certainly the politicians are being pushed into action. Certainly something has been done to limit the logging and the ranching. But in this last factor lies a danger, strange as it may sound. While we acclaim ourselves for tackling the two most obvious agents of deforestation, we are aiming at the less important targets. All too little is being done to broach the super-problem of the shifted cultivator, even though he causes more deforestation than the other two put together. To figure out why this pre-eminent issue is being neglected, let us take a closer look at what he is about, why he is in the forests at all.

THE SHIFTED CULTIVATOR: A FORGOTTEN FIGURE

Not only is the shifted cultivator now accounting for well over half of all deforestation, the proportion is expanding rapidly. So even if we call a complete halt to commercial logging and cattle ranching forthwith, we shall have fixed the lesser part of the problem. Moreover these two agents of deforestation can readily gain their livelihood without felling a single further tree in primary or mature forests. The logger should be required to take his harvest from secondary or disturbed forests, and from tree farms established on already deforested lands. If recent deforestation rates persist, he will shortly have no option anyway. Similarly the cattle rancher should be persuaded to engage in the modicum of management that would enable him to produce an increased flow of beef from established pasturelands, thus eliminating the urge to clear more forest for new ranches. But all this potential progress would be overtaken within just a few years by fast-spreading deforestation at the hands of the shifted cultivator—who presents a far less tractable problem.

The shifted cultivator is the peasant farmer who finds himself squeezed out of traditional farmlands in areas many horizons away from his country's forests. So he feels compelled to head for the only unoccupied lands available, the forests. He advances upon the forest fringe in ever-growing throngs, pushing deeper into the forest year by year. Behind him come more multitudes of displaced peasants, allowing the forest no chance to restore itself.[1]

There are various factors driving the shifted cultivator into the forests.[2] They include population growth, maldistribution of farmlands in established agricultural territories, and lack of government attention to subsistence agriculture (as opposed to Green Revolution agriculture) in order to help the impoverished peasant make more intensive and permanent use of his outside-the-forests smallholding. Worst of all, there is generally an inadequate policy emphasis on rural development overall, favoring cities over the countryside.

As a measure of how difficult it has been for policymakers to tackle the shifted-cultivator problem, note that the issue has been recognized for at least a decade, yet it has been largely ignored. Throughout the 1980s the deforestation focus has been almost entirely confined to the logger and the rancher. We do not even have a basic idea of how numerous the shifted cultivators have become; we have only broad estimates that range from 200 to 500 million people. If the latter estimate is correct, they account for almost one in ten of humankind. Yet the impoverished peasant remains a forgotten figure.

In short, the survival of tropical forests depends on a range of policy initiatives to help the subsistence farmer make an acceptable living in established farmlands way outside the forests. Nothing less will work to forestall his compulsion to head for the forests. So we need to think in terms of measures to promote efficient agriculture and agrarian reform (or fairer sharing-out of existing farmlands), plus a spectrum of support factors such as extension services, marketing networks, and credit facilities. In short, the shifted cultivator constitutes a challenge that can be confronted only by a seismic shift in development strategies by tropical-forest governments—all of which constitutes a far cry from current approaches to the deforestation problem. To certain conservationists, it may sound strange to invoke esoteric-seeming measures such as agrarian reform, credit facilities, and general rural development as a strategy to combat deforestation in lands far away from the scene of critical action. But however many horizons away they may be, are they any further removed than from the horizons of conventional conservation in our heads?

All in all, then, we need to broaden our understanding of what is needed to save remaining forests. Primarily we should recognize that at its heart it is no longer a forestry problem. While much can still be achieved through traditional forestry practices (also through more protected areas within forests, plus other "defensive" measures), these measures often do no more than tackle symptoms of deeper problems. In turn this means that the fate of tropical forests can no longer be left to foresters, splendid job though many of them have done for decades. Front and center, we must address the source problem of the shifted cultivator. Anything less is akin to building a fence around tropical forests (which would take an awful lot of timber)—a fence that would be speedily overrun by multitudes of land-hungry farmers.

A GLOBAL FORESTS CONVENTION?

Who will take on the seismic shift of strategy? Who can deploy the visionary approach right across the board, to turn a profound problem into a splendid opportunity? Since the problem is intrinsically international, affecting and affected by people in many lands, where shall we find an international agency that can measure up to the job?

The United Nations outfit with prime responsibility for tropical forestry is the Food and Agriculture Organization. But for all its professional skills and decades-long experience in the field, the Organization is hardly equal to the challenge. It has entirely failed to respond to the opportunity presented by the Tropical Forestry Action Plan.

There has been some talk of incorporating the tropical forestry issue into the Global Climate Convention that is being hammered out by the United Nations, through a Convention Protocol on World Forest Conservation. This makes some sense in light of the forests' role in planetary climate systems generally and in global-warming processes in particular. But there are several drawbacks. It is far from certain that the Climate Convention will be concluded within the next few years. A number of tropical-forest nations, notably Brazil, protest they would not want to support a forests protocol unless the main agents of global warming, the industrialized nations, commit themselves to doing much more to limit their fossil-fuel emissions of greenhouse gases. Most important of all, a protocol is essentially a statement of general intent. It would not offer mechanisms to get something done for tropical forests, especially as concerns the shifted cultivator. It would likely turn out to be no more than a pious expression of intent.

So the front-runner option now seems to be a Global Forests Convention, leading to a World Forests Organization backed by a Tropical Forests Fund. At the July 1990 Western leaders' conference in Houston, Texas, the communiqué declared, "We are ready to begin negotiations . . . on a global forests convention or agreement, which is needed to curb deforestation, protect biodiversity, stimulate positive forestry actions, and address threats to the world's forests." So the basic intent would be to reflect the common global interest in tropical forests (in all other forests too, though others are not so widely and immediately threatened as are tropical forests). This would not mean of course that tropical-forest nations would lose their sovereign rights to use their forest resources for their own purposes. But tropical forests are now seen to serve global purposes too, hence there is need to express this wider dimension through an initiative reflecting the "universal heritage" in tropical forests.

Fortunately there is now some consensus, among tropical-forest nations and other nations alike, on the nature and urgency of the issue. The stronger this consensus becomes, the better the prospect of articulating a forest-safeguard strategy as an authoritative goal of all nations. In response to this intent, many tropical-forest nations would be more ready to conform to collectively perceived needs and opportunities. Rather than feeling pressured in a manner they feel offends their sovereignty, they could find their prestige and political profile enhanced among the community of nations.

The global-convention concept is still being explored. As currently envisaged, it would promote the entire array of the forests' benefits for people everywhere and forever. It would supply a conceptual framework to deal with all aspects of the management, conservation, and development of the forests, including threats from agricultural encroachment. It would mobilize a critical mass of intersectoral expertise in all areas related to tropical forests; it would be much more than an assemblage of foresters. It would offer a forum for continuous contributions from nongovernmental bodies, especially those that promote environmental values and grassroots interests such as indigenous peoples.

Conversely, there is a marked disadvantage in that the convention would produce yet another international agency. Many observers protest we have enough of them already: why not spend the same effort on making the existing ones work better? But as we have seen, the main existing agency, the Food and Agriculture Organization, seems constitutionally unable to take on the job. In fact its operational makeup fosters precisely those sectoral divisions that have helped generate the deforestation crisis in the first place. The foresters assert that the source of the shifted-cultivator problem lies with the agricultural sector, so it's not their pigeon. The agriculturalists respond that when the shifted cultivator disappears into the forests, he ceases to be an agricultural issue, and it's up to the foresters to sort him out. Result: the problem falls through the famous institutional cracks—but doesn't become any the less of a super-problem.

WE'RE ALL IN IT TOGETHER

As we have seen, tropical forests are a concern for all of us, including those of who live far outside the tropical-forest countries. This is partly because we all contribute to setting up the problem and to blocking its resolution. For instance, we could all urge our governments to do a better job of sorting out the North-South issues that help precipitate the plight of the shifted cultivator. Partly too it is a problem for all of us because, as this book has repeatedly emphasized, we shall all lose through the demise of the forests. We're all in it together, and deeply so, whether we recognize it or not.

So that we can get a grasp of what is at stake for us, let's consider the prospect in ultimate big-picture terms. Supposing we—meaning the world at large—were to take the decisive step of offering to compensate tropical-forest countries for what they would lose were all remaining tropical forests to be declared effectively off limits to development. What would it cost?

For sure, the question is not altogether a true one. Were the forests to be "fenced off," that would not stop the shifted cultivator from breaking down the fences. Then there is the sovereignty factor: would Brazil, for instance, agree to international "intrusion" of this sort? Moreover, protected forests are not useless forests: they supply a host of valuable outputs through their mere existence, notably in the form of climate regulation. So the compensation idea is speculative for various reasons. But it serves to illuminate the responsibility of outside nations, which will surely have to make a hefty contribution to conservation of the forests if we are to have many forests left within another few decades. It also helps to give a sense of the scale of costs entailed in a worldwide support effort.

Fortunately, some preliminary arithmetic has been put together by one of our more creative economists, Professor David Pearce of the University of London. He has looked at the outlook if Brazilian Amazonia were to be relieved of most of the exploitation that has resulted in much deforestation during the 1980s.[3] He postulates that the annual "opportunity costs," or revenues forgone through forest preservation, would amount to around $7 billion a year. A parallel calculation by Dr. Martin Katzman of the Oak Ridge National Laboratory and Dr. William Cale of Indiana University in Pennsylvania,[4] employing a different form of analysis, concludes that a conservation easement for Brazilian Amazonia would cost around $6 billion a year. These assessments are no more than exploratory, and the totals could be off target quite a bit either way. But as ballpark estimates, they help to clarify the challenge to the outside world were we to undertake the effort.

Suppose too we were to engage in a "heroic extrapolation," and, considering that Brazilian Amazonia amounts to roughly one-third of all remaining tropical forests, extend the calculation to the entire forest realm. Of course the "ifs" and "buts" start to multiply apace, since there is much more logging in Africa and Asia, while less ranching. Local circumstances make the arithmetic an exceptionally dodgy exercise. But purely to help us get a handle on the scale of costs involved, and in the absence of any better calculation, it is worthwhile for present purposes. We should anticipate that to safeguard all tropical forests through a similar initiative could cost us in the rough region of $20 billion a year—perhaps rather less, perhaps rather more. But plainly not hundreds of billions of dollars.

Whatever the true figure, it should surely not prove beyond the means of the global community. If we are sincere in our proclaimed readiness to safeguard remaining forests, our reaction should not be, "How could we afford to do it?," rather it should be, "How can we afford not to do it?" Or, to put it another way, we should not only consider the costs of action but the concealed costs of inaction. Let's bear in mind, for instance, some of the eventual price we shall have to pay for global warming, to which deforestation already contributes 30 percent of the principal greenhouse gas, carbon dioxide. To rejig the United States' irrigation networks and hydropower installations in the wake of the greenhouse effect would, the Environmental Protection Agency has estimated, carry an eventual bill of $200 billion, and to safeguard just the eastern seaboard against sea-level rise would cost between $75 billion and $111 billion. So from the standpoint of the United States alone, never mind the many other outsider nations involved, a compensation plan of $20 billion a year should prove a splendidly sound investment.

Above all, the calculation highlights our responsibility in the tropical-forest crisis. To date we have been inclined to assert we would do anything to save the forests, yet with scant scope to do so in clear-cut terms. Here would be a full-scale opportunity: it would serve as a measure of the readiness of the world at large to pay for what it often says it wants. At long splendid last we would have a chance to put up or shut up—a response that has not hitherto been available with our limited opportunities to assist the forests.

Nor need the prospect be an all-or-nothing affair. We could start with trial efforts directed at those tropical-forest countries that wish to make common cause with the global community, before expanding the exercise to include those other countries that might at first hesitate until they see how things work out. With refinement through experience, the compensation idea could eventually operate as a kind of proxy "pricing system" to express out evaluation of tropical forests, supplying a framework that reflects costs and benefits as perceived by all participants.

For sure, there could be all manner of difficult negotiations before the idea becomes practice. But would that be any more difficult than trying to live in a world bereft of its tropical forests?

GRANDSCALE REFORESTATION

In addition we may soon find new hope, exceptionally large hope too, for tropical forests by virtue of their role in the global carbon cycle and hence their potential contribution to a solution for global warming. Just as a tree burning releases its carbon, so a tree growing absorbs carbon. Through photosynthesis it absorbs sizeable quantities from the atmosphere: a tree is half carbon. If we were to plant enough trees, we could soak up a good part of the carbon dioxide that is bringing on the greenhouse effect.[5]

The key question is, How many trees? That depends on how fast they grow. Using the fastest-growing species for tropical tree plantations, we can reckon that one hectare generates twenty metric tons of additional wood each year, hence it locks away ten tons of carbon. It keeps on doing this throughout its period of major growth, which can last thirty or forty years. The rate is way ahead of what can be achieved in the most favorable tree-growing parts of the United States such as Georgia and Louisiana, and it's several times better than the Pacific Northwest. Of course, the ten-tons figure is a rough and ready reckoning, with much variation according to different sectors of the tropics. But it is conservative too. In southern Brazil, the best types of eucalyptus plantations rack up a growth rate three and a half times as high. Moreover, and by courtesy of genetic engineering, we could soon have tree varieties that will sprout ultra-rapidly—a case of "plant your seedling and jump aside."

With these base-line figures, we can calculate that to soak up one billion tons of carbon we'd have to plant one million square kilometers of fast-growing trees. One million square kilometers is a lot of land; it is one and a half times the size of Texas. Fortunately there are sufficient lands available in the humid tropics, out of the eight million square kilometers already deforested and that have not been taken for permanent agriculture or other purposes. In any case, an area almost twice the size needs to be reforested to meet other needs, such as rehabilitation of watersheds. No doubt the initial expanse could be extended eventually, as the idea caught on and the multiple payoffs became plain. But let's stick with just one million square kilometers—enough to knock quite a hole in the carbon dioxide problem.

What would it cost? Suppose that land and labor are provided free, in light of the benefits that local people would gain from all those trees. To use a crude average figure again, we can reckon that a one-hectare plantation costs $400 to establish and maintain. So one million square kilometers would cost $40 billion. Again, we are dealing with super-size figures. But as we have seen above, these costs need to be compared with the concealed costs of inaction on the global-warming front.

In any case, the effort would presumably be spaced out over, say, ten years. This would mean an annual budget of $4 billion—or almost three times more than the most ambitious save-the-forests effort to date, the United Nations' Tropical Forestry Action Plan.

Who would pay? The main share should be stumped up by those nations, the rich nations, that have caused most of the greenhouse effect through their profligate burning of fossil fuels, *and* that could well suffer unduly through global warming. Nobody has worked out the proportional figures for each nation yet, since this would require hard-nosed negotiations and the most delicate diplomacy. In any case, the tree-planting idea is still an idea. It is a carefully analyzed idea, with lots of supporting evidence and detailed documentation. But these fine concepts are not turned into workable programs overnight. All we can say right now about "Who pays what?" is that the United States should surely undertake a hefty share of the bill, since it is the world's most profligate consumer of fossil fuels. Per person, Americans contribute more to the greenhouse effect than any other nation, and five times more than the global average. The United States and Canada combined account for more than one quarter of the greenhouse effect worldwide.

Who would actually do the job? There is a growing awareness

that we must find who could best undertake the work at local level—that is, who would actually stick the seedlings in the ground and make sure they were protected during their early years of growth. While we face a single problem, we should not seek a single solution. We would need thousands of mini-solutions, each tailored to local circumstances. It would be a strategic error to envisage a small number of horizon-to-horizon plantations of eucalyptus, pines, and a handful of other species. Rather we should plan on a smorgasbord of possibilities—village woodlots, farm plantings, agroforestry blocks, urban forests—each utilizing a broad variety of trees, whatever meets the needs and capacities of local communities. To a central-government planner it might seem a "messy" response. All the more, then, the best organizations for the job will be the ones that delight in on-the-spot activity, messiness included.

To indicate what can be achieved by grass-rooters, consider the case of Haiti. Until recently, the country's forests have been reckoned a matter of history. Deforestation has been causing widespread soil erosion and loss of watershed services, on top of massive shortages of forest products such as construction timber and fuelwood. It has also become plain that, if over-cutting persists, fuelwood supplies will be finished by the mid-1990s. This "impossible" situation has been largely ignored by the Haitian government, which has launched only a handful of reforestation projects.

Enter an alliance of nongovernmental organizations, made up of American bodies such as CARE, Oxfam USA, and Catholic Relief Services, working in conjunction with a full 200 counterpart groups in Haiti. During the second half of the 1980s, they have enabled 135,000 farmers to plant 35 million trees, for an average of 270 trees per smallholding. The key to the project's success? From-the-start involvement of local people in every aspect of operations.

Whatever the success of on-the-ground activists, it is governments that will have to mobilize the funding. Fortunately the tree-planting idea seems to be catching on as a promising measure to tackle the greenhouse effect. A United Nations outfit, the Intergovernmental Panel on Climate Change, has recommended that we plant 2.4 million square kilometers of trees in various parts of the world over a period of twenty years as a counter-greenhouse measure. It is this sort of thinking too that partly lies behind President Bush's proposal to plant one billion trees a year in America until the year 2000, and behind Prime Minister Robert Hawke's plan to plant one billion trees across Australia during the next three years. But the campaign we're talking about would entail several hundred billion trees.

Of course, tree planting can be no more than a stopgap solution to the greenhouse problem (and only part of the problem at that). When the trees reach maturity and stop growing, they halt their carbon-absorbing service. But the effort would gain us several decades of breathing space while we devise better ways to meet our energy needs than by burning carbon-producing fuels.

By the same token, of course, we would have to cut back on carbon-producing deforestation. All our best efforts to date have failed. The tree-planting strategy may work the trick. After all, it offers megabucks to countries that want to engage in massive reforestation—and thus supply incentive to get to final grips with deforestation. Could we one day look back and find that one problem, runaway deforestation, has been combined with another problem, the greenhouse effect, to produce a startling solution?

SIGNING OFF AND SIGNING ON

So as we come to the end of this book, we could be witnessing a new start for tropical forests. Especially will this be the case if each one of us remembers our role. I often have cause to think of Edmund Burke's remark that I put on the frontispiece: "Nobody made a greater mistake than he who did nothing because he could do only a little." Friends have reminded me of another remark, this one by Margaret Mead, equally pertinent and more recent: "Never doubt that a small group of thoughtful committed citizens can change the world; indeed it's the only thing that ever has."

So this is a time of signing on. Our task will not be easy, nor will it be short. But it will be a great endeavor, one of the most exhilarating we could envisage. True, there have been many times during the past twenty years when I have felt the problem has been growing more problematical than ever, even that it seems to be getting away from us. On those occasions I have reminded myself that far from being disheartened, I should feel exhilarated by the very essence of what we are taking on. After all, we are trying to save the greatest celebration of nature that has ever graced the face of the planet. So should we not feel privileged that we, alone of all human generations, should face this challenge? No past generation has ever had the chance we enjoy: the problem was simply not there. No future generation will have our chance either: if we don't convert this profound problem into a glorious opportunity, there will be nothing left for our successors to do but pick up the pieces. It is in our hands. And as we measure up to the challenge, won't it cause us to feel ten feet tall? In the eyes of future generations, won't we have made ourselves giants of the human condition?

Those tropical forests are waiting to hear from us. Let's give them a message.

APPENDIX

REFERENCES

INDEX

Appendix

COUNTRY-BY-COUNTRY REVIEW OF CONVERSION RATES IN TROPICAL FORESTS

Areas Undergoing Broad-Scale Conversion at Rapid Rates

Australia. Most lowland tropical forests, both rainforests and seasonal forests, due to timber exploitation and planned agriculture; could be little left by 1990 if not earlier.

Bangladesh. Most forests, both lowland and upland, predominantly rainforests, due to timber exploitation, forest farming, and population pressure; could be little left by 1990 if not earlier.

India. Much of the forests, predominantly seasonal forests, mainly upland, due to forest farming and population pressure; could be little left by 1990.

Indonesia. Much if not most of the lowland forest, predominantly rainforests, due to timber exploitation, forest farming, and transmigration programs; could be little left in Sumatra and Sulawesi by 1990, in Kalimantan and most of the smaller islands by 1995, and in Irian Jaya by the year 2000.

Malaysia. Most of the Peninsula's lowland forests, almost all rainforests, due to timber exploitation and planned agriculture; could be little left by 1990 if not earlier. Much of Sarawak and Sabah lowland forests, almost all rainforests, due to timber exploitation; could be little left by the year 2000 if not earlier.

Melanesia. Much if not most of the lowland forests, due to timber exploitation and planned agriculture; could be little left by 1990.

Philippines. Most lowland forests, predominantly rainforests, due to timber exploitation and forest farming; could be little left by 1990 if not earlier.

Sri Lanka. Much if not most of the forests, predominantly rainforests, mostly upland, due to timber exploitation and forest farming; could be little left by 1990.

Thailand. Much if not most of the forests, almost all seasonal forests, both lowland and upland, due to timber exploitation (especially illegal felling) and forest farming; could be little left by 1990 if not earlier.

Vietnam. Much of the forests, almost all seasonal forests, both lowland and upland, especially in the south, due to forest farming, timber exploitation, and immigration from the north; could be little left by 1990.

Brazil. Parts of the southern and eastern sectors of Amazonia, lowland rainforests, notably in Rondônia, Pará, and Mato Grosso; due to cattle raising and colonist settlement; appreciable tracts could be converted by 1990 and much more by the year 2000. Most if not virtually all of this country's Atlantic coast strip of moist forest, due to timber exploitation and cash-crop agriculture, notably sugarcane plantations; could be little left by 1990 if not a good deal earlier.

Central America. Much if not most of the forests, notably rainforests, both lowland and upland, due to forest farming, cattle raising, and timber exploitation; could be little left by 1990 if not earlier.

Colombia. Parts of the lowland rainforests on the borders of

Amazonia, especially in Caqueta and Putumayo, due to colonist settlement and cattle raising; extensive tracts could be converted by 1990, and much more by the year 2000.

Ecuador. Much of this country's Pacific coast forests, mostly very wet and very rich rainforests, both lowland and upland, due to plantation agriculture and some timber exploitation; could be almost entirely converted by 1990.

Madagascar. Much if not most of the forests, especially rainforests, both lowland and upland, due to forest farming and timber exploitation; could be little left by 1990 if not earlier.

East Africa. Much if not most of East Africa's relict montane forests, especially in northern Tanzania, mostly seasonal forests, due to timber exploitation, firewood cutting, and forest farming; could be little left by 1990.

West Africa. Much if not most of West Africa's forests, mainly seasonal forests, due to timber exploitation and forest farming; could be little left by 1990 if not earlier.

Areas Undergoing Moderate Conversion at Intermediate Rates

These areas cannot be so readily listed as those under *Areas Undergoing Broad-Scale Conversion at Rapid Rates,* since less is known about their present status and future prospects. The listing is deliberately conservative.

Burma. Parts of the lowland forests, almost all seasonal, due to forest farming and some timber exploitation; appreciable areas could be converted by the year 2000 if not earlier.

Papua New Guinea. Parts of the forests, mostly seasonal, both lowland and upland, due to timber exploitation and forest farming; extensive areas could be converted by the year 2000 if not earlier.

Brazil. Parts of Brazil's Amazonia forests, lowland rainforests, notably in Amapa, Acre, sections of the TransAmazonia Highway system, and areas selected for timber exploitation, e.g., Tapajos River area; due to colonist settlement, forest farming, cattle raising, and timber exploitation; appreciable tracts could be converted by 1990.

Colombia. Parts of Colombia's Pacific coast forests, very wet and very rich rainforests, both lowland and upland, due to timber exploitation; extensive sectors could become converted by 1990.

Ecuador. Much of Ecuador's Amazonia forests, almost all rainforests, both lowland and upland, due to colonist settlement, forest farming, some planned agriculture, and also oil exploitation; appreciable areas could be converted by 1990, and much more by the year 2000.

Peru. Much of Peru's Amazonia forests, almost all rainforests, both lowland and upland, due to colonist settlement, forest farming, and some planned agriculture; appreciable areas could be converted by 1990, and much more by the year 2000.

Cameroon. Parts of the forests, both seasonal and rainforests, both lowland and upland, due to timber exploitation and forest farming; extensive areas could be converted by 1990, and much more by the year 2000.

Areas Apparently Undergoing Little Change

Like *Areas Undergoing Moderate Conversion,* these areas cannot be so readily listed as those under *Areas Rapidly Undergoing Broad-Scale Conversion,* since less is known about their present status and future prospects. The listing is deliberately conservative, especially as concerns the long term.

Brazil. Much of Brazil's western Amazonia, lowland rainforests, generally wetter and richer than eastern Amazonia; except for some timber extraction in limited areas (and some cultivation of *varzea* floodplains), exploitation of this huge zone could prove difficult in view of its unusually wet climate and distance from markets; it is reasonable to anticipate—so far as can be ascertained, and the point is stressed—that much of this vast tract of lowland rainforest could remain little changed for some time, probably until the year 2000, even for a time thereafter.

French Guiana, Guyana, and Surinam. Much of the forests

of these countries, almost all rainforests, both lowland and upland; timber exploitation, at present very limited, may expand, but because population pressures are low, there is little likelihood of widespread colonist settlement and forest farming. So it is reasonable to anticipate—with the caveat as applied to Brazil, above—that large areas may remain little changed for a good while to come, probably until the year 2000, conceivably for a time thereafter.

Zaire Basin. Much of the Zaire basin, comprising Congo, Gabon, and Zaire; some rainforest in Gabon, remainder mainly seasonal, almost entirely lowland; population pressures are low, and there are abundant mineral resources on which to base national economic development; timber exploitation, primarily limited to northern Congo and to Gabon, could expand; but in the main, it is reasonable to anticipate—with caveat as applied to Brazil, above—that large areas may remain little changed for a good while to come, probably until the year 2000, even for a time thereafter.

In sum, the situation is highly differentiated, between and within the three main regions, as well as within some individual countries.

With respect to future trends, it seems clear that human population growth and economic aspirations, already factors of importance, will exert progressive pressures on tropical forests, until their impact by the year 2000 could represent a profound change from the present position. At the same time, it is necessary to bear in mind that exploitation patterns can alter. It would be a mistake to suppose that the future will amount to a simple extrapolation of the present situation. Some exploitation trends could reveal a geometric rather than an arithmetic progression, while others could decline in significance or even fade away as they are supplanted by innovative forms of exploitation.

References

Prologue to the 1992 Edition

1. J.-P. Malingreau and C.J. Tucker, 1988, Large-Scale Deforestation in the Southern Amazon Basin of Brazil, *Ambio* 17:49–55.
2. N. Myers, 1990, Mass Extinctions: What Can the Past Tell Us About the Present and the Future?, *Palaeoecology, Palaeoclimatology, Palaeoecology (Global and Planetary Change Section)* 82:175–85; P.H. Raven, 1987, *We're Killing Our World: The Global Ecosystem in Crisis*, The MacArthur Foundation, Chicago, Illinois; E.O. Wilson, 1989, Threats to Biodiversity, *Scientific American* 261:108–16.
3. D.J. Greathead, 1983, The Multi-Million Dollar Weevil That Pollinates Oil Palm, *Antenna* (Royal Entomological Society of London) 7:105–7.
4. N. Myers, 1988, Threatened Biotas: "Hot Spots" in Tropical Forests, *The Environmentalist* 8:187–208; N. Myers, 1990, The Biodiversity Challenge: Expanded Hot-Spots Analysis, *The Environmentalist* 10:243–56.
5. N. Myers, 1989, Tropical Deforestation and Climatic Change, *Environmental Conservation* 15:293–98.
6. R.A. Houghton, 1990, The Future Role of Tropical Forests in Affecting the Carbon Dioxide Concentration of the Atmosphere, *Ambio* 19:204–9.
7. N. Myers, 1986, The Environmental Dimension to Security Issues, *The Environmentalist* 6:251–57.
8. *Ibid.*
9. N. Myers, 1989, Environmental Security: The Case of South Asia, *International Environmental Affairs* 1:138–54.
10. R. Repetto and M. Gillis, 1988, *Public Policies and the Misuse of Forest Resources*, Cambridge University Press, Cambridge, U.K.
11. C.M. Peters, A.H. Gentry, and R.O. Mendelsohn, 1989, Valuation of an Amazonian Rain Forest, *Nature* 339: 655–56.

12. S. Schwartzman, 1989, Extractive Reserves: The Rubber Tappers' Strategy for Sustainable Use of the Amazon Rainforest, in J.O. Browder, editor, *Fragile Lands of Latin America: Strategies for Sustainable Development:* 150–65, Westview Press, Boulder, Colorado.

1. Nature's Powerhouse

1. P. S. Ashton, 1964, Ecological Studies in the Mixed Dipterocarp Forest of Brunei State, *Oxford Forestry Memoirs* 25:1–75; and 1981, Forest Conditions in the Tropics of Asia and the Far East, *Studies in Third World Societies* 13: 169–79. For general information on dipterocarp forests of Indonesia and Southeast Asia, see M. Jacobs and T. J. J. de Boo, 1980, *Conservation Literature on Indonesia: Selected Annotated Bibliography*, Rijksherbarium, Leiden, Netherlands.
2. G. T. Prance, 1977, The Phytogeographic Subdivisions of Amazonia and Their Influence on the Selection of Biological Reserves, in G. T. Prance and T. S. Elias, editors, *Extinction is Forever:* 195–213, New York Botanical Garden, Bronx, New York.
3. D. H. Janzen, 1977, Promising Directions in Tropical Animal-Plant Interactions, *Annals of Missouri Botanical Garden* 64:706–36; see also same author, 1975, *Ecology of Plants in the Tropics*, Edward Arnold Publishers Ltd., London, U.K.
4. N. M. Collins, 1983, Rain Forests of Gunung Mulu, in S. M. Wells, R. M. Pyle, and N. M. Collins, editors, *The IUCN Invertebrate Red Data Book:* 567–76, International Union for Conservation of Nature and Natural Resources, Gland, Switzerland.

2. Forests and Forests

1. A. H. Gentry, 1977, Endangered Plant Species and Habitats of Ecuador and Amazonian Peru, in G. T. Prance and T. S. Elias, editors, *Extinction is Forever:* 136–49, New York Botanical Garden, Bronx, New York.
2. C. G. van Steenis, 1950, The Delimitation of Malaysia and its Main Plant Geographical Divisions, *Flora Malesiana* Ser. I (1):70–75; see also M. J. Jacobs and T. J. J. de Boo, 1980, reference 1 of chapter 2.
3. N. Myers, 1980, *Conversion of Tropical Moist Forests*, National Research Council, Washington D. C. (a scientific report listing some 400 references that deal with tropical forests and forestry generally, covering biology and ecology, exploitation agents, and remote-sensing surveys); see also UNESCO, 1978, *Tropical Forest Ecosystems*, UNESCO, Paris, France; F. B. Golley, editor, 1983, *Tropical Rain Forest Ecosystems*, Elsevier Scientific Publishing Company, New York; and S.L. Sutton, T.C. Whitmore and A.C. Chadwick, editors, 1983, *Tropical Rain Forests: Ecology and Management*, Blackwell Scientific Publications, Oxford, U.K.
4. See references in Myers, above; also P. W. Richards, 1952, *The Tropical Rain Forest*, Cambridge University Press, Cambridge, U. K., and T. C. Whitmore, 1984, *Tropical Rain Forests of the Far East*, second edition, revised, Clarendon Press, Oxford, U. K.
5. P. A. Opler, H. G. Baker, and G. W. Frankie, 1977, Recovery of Tropical Lowland Forest Ecosystems, in J. Cairns and K. Dickson, editors, *Recovery and Management of Damaged Ecosystems:* 379–421, University of Virginia Press, Charlottesville, Virginia; see also P. S. Ashton, cited under reference 1, chapter 2, and L. E. Gilbert, 1980, Food Web Organization and Conservation of Neotropical Diversity, in M. E. Soule and B. A. Wilcox, editors, *Conservation Biology:* 11–34, Sinauer Associates, Sunderland, Massachusetts.

6. M. Goulding, 1980, *The Fishes and the Forest,* University of California Press, Berkeley, California.
7. Asian Development Bank, 1980, *Sector Paper on Forestry and Forest Industry,* Asian Development Bank, Manila, Philippines.

3. Biological Richness

1. R. d'O. Good, 1974, *The Geography of the Flowering Plants,* Longman, London, U. K.; A. H. Gentry, 1982, "Patterns of Neo-Tropical Plant Species Diversity," in M. K. Hecht, B. Wallace, and G. T. Prince, eds., *Evolutionary Biology* 15:1–84, Plenum Press, New York; V. H. Heywood, 1978, *Flowering Plants of the World,* Oxford University Press, Oxford, U. K.; and Perspectives in Tropical Botany, *Annals of the Missouri Botanical Garden* 64 (4):1977.
2. A. H. Gentry, 1979, Extinction and Conservation of Plant Species in Tropical America: A Phytogeographical Perspective, in I. Hedberg, editor, *Systematic Botany: Plant Utilization and Biosphere Conservation* : 115–21, Almqvist and Wiksell, Uppsala, Sweden; A. H. Gentry, 1982, Patterns of Neotropical Plant Species Diversity, in M. K. Hecht, B. Wallace, and G. T. Prance, editors, *Evolutionary Biology,* vol. 15:1–84; and C. H. Dodson and A. H. Gentry, 1978, Flora of the Rio Palenque Science Center, *Selbyana:* 4(1–6).
3. A. H. Gentry, ibid.
4. See references for P. S. Ashton, under reference 1 for chapter 1 above; see also M. J. Jacobs and T. J. J. de Boo, 1980, reference 1, chapter 1.
5. A. C. Diamond, 1983, *Conservation of Tropical Forest Birds,* Technical Publication of International Council for Protection of Birds, Cambridge, U. K.
6. A. Keast and E. S. Morton, 1980, *Migrant Birds in the Neotropics: Ecology, Behavior, Distribution and Conservation,* Smithsonian Institution Press, Washington D. C.
7. R. A. Mittermier, H. de Macedo Ruiz, and L. Luscombe, 1975, A Woolly Monkey Rediscovered in Peru, *Oryx* 13:41–46.
8. D. F. Owen, 1971, *Tropical Butterflies,* Clarendon Press, Oxford, U. K.
9. J. A. R. Anderson, A. C. Jermy and the Earl of Cranbrook, 1982, *Gunung Mulu National Park: A Management and Development Plan,* Royal Geographical Society, London, U. K.
10. T. L. Erwin, 1982, Tropical Forests: Their Richness in Coleoptera and Other Arthropod Species, *Coleopterists Bulletin,* 36:74–75.
11. A. H. Gentry, 1979, reference 2 above.
12. J. Steyermark, 1977, Future Outlook for Threatened and Endangered Species in Venezuela, in G. T. Prance and T. S. Elias, editors, *Extinction is Forever:* 128–35, New York Botanical Garden, Bronx, New York.
13. A. H. Gentry, 1982, reference 2 above.
14. P. H. Raven, 1980, *Research Priorities in Tropical Biology,* National Research Council, Washington D. C.

4. Ecological Complexity

1. P. S. Ashton, 1977, The Contribution of Rain Forest Research to Evolutionary Theory, *Annals of Missouri Botanical Garden* 64(4):694–705; H. G. Baker, 1970, Evolution in the Tropics, *Biotropica* 2 (2):101–11; J. R. Flenley, 1979, *The Equatorial Rain Forests: A Geological History,* Butterworths, Boston, Mass.; D. H. Janzen, 1975, *Ecology of Plants in the Tropics,* Edward Arnold Publishers Ltd., London, U. K.; R. H. Lowe-McConnel, editor, 1969, *Speciation in Tropical Environments,* Academic Press, New York; R. M. May, 1975, The Tropical Rainforest, *Nature* 257:737–38; and P. H. Raven, 1980, reference 14 for Chapter 3. For much generalized information

on the ecology of tropical forests, see National Research Council, 1982, *Ecological Aspects of Development in the Humid Tropics*, National Academy Press, Washington D.C.; and for a first-rate review of forest ecology in a salient sector of the biome, that is, Indonesia, see M. J. Jacobs, see T. J. J. de Boo, 1980, reference 1, Chapter 1.

2. See a whole series of papers on this topic in G. T. Prance, editor, 1982, *Biological Diversification in the Tropics*, Colombia University Press, New York.

3. A. Whitten, 1982, *The Gibbons of Siberut*, J. M. Dent, London, U. K.

4. F. Smiet, 1982, Threats to the Spice Islands, *Oryx* 16(4):323–28.

5. P. A. Sanchez, 1976, *Properties and Management of Soils in the Tropics*, John Wiley, New York; P. A. Sanchez, D. E. Bandy, J. H. Villachica, and J. J. Nicholaides, 1982, Soils of the Amazon Basin and Their Management for Continuous Crop Production, *Science*, 216:821–27; T. T. Cochrane and P. A. Sanchez, 1980, Land Resources, Soil Properties and Their Management in the Amazon Region: A State of Knowledge Report, in S. B. Hecht, editor, *Land-Use and Agricultural Research in the Amazon Basin*, CIAT, Cali, Colombia.

6. J. Posner, 1981, *The Densely Populated Steep Slopes of Tropical America: Profile of a Fragile Environment*, Report to the Rockefeller Foundation, New York.

7. F. B. Golley, 1973, Nutrient Cycling and Nutrient Conservation, in F. B. Golley, editor, *Tropical Rain Forest Ecosystems:* 137–56, Elsevier Scientific Publishing Company, New York; R. Herrera, C. F. Jordan, H. Klinge, and E. Medina, 1978, Amazon Ecosystems: Their Structure and Functioning, with Particular Emphasis on Nutrients, *Interciencia* 3(4):223–32; D. H. Janzen, 1973, Tropical Agroecosystems, *Science* 182:1212–19; C. F. Jordan, 1982, Amazon Rain Forests, *American Scientist* 70:394–410; C. F. Jordan and R. Herrera, 1981, Tropical Rain Forests: Are Nutrients Really Critical? *American Naturalist* 117(2):167–80; N. Stark and C. F. Jordan, 1978, Nutrient Retention by the Root Mat of an Amazonian Rainforest, *Ecology* 59:434–37; and C. Uhl, 1983, You Can Keep a Good Forest Down, *Natural History* 92(4):70–79.

8. M. Falkenmark and G. Lindh, 1976, *Water for a Starving World*, Westview Press, Boulder, Colorado.

9. See N. Stark and C. F. Jordan, 1978, reference 7 above.

10. See Herrera *et al.*, 1978, reference 7 above; D. P. Janos, 1980, Vesicular-Arbuscular Mycorrhizae Affect Lowland Tropical Rainforest Plant Growth, *Ecology* 61(1):151–62; D. W. Malloch, K. A. Pirozynski, and P. H. Raven, 1980, Ecological and Evolutionary Significance of Mycorrhizal Symbioses in Vascular Plants (A Review), *Proceedings of National Academy of Sciences* 77(4):2113–18; F. S. P. Ng, 1977, Gregarious Flowering of Dipterocarps in Kepong 1976, *Malaysian Forester* 40:126–37.

11. C. F. Jordan, F. B. Golley, J. D. Hall, and J. Hall, 1980, Nutrient Scavenging of Rainfall by the Canopy of an Amazonian Rain Forest, *Biotropica* 12(1):61–66.

12. E. F. Brünig, 1977, The Tropical Rain Forests—A Wasted Asset or an Essential Biospheric Resource? *Ambio* 6(4):187–91.

13. P. S. Ashton, 1981, reference 1, chapter 1, above; also J. Burley and B. T. Styles, editors, 1976, *Tropical Trees: Variation, Breeding and Conservation*, Academic Press, London, U. K.

14. G. S. Hartshorn and G. H. Orians, 1978, *The Influence of Gaps in Tropical Forests on Tree Species' Richness*, Tropical Science Center, San Jose, Costa Rica; G. H. Orians, 1983, The Influence of Tree Falls in Tropical Forests on Tree Species Richness, *Tropical Ecology* 23:255–79.

15. P. S. Ashton, 1980, Techniques for the Identification and Conservation of Threatened Species in Tropical Forests, in H. Synge, editor, 1981, *The Biological Aspects of Rare Plant Conservation:* 155–64. John Wiley, London, U. K.

16. M. Jacobs, 1982, The Study of Minor Forest Products, *Flora Malesiana Bulletin* 35:3768–82.

17. L. E. Gilbert, 1980, reference 5, chapter 2, above; S. T. A. Pickett and J. N. Thompson, 1978, Patch Dynamics and Design of Nature Reserves, *Biological Conser-*

vation 13:27–37; and J. Roughgarden, 1979, *Theory of Population Genetics and Evolutionary Ecology: An Introduction*, McMillan, New York.

18. A. N. Start and A. G. Marshall, 1976, Nectarivorous Bats as Pollinators of Trees in West Malaysia, in J. Burley and B. T. Styles, editors, *Tropical Trees: Variation, Breeding and Conservation*: 141–50, Academic Press, London, U. K.

19. D. H. Janzen, 1979, How to be a Fig, *Annual Review of Ecology and Systematics* 10:13–51; B. W. Ramirez, 1970, Host Specificity of Fig Wasps, *Evolution* 24: 680–91, and 1974, Coevolution of *Ficus and Agaonidae*, *Annals of Missouri Botanical Garden*, 61:770–80

20. L. E. Gilbert, 1980, reference 5, chapter 2, above.

21. L. E. Gilbert and P. H. Raven, editors, 1975, *Coevolution of Animals and Plants*, University of Texas Press, Austin, Texas.

22. C. W. Elton, 1973, The Structure of Invertebrate Populations Inside Tropical Rainforest, *Journal of Animal Ecology* 42:55–104; and L. E. Gilbert, 1980, reference 5, chapter 2, above; also D. H. Janzen, 1975, reference 1, above.

5. The Commercial Logger

1. These pages, and this chapter generally, draw from a number of background documents, notably Food and Agriculture Organization, 1983, *Forest Products Yearbook*, Food and Agriculture Organization, Rome, Italy; N. Myers, 1980, reference 3, chapter 2; and U.S. Forest Service, 1982, *An Analysis of the Timber Situation in the United States 1952–2030*, U. S. Forest Service, Washington D. C.

2. Economic and Social Commission for Asia and the Pacific, 1981, *Transnational Corporations and the Tropical Hardwood Industry of Indonesia*, Economic and Social Commission for Asia and the Pacific, Bangkok, Thailand; G. Sacerdoti, 1979, Indonesia at the Crossroads, *Far Eastern Economic Review* 106(48):63–66; and K. Takeuchi, 1979, *Market Prospects for Tropical Hardwood*, Commodities and Export Projections Division, The World Bank, Washington D. C.

3. L. Daryadi, 1980, The Management and Conservation of Dipterocarps in Indonesia, in *Proceedings of Roundtable on Dipterocarps*, Forest Research Institute, Kepong, Malaysia.

4. C. W. Clark, 1977, *Mathematical Bioeconomics: The Optimal Management of Renewable Resources*, John Wiley and Sons, New York; and N. Myers, 1977, Discounting and Depletion: The Case of Tropical Moist Forests, *Futures* 9(6):502–9.

5. K. Imamura, 1982, Human Resources in Japanese Forestry, *Unasylva* 34(135); and M. Matsui, 1980 Japan's Forest Resources, *Unasylva* 32(128):19–20.

6. R. B. Phelps, 1981, *Outlook for Timber Products*, USDA Forest Service, Washington D.C.; and U.S. Forest Service, 1982, reference 1, above.

7. S. H. Spurr, 1979, Silviculture, *Scientific American* 240(1):76–91.

8. Food and Agriculture Organization, 1983, reference 1, above; and U.S. Forest Service, 1982, reference 1, above.

9. E. Shimokawa, 1977, Japan's Dependence Upon Wood Chips Pulp, *Unasylva* 29:26–27; see also K. Imamura, 1982, and M. Matsui, 1980, reference 5, above.

10. M. Chudnoff, 1980, *Tropical Timbers of the World*, U. S. Government Printing Office, Washington D. C.; and S. D. Richardson, 1978, Foresters and the Faustian Bargain, in *Proceedings of Conference on Improved Utilization of Tropical Forests*: 151–61 U. S. Forest Service, Forest Products Laboratory, Madison, Wisconsin.

11. E. O. Chijicke, 1980, *Impact on Soils of Fast-Growing Species in Lowland Humid Tropics*, Food and Agriculture Organization, Rome, Italy; J. Ewel and L. Conde, 1978, Environmental Implications of Any-Species Utilization in the Moist Tropics, in *Proceedings of Conference on Improved Utilization of Tropical Forests*: 106–23, Forest Products Laboratory, U. S. Forest Service, Madison, Wisconsin.

12. R. Abdulhadi, K. Kartawinata, and S. Sukardjo, 1981, Effects of Mechanized Logging in the Lowland Dipterocarp Forest at Lempake, East Kalimantan, *Malaysian Forester* 44 (2 and 3):407–18; J. Ewel and L. F. Conde, 1980, *Potential Ecological Impact of Increased Intensity of Tropical Forest Utilization*, BIOTROP (Regional Centre For Tropical Biology), Bogor, Indonesia; A. D. Johns, 1983, *Effects of Selective Logging on a West Malaysian Hill Forest Community*, unpublished Ph.D. Thesis, University of Cambridge, U. K.; K. Kartawinata, S. Adisoemarto, S. Riswan, and A. P. Vayda, 1981, The Impact of Man on a Tropical Forest in Indonesia, *Ambio* 10(2–3):115–19; D. I. Nicholson, 1979, *The Effects of Logging and Treatment on the Mixed Dipterocarp Forests of Southeast Asia*, Food and Agriculture Organization, Rome, Italy; R. S. Suparto et al., 1978, *Proceedings of Symposium on the Long-Term Effects of Logging in South East Asia*, BIOTROP, Bogor, Indonesia; U. Tinal and J. L. Palenewan, 1974, *A Study of Mechanical Logging Damage After Selective Cutting in the Lowland Dipterocarp Forest of East Kalimantan*, BIOTROP, Bogor Indonesia.
13. C. Chandrasekharan, 1980, *Forests and Forestry in Thailand*, Regional Office for Asia and the Pacific, Food and Agriculture Organization, Bankok, Thailand.
14. H. Olofson, editor, 1982, *Adaptive Strategies and Change in Philippine Swidden-Based Societies*, Forestry Organization Research Institute, Laguna, Philippines.
15. J. Ewel, editor, 1980, Tropical Succession, *Biotropica, Supplement* to vol. 12 (2); and A. Sasson, 1980, Development of Forest Resources in Tropical Regions, *Impact of Science on Society* 30(3):211–16.
16. N. Johnson and G. Dykstra, 1978, *Multinational Timber Corporations and Forest Harvesting in East Kalimantan*, Weyerhaeuser Corporation, Jakarta, Indonesia, and Tacoma, Washington.
17. H. N. Gregersen and A. Contreras, 1975, *U.S. Investment in the Forest-Based Sector in Latin America*, Johns Hopkins University Press, Baltimore, Maryland.
18. M. Gillis, 1980, *Fiscal and Financial Issues in Tropical Hardwood Concessions*, Harvard Institute for International Development, Harvard University, Cambridge, Mass.
19. R. C. Stadelman, 1979, *The United States Market for Tropical Hardwoods*, Food and Agriculture Organization, Rome, Italy.

6. The Fuelwood Gatherer

1. J. E. M. Arnold, 1979, Wood Energy and Rural Communities, *Natural Resources Forum* 3:229–52; J. E. M. Arnold and J. Jongma, 1978, Fuelwood and Charcoal in Developing Countries, *Unasylva* 29(118):2–9; M. R. Montalembert, 1981, *Fuelwood in Asia: An Identification of Critical Situations*, Food and Agriculture Organization, Rome, Italy; R. P. Moss and W. B. Morgan, 1982, *Fuelwood and Rural Energy Production and Supply in the Humid Tropics*, Tycooly Press, Dublin, Republic of Ireland; K. Openshaw, 1978, Woodfuel—A Time for Re-Assessment, *Natural Resources Forum* 3:35–51; J. S. Spears, 1978, *Wood as an Energy Source: The Situation in the Developing World*, Forestry Division, The World Bank, Washington D. C.; and United Nations, 1981, *Proceedings of Conference on New and Renewable Sources of Energy*, Conference Secretariat, United Nations, New York.
2. B. Lekagul and J. McNeely, 1978, Thailand Launches Extensive Reforestation Program, *Tigerpaper* 5(1):9–13.
3. M. M. Pant, 1980, The Impact of Social Forestry on the National Economy of India, *The International Tree Crops Journal*, 1:69–92; and J. S. Spears, 1978, reference 1, above.
4. Volunteers in Technical Assistance, 1979, *Woodstove Design Manual*, Volunteers in Technical Assistance, Mount Rainier, Maryland.
5. J. Evans, 1982, *Plantation Forestry in the Tropics*, Clarendon Press, Oxford, U. K.; M. M. Pant, 1979, Social Forestry in India, *Unasylva* 31(125):19–26; J. Romm, 1982,

A Research Agenda for Social Forestry, *The International Tree Crops Journal* 2:25–59; and World Bank, 1981, *Fuelwood Needs and Plantation Forestry in the Developing World*, World Bank, Washington D. C.

6. National Academy of Sciences, 1980, *Firewood Crops: Shrub and Tree Species for Energy Production*, National Research Council, Washington D. C.

7. H. Chouvin, 1980, *Comparative Evaluation of the Possibilities for Supplying Fuelwood Deficit Areas from Distant Available Resources*, Proceedings of United Nations Conference on New and Renewable Sources of Energy, United Nations, New York; J. Fritz, J. J. Gordon and V. T. Nguyen, 1979, *Status Review of Wood Biomass*, The MITRE Corporation, McLean, Virginia; and J. Janczak, 1981, Simple Techniques for Basic Bio-Fuels, *Unasylva*, 33(131):30–36.

7. The Cattle Raiser

1. D. B. Agnew, 1979, The Outlook for Hamburger, *Livestock and Meat Situation* (publication of U.S. Department of Agriculture), April 1979: 26–28; B. R. DeWalt, 1983, The Cattle are Eating the Forest, *Bulletin of the Atomic Scientists* 39(1):18–23; N. Myers, 1981, The Hamburger Connection, *Ambio* 10(1):3–8; J. D. Nations and D. I. Komer, 1982, *Rainforests, Cattle and the Hamburger Society*, Center for Human Ecology, Austin, Texas; J. J. Parsons, 1976, Forest to Pasture: Development or Destruction? *Revista Biologica Tropicale* 24:121–38; D.R. Shane, 1980, *Hoofprints on the Forest: An Inquiry Into the Beef Cattle Industry in the Tropical Forest Areas of Latin America*, Report to Office of Environmental Affairs, Department of State, Washington D.C.; J. R. Simpson and D. E. Farris, 1982, *The World's Beef Business*, Iowa State University Press, Ames, Iowa; and R. Wasserstrom, 1978, Population Growth and Economic Development in Chiapas, 1524–1975, *Human Ecology* 6(2):127–43.

2. D. Pimentel and M. Pimentel, 1979, *Food, Energy and Society*, Edward Arnold Publishers Ltd., London, U. K.

3. This section is based on the author's on-the-ground researches during repeated visits to Central America from 1976 onwards, and to discussions with 38 colleagues who have worked in the cattle raising/tropical forest sector in Central America.

4. James D. Nations, Center for Human Ecology, University of Texas, Austin, Texas, personal communication, 1983.

5. M. J. Esman, 1979, *Landlessness and Near Landlessness in Developing Countries*, Cornell University, Ithaca, New York; C. A. Lassen, 1980, *Landlessness and Rural Poverty in Latin America*, Center for International Studies, Cornell University, Ithaca, New York; and see Nations and Komer, 1982, reference 1, above.

6. P. M. Fearnside, 1980, The Effects of Cattle Pasture on Soil Fertility in the Brazilian Amazon: Consequences for Beef Production Sustainability, *Tropical Ecology*, 21(1):125–37; S. B. Hecht, editor, 1980, *Land-Use and Agricultural Research in the Amazon Basin*, CIAT, Cali, Colombia, and 1981, Deforestation in the Amazon Basin: Magnitude, Dynamics and Soil Resource Effects, *Studies in Third World Societies* 13:61–100; J. M. Kirby, 1976, Agricultural Land Use and Settlement of Amazonia, *Pacific Viewpoint* 17(2):105–32; D. F. Osborn, 1975, Beef Production from Improved Pastures in the Tropics, *World Review of Animal Production* 11(4):23–31.

7. D. J. Mahar, 1979, *Frontier Development Policy in Brazil: A Study of Amazonia*, Praeger Publishers, New York, U. S. A.

8. The Forest Farmer

1. S. C. Chin, 1977, Shifting Cultivation: A Need for Greater Understanding, *Sarawak Museum Journal* 25(46):107–28; W. C. Clarke, 1976, Maintenance of Agriculture

and Human Habitats Within the Tropical Forest Ecosystem, *Human Ecology* 4:247–59; W. M. Denevan, 1980, Latin America, in G. A. Klee, editor, *World Systems of Traditional Resource Management*:216–17, Halsted Press, New York; T. B. Grandstaff, 1981, Shifting Cultivation: A Reassessment of Strategies, *Ceres* 14(4):28–30; D. J. Greenland and R. Herrera, 1977, Shifting Cultivation and Other Agricultural Practices, in *Patterns of Use of Tropical Forest Ecosystems*, Soil Sciences Department, University of Reading, U. K.; R. R. Harwood, 1979, *Small Farm Development: Understanding and Improving Farming Systems in the Humid Tropics*, Westview Press, Boulder, Colorado.

2. C. Chandrasekharan, 1980, reference 13, Chapter 5; S. H. Cochrane, 1979, *The Population of Thailand: Its Growth and Welfare*, The World Bank, Washington D.C.; W.J. Van Liere, 1977, Southeast Asia—Potential World Food Basket, *Netherlands Journal of Agricultural Science* 25:14–25.

3. E. Eckholm, 1979, *The Dispossessed of the Earth: Land Reform and Sustainable Development*, Worldwatch Institute, Washington D. C.; M. J. Esman, 1979, reference 5, Chapter 7; Food and Agriculture Organization, 1980, *Proceedings of World Conference on Agrarian Reform and Rural Development*, Food and Agriculture Organization, Rome, Italy; International Labour Organization, 1977, *Poverty and Landlessness in Rural Asia*, International Labour Office, Geneva, Switzerland; C.L.J. van der Meer, 1981, *Rural Development in Northern Thailand*, State University of Groningen, Groningen, Netherlands.

4. M. J. Dourojeanni, 1979, *Desarrollo Rural Integral en la Amazonia Peruana, con Especial Referencia a las Actividades Forestales*, Department of Forest Management, Universidad Nacional Agraria, Lima, Peru; K. F. Durham, 1979, *The Forest Frontier: Environment and Development in Eastern Peru*, Ph.D. Dissertation, University of Michigan, Ann Arbor, Michigan; W. Mertens, 1980, *Amazonic Population Ideologies in Peru*, Ford Foundation; Lima, Peru.

5. A. Duldulao, 1981, The Implications of Forest Occupancy Management on Natural Resources Conservation in the Philippines, *Journal of the Natural Resources Management Forum* 3:32–39; P. B. Durst, 1981, Problems Facing Reforestation in the Philippines, *Journal of Forestry*, October 1981: 686–88, 699; O. Mehmet, editor, 1979, *Poverty and Social Change in Southeast Asia*, University of Ottawa Press, Ottawa, Canada; and P. E. Sajise, 1980, *Our Degrading Uplands and What We Can Do About It*, San Carlos University, Cabu City, Philippines. For information on migrations of displaced peasants and other landless people in Southeast Asia generally, see G. W. Jones and H. V. Richter, 1981, *Population Mobility and Development: Southeast Asia and the Pacific*, Development Studies Centre, Canberra, Australia, and R. J. Pryor editor, 1979, *Migration and Development in Southeast Asia*, Oxford University Press, Oxford, U.K.

6. See International Labour Organization, 1977, reference 3, above.

7. See, for example, Duldulao, 1981, reference 5, above; and Population Center Foundation, 1978, *Images From the Future: The Philippines in the Year 2000*, Population Center Foundation, Manila, Philippines.

8. K. F. S. King and T. Chandler, 1978, *Wasted Lands*, International Council for Research in Agroforestry, Nairobi, Kenya; N. Myers, 1980, reference 3, chapter 2, above; R. Persson, 1974, *World Forest Resources: Review of the World's Forest Resources in the Early 1970s*, Royal College of Forestry, Stockholm, Sweden; W. Schulte, 1981, *Report on Population Data in Slash-and-Burn Forestry Communities in Asia*, Population Programme of Economic and Social Policy Division, Food and Agriculture Organization, Rome, Italy; Food and Agriculture Organization, 1980, *Agriculture: Toward 2000*, Food and Agriculture Organization, Rome, Italy. But note that at least one noted authority, R. F. Watters, proposed a figure of 200 million for as far back as the late 1950s: 1960, The Nature of Shifting Cultivation, *Pacific Viewpoint* 1:59–64, and 1971, *Shifting Cultivation in Latin America*, Food and Agriculture Organization, Rome, Italy. A recent United Nations publication (UNESCO, 1978, *Tropical Forest Ecosystems*, UNESCO, Paris, France) suggested

a total of 240 million for the mid-1970s. An expert in this field, W. M. Denevan, reference 1 above, believes the figure could be as high as 300 million.

9. C. Chandrasekharan, editor, 1978, Shifting Cultivation, *Forest News* 2(2):1–25; K. Kartawinata *et al.*, 1981, reference 12, Chapter 5; Food and Agriculture Organization, 1980, reference 8, above; and see references for Thailand, reference 2, above, and for the Philippines, reference 5, above.

10. H. Braun, 1974, *Shifting Cultivation in Africa,* Food and Agriculture Organization, Rome, Italy; F. W. Hauck, 1974, *Shifting Cultivation and Soil Conservation in Africa,* Food and Agriculture Organization, Rome, Italy; R. Persson, 1977, *Forest Resources of Africa,* Royal College of Forestry, Stockholm, Sweden.

11. W. M. Denevan, 1980, reference 1, above; and R. F. Watters, 1971, *Shifting Cultivation in Latin America,* Food and Agriculture Organization, Rome, Italy.

12. This figure has been established by the author during discussions between 1976 and 1983 with field researchers throughout the biome; and confirmed by the author during consultations with officials of the Food and Agriculture Organization in preparation for the early 1985 World Forestry Congress, and during the course of his background work with colleagues at the U.N. Fund for Population Activities in preparation for the August 1984 Conference on Population.

13. This average figure can range from well below half a hectare to almost two hectares, depending upon population densities, soil types, and the like. For some background documentation, see references by Chandrasekharan, Denevan, Esman, Duldulao, Greenland, King and Chandler, and Persson, above; and see C. Uhl and P. Murphy, 1981, A Comparison of Productivities and Energy Values Between Slash-and-Burn Agriculture and Secondary Succession in the Upper Rio Negro Region of the Amazon Basin, *Agro-Ecosystems* 7:63–83.

14. P. Guiness, editor, 1977, *Transmigrants in South Kalimantan and South Sulawesi,* Population Institute, Gadjah Mada University, Yogyakarta, Indonesia; G. E. Hansen, editor, 1981, *Agricultural and Rural Development in Indonesia,* Westview Press, Boulder, Colorado; P. A. Meyer and C. MacAndrews, 1978, *Transmigration in Indonesia:An Annotated Bibliography,* Gadjah Mada University Press, Yogyakarta, Indonesia; S. Mustaffa, 1978, Indonesia Steps Up Transmigration, *Populi* 5(3):7–13; M. R. Ross, 1980, The Role of Land Clearing in Indonesia's Transmigration Program, *Bulletin of Indonesian Economic Studies* 16(1):75–87.

15. F. Barbira-Scazzocchio, editor, 1980, *Land, People and Planning in Contemporary Amazonia,* Centre of Latin American Studies, Cambridge University, Cambridge, U. K.; V. Cox, 1977, Brazil: The Amazon Gamble, *Development Digest* 15(1):82–86; R. J. A. Goodland and H. S. Irwin, 1975, *Amazon Jungle: Green Hell to Red Desert?* Elsevier Scientific Publishing Company, New York; E. F. Moran, editor, 1979, Changing Agricultural Systems in Africa, *Studies in Third World Society,* no. 8, College of William and Mary, Williamsburg, Virginia.

16. N. J. H. Smith, 1982, *Rainforest Corridors: The TransAmazon Colonization Scheme,* University of California Press, Berkeley, California.

17. P. Röben, 1980, Ende des Regenwaldes in Sicht? *UmSchau* 80(15):459–62.

18. World Bank and Organization for Economic Cooperation and Development, 1981, *Migration in West Africa,* Oxford University Press, Oxford, U. K.

19. S. Atmosoedarjo and S. G. Banyard, 1978, The Prosperity Approach to Forest Community Development in Java, *Commonwealth Forestry Review* 57(2); J. G. Bene, H. W. Beall, and A. Cote, 1977, *Trees, Food and People,* International Development Research Centre, Ottawa, Canada; G. Budowski, 1978, *Agro-Forestry: A Bibliography,* Centro Agronomico Tropical de Investigacion y Ensenanza (CATIE), Turrialba, Costa Rica; J. Combe, H. J. Saa and C. Monge, 1981, *Bibliography on Tropical Agroforestry,* CATIE, Turrialba, Costa Rica; H. O. Mongi and P. A. Huxley, editors, 1979, *Soils Research in Agroforestry,* International Council for Research in Agroforestry, Nairobi, Kenya; G. De Las Salas, 1979, *Agroforestry Systems in Latin America,* CATIE, Turrialba, Costa Rica.

20. J. L. Brewbaker and Ta Wei Hu, 1981, *Nitrogen Fixing Trees of Importance in the*

Tropics, University of Hawaii, Honolulu, Hawaii; J. A. Duke, 1981, *Handbook of Legumes of World Economic Importance*, Plenum Press, New York; P. Felker and L. S. Bandurski, 1979, Uses and Potential Uses of Leguminous Trees for Minimal-Energy Input Agriculture, *Economic Botany* 33:172–84; R. M. R. Harun, editor, 1981, *Legumes in the Tropics*, Faculty of Agriculture, Universiti Pertanian Malaysia, Serdang, Selangor, Malaysia; D. Isely, 1982, Leguminosae and *Homo sapiens*, 1982, *Economic Botany* 36:46–69; R. M. Polhill and P. H. Raven, editors, 1981, *Advances in Legume Systematics*, Royal Botanic Gardens, Kew, U. K.

21. K. F. S. King, 1979, *Concepts of Agroforestry*, International Council for Research in Agroforestry, Nairobi, Kenya.

22. T. B. Grandstaff, 1980, Shifting Cultivation in North Thailand: Possibilities for Development, *United Nations University Resource Systems, Theory and Methodology Series* no. 3, United Nations University, Tokyo, Japan; P. Johnson, 1980, Tackling Drug Traffic at its Source, *Ceres* 13(4):27–32.

23. C. P. Diaz, 1976, *Industrial Forest Plantations: An Economic-Oriented Approach to Reforestation in the Philippines*, Forest Research Institute, Laguna, Philippines; H. Gregersen and J. Haygreen, 1979, *Planning for a Large-Scale Integrated Forestry and Forest Industries Project: A Case Study of the Paper Industry Corporation of the Philippines*, Food and Agriculture Organization, Rome, Italy; E. L. Hyman, 1983, Smallholder Tree Farming in the Philippines, *Unasylva* 35(139):25–31.

24. W. C. Clarke, 1976, reference 1, above; J. C. Dickenson III, 1972, Alternatives to Monoculture in the Humid Tropics of Latin America, *Professional Geographer* 24:-217–22; S. R. Gliessman, R. Garcia, and M. Amador, 1981, The Ecological Basis for the Application of Traditional Agricultural Technology in the Management of Tropical Agroecosystems, *Agro-Ecosystems* 7:173–85; T. B. Grandstaff, 1980, reference 22, above; D. J. Greenland, 1975, Bringing the Green Revolution to the Shifting Cultivator, *Science* 190:841–44; D. H. Janzen, 1973, Tropical Agroecosystems, *Science* 182:1212–19; P. Kundstadter, E. C. Chapman, and S. Sabhasri, editors, 1978, *Farmers in the Forest: Economic Development and Marginal Agriculture in Northern Thailand*, University Press of Hawaii, Honolulu, Hawaii; J. D. Nations and R. B. Nigh, 1978, Cattle, Cash, Food and Forest: The Destruction of the American Tropics and the Lacandon Maya Alternative, *Culture and Agriculture* 6:1–5; H. Ruthenberg, 1983, *Farming Systems in the Tropics* (third edition), Oxford University Press, London, U.K.

25. P. Kunstadter *et al.*, 1978, reference 24, above.

26. D. Lee, 1980, *The Sinking Ark*, Heinemann Educational Books Ltd., Kuala Lumpur, Malaysia.

27. S. R. Gliessman et al., 1981, reference 24, above.

28. J. D. Nations and R. B. Nigh, 1980, reference 24, above.

29. D. Lee, 1980, reference 26, above.

30. C. De'Aeth, 1980, *The Throwaway People: Social Impact of the Gogol Timber Project, Madang Province*, Institute of Applied Social and Economic Research, Boroko, Papua New Guinea; D. S. Liem, 1978, *Ethnozoological Survey of the Gogol Timber Area*, Institute of Applied Social and Economic Research, Boroko, Papua New Guinea.

31. G. C. Wilken, 1977, Integrating Forest and Small-Scale Farm Systems in Middle America, *Forest Ecology and Management* 1:223–35.

9. Bottom-Line Reckoning

1. C. Uhl, 1983, You Can Keep a Good Forest Down, *Natural History* 92(4):70–79.

2. T. E. Lovejoy, 1979, A Projection of Species Extinctions, in *The Global 2000 Report*: 328–29, Council on Environmental Quality, Washington D. C.

3. A. Sommer, 1976, Attempt at an Assessment of the World's Tropical Moist Forests, *Unasylva*, 28(112 and 113):5–24; R. Persson, 1974, reference 8, chapter 8.
4. J.-P. Lanly, 1982, *Tropical Forest Resources*, Food and Agriculture Organization, Rome, Italy, in conjunction with the United Nations Environment Programme, Nairobi, Kenya.
5. R. A. Houghton, N. Myers *et al.*, 1983, *Current Estimates of Deforestation in the Tropics: A Synthesis* (in press); J. M. Melillo, N. Myers *et al.*, 1983, *A Comparison of Recent Estimates of Disturbance in Tropical Forest* (in press); see also M. Hadley and J. P. Lanly, 1983, Tropical Forest Ecosystems: Identifying Differences, Seeing Similarities, *Nature and Resources* 19(1):2–19.

10. A Cornucopia of Foods

1. P. Carlson, editor, 1980, *Biology of Crop Productivity*, Academic Press, New York; J. R. Harlan, 1975, *Crops and Man*, American Society of Agronomy, Madison, Wisconsin; N. Myers, 1983, *A Wealth of Wild Species*, Westview Press, Boulder, Colorado; M. L. Oldfield, 1981, Tropical Deforestation and Genetic Resources Conservation, *Studies in Third World Societies* 14:277–346.
2. A. H. Gentry, 1982, reference 2, chapter 3.
3. J. R. Harlan, 1975, reference 1, above.
4. F. P. Ferwerda, 1976, Coffees, in N. W. Simmons, editor, *Evolution of Crop Plants*, 252–60, Longman, New York.
5. H. H. Iltis, J. F. Doebley, R. M. Guzman, and B. Pazy, 1979, *Zea diploperennis* (Gramineae), a New Teosinte From Mexico, *Science* 203:186–88.
6. L. R. Nault and W. R. Findley, 1981, Primitive Relative Offers New Traits for Corn Improvement, *Ohio Report* 66(6):90–92.
7. A. C. Fisher, 1982, *Economic Analysis and the Extinction of Species*, Department of Agriculture and Resource Economics, University of California at Berkeley; N. Myers, 1983, reference 1, above.
8. M. Chai, E. Soepadmo, and H. S. Yong, editors, 1979, Proceedings of Workshop on Genetic Resources of Plants, Animals and Microorganisms in Malaysia, *Malaysian Applied Biology, Special Issue* 8(1); M. A. Rifai, editor, 1976, *ASEAN Grain Legumes*, Central Research Institute of Agriculture, Bogor, Indonesia; J. T. Williams, C. H. Lamoureux, and N. Wulijarni-Soetjipto, editors, 1975, *Southeast Asian Plant Genetic Resources*, BIOTROP, Bogor, Indonesia.
9. National Academy of Sciences, 1975, *The Winged Bean: A High Protein Crop for the Tropics*, National Academy of Sciences, Washington D. C.
10. K. Jong, editor, 1979, *Biological Aspects of Plant Genetic Resource Conservation in Southeast Asia*, Institute of Southeast Asian Biology, University of Aberdeen, Scotland, U. K.
11. S. Nagy and P. E. Shaw, editors, 1980, *Tropical and Subtropical Fruits*, AVI Publishing Company, Westport, Connecticut; J. A. Samson, 1980, *Tropical Fruits*, Longman, New York; and for several illustrative examples, see National Academy of Sciences, 1975, *Underexploited Tropical Plants with Promising Economic Value*, National Academy of Sciences, Washington D. C.
12. National Academy of Sciences, 1979, *Tropical Legumes: Resources for the Future*, National Academy of Sciences, Washington D. C.
13. G. A. C. Herklots, 1972, *Vegetables in Southeast Asia*, George Allen and Unwin, London, U. K.; F. W. Martin and R. M. Ruberte, 1979, *Edible Leaves of the Tropics*, Institute of Tropical Agriculture, Mayaguez, Puerto Rico; J. J. Ochse, 1977, *Vegetables of the Dutch East Indies*, A. Asher and Company Publishers Ltd., Amsterdam, Netherlands; J. C. Okafor, 1980, Edible Indigenous Woody Plants in the Rural Economy of the Nigerian Zone, *Forest Ecology and Management* 1:235–47; H. A. P. C. Oomen and G. J. H. Grubben, 1977, Tropical Leaf Vegetables in Human

Nutrition, in *Communication* 69:24–41, 51–55, Department of Agricultural Research, Koninklijk Instituut voor de Tropen, Amsterdam, Netherlands; and R. L. Villareal and R. T. Opena, 1976, The Wild Vegetables of Southeast Asia, *American Horticulturalist* 55(3):1–4.

14. J. A. Duke, 1981, *The Gene Revolution,* Office of Technology Assessment, U.S. Congress, Washington D.C.; N. Mohan and G. P. Srivastava, 1980, Studies on the Extractability and Chemical Composition of Leaf Proteins from Certain Trees, *Journal of Food Science and Technology* 18:48–50; N. W. Pirie, editor, 1975, *Food Protein Sources,* Cambridge University Press, London, U. K.;

15. J. A. Duke, 1983, *Ecological Amplitudes of Crops Used as Sweeteners,* Plant Genetics and Germplasm Institute, Agricultural Research Service, Beltsville, Maryland; J. D. Higginbotham, 1979, Protein Sweeteners, in C. A. M. Hough, K. J. Parker, and A. J. Vletof, editors, *Developments in Sweeteners,* Applied Science Publishers, London, U. K.; G. E. Inglett, 1981, Sweeteners—A Review, *Food Technology,* March 1981, 37–41.

16. L. E. Gilbert and P. H. Raven, editors, 1975, *Coevolution of Animals and Plants,* revised edition, University of Texas Press, Austin, Texas; D. H. Janzen, 1975, *Ecology of Plants in the Tropics,* Arnold, London, U. K.; R. L. Metcalfe, 1977, Plant Derivatives for Insect Control, in R. S. Seigler, editor, *Crop Resources:* 165–78, Academic Press, New York.

17. W. Kraus *et al.,* 1980, New Insect Antifeedants from *Toona* Species *(Meliaceae),* in *Proceedings of Fourth Asian Symposium on Medicinal Plants and Species:* 127–35, Government of Thailand in Conjunction with UNESCO and Faculty of Science, Mahidol University, Bangkok, Thailand.

18. S. W. T. Batra, 1982, Biological Control in Agroecosystems, *Science* 215:135–39; C. P. Clausen, editor, 1978, *Introduced Parasites and Predators of Arthropod Pests and Weeds: A World Review,* Agricultural Research Service of U.S. Department of Agriculture, Washington D. C.; C. B. Huffaker, 1980, Use of Predators and Parasitoids in Biological Control, in R. C. Staples and R. J. Kuhr, editors, *Linking Research to Crop Production:* 173–98, Plenum Publishing, New York; J. S. Marsden, G. E. Martin, D. J. Parham, T. J. Ridsdill-Smith, and B. G. Johnston, 1980, *Returns on Australian Agricultural Research,* Division of Entomology, Commonwealth Scientific and Industrial Research Organization, Canberra, Australia; R. I. Sailer, 1981, Progress Report on Importation of Natural Enemies of Insect Pests in the U.S.A., in J. R. Coulson, editor, *Use of Beneficial Organisms in the Control of Crop Pests:* 20–26, Entomological Society of America, College Park, Maryland; R. Van den Bosch, P. S. Messenger, and A. P. Gutierrez, 1982, *An Introduction to Biological Control,* Plenum Publishing, New York.

19. V. L. Delucchi, editor, 1976, *Studies in Biological Control,* Cambridge University Press, Cambridge, U. K.

20. D. M. Maddox, L. A. Andres, R. D. Hennessey, R. D. Blackburn, and N. R. Spencer, 1971, Insects to Control Alligatorweed, *BioScience* 21:985–91.

21. National Academy of Sciences, 1976, *Making Aquatic Weeds Useful,* National Academy of Sciences, Washington D. C.

22. C. B. Huffaker, 1980, and R. Van den Bosch *et al.,* 1982, reference 18, above.

23. R. I. Sailer, 1981, reference 18, above.

24. M. L. Oldfield, 1981, *The Value of the Conservation of Genetic Resources,* Department of Agriculture, University of Texas, Austin, Texas; N. D. Vietmeyer, 1983, *Little-Known Asian Animals with a Promising Economic Future,* National Academy Press, Washington, D. C.

25. S. S. Ajayi, 1979, *Utilization of Forest Wildlife in West Africa,* Food and Agriculture Organization, Rome, Italy; A. DeVos, 1977, Game as Food, *Unasylva* 29:2–12; W. Krostitz, 1979, The New International Market for Game Meat, *Unasylva* 31(123):32–36; J. B. Sale, 1983, *The Importance and Values of Wild Plants and Animals in Africa,* International Union for Conservation of Nature and Natural Resources, Gland, Switzerland.

26. R. A. Mittermeier, 1978, South America's River Turtles: Saving Them by Use, *Oryx* 14(3):222–30.
27. A. Paucar and A. Gardner, 1981, *Establishment of a Scientific Research Station in the Yasuni National Park of the Republic of Ecuador,* National Zoo, Washington D. C.

11. Pharmaceutical Factories

1. R. E. Schultes, 1980, The Amazonia as a Source of New Economic Plants, *Economic Botany* 33(3):259–66.
2. L. M. Perry, 1980, *Medicinal Plants of East and Southeast Asia,* MIT Press, Cambridge, Mass.; S. Soedigdo *et al.,* 1980, Studies on the Chemistry and Pharmacology of Indonesian Medicinal Plants, in *Proceedings of Fourth Asian Symposium on Medicinal Plants and Spices:* 112–19, UNESCO and Faculty of Science, Mahidol University, Bangkok, Thailand.
3. A. Whitten, 1982, *The Gibbons of Siberut,* J. M. Dent, London, I. K.
4. J. D. Douros and M. Suffness, 1980, The National Cancer Institute's Natural Products Antineoplastic Development Program, in S. K. Carter and Y. Sakurai, editors, *Recent Results in Cancer Research* 70:21–44; J. A. Duke, 1981, reference 14, chapter 10; J. A. Duke and K. K. Wain, 1981, *Medicinal Plants of the World: Computer Index with More Than 85,000 Entries,* 3 vols., Plant Genetics and Germplasm Institute, Agricultural Research Service, Beltsville, Maryland; see L. M. Perry, 1980, reference 2, above; 5. Von Reis and F. J. Lipp, 1982, *New Plant Sources for Drugs and Foods,* Harvard University Press, Cambridge, Mass.; see also H. Wagner and P. Wolff, 1977, *New Natural Products and Plant Drugs,* Springer Verlag, New York.
5. N. R. Farnsworth, 1982, *The Consequences of Plant Extinction on the Current and Future Availability of Drugs,* University of Illinois Medical Center, Chicago, Illinois; N. Myers, 1983, reference 7, chapter 10.
6. D. A. Levin, 1976, Alkaloid-Bearing Plants: An Ecogeographic Perspective, *American Naturalist* 110:261–84; S. Moody, 1978, Latitude, Continental Drift, and the Percentage of Alkaloid-Bearing Plants in Floras, *American Naturalist* 112:965–68; R. F. Raffauf, 1970, *Handbook of Alkaloids and Alkaloid-Containing Plants,* John Wiley and Sons, New York; T. Robinson, 1981, *The Biochemistry of Alkaloids,* Springer-Verlag, New York; G. R. Waller and E. K. Nowacki, 1978, *Alkaloid Biology and Metabolism in Plants,* Plenum Press, New York.
7. G. Staritsky, 1980, Is There a Future for Medicinal Crops? *Span* 23(2):80–82.
8. M. L. Oldfield, 1981, reference 1, chapter 10.
9. P. Brooke, 1978, *Cancer: Drug Utilization and Drug Prospects,* Cyrus J. Laurence Inc., New York; W. H. Lewis and M. P. F. Elvin-Lewis, 1977, *Medical Botany,* John Wiley and Sons, New York.
10. S. K. Carter and R. B. Livinston, 1976, Plant Products in Cancer Chemotherapy, *Cancer Treatment Reports* 60(8):1141–56; W. I. Taylor and N. R. Farnsworth, editors, 1975, *The Catharanthus Alkaloids,* Marcel Dekker Inc., New York.
11. International Marketing Statistics, 1980, *National Prescriptions Audit,* Ambler, Pennsylvania.
12. A. S. Barclay and R. E. Perdue, 1976, Distribution of Anti-Cancer Activity in Higher Plants, *Cancer Treatment Reports* 60(8):1081–1113; G. A. Cordell, 1978, Alkaloids, in *Encyclopedia of Chemical Technology* vol. 1, third edition: 883–943, John Wiley and Sons, New York; J. D. Douros and M. Suffness, 1980, reference 4, above; R. W. Spjut and R. E. Perdue, 1976, Plant Folklore: A Tool for Predicting Sources of Anti-Tumor Activity? *Cancer Treatment Reports* 60(8):979–85; M. Suffness and J. Douros, 1979, Drugs of Plant Origin, in V. T. DeVita and H. Busal, editors, *Methods in Cancer Research,* vol. XVI, Part A:73–126, National Cancer Institute, Bethesda, Maryland.

13. J. A. Duke, 1982, *Contributions of Neotropical Forests to Cancer Research*, Economic Botany Laboratory, U. S. Department of Agriculture, Beltsville, Maryland.

14. L. E. Gilbert and P. H. Raven, 1975, reference 16, chapter 10; D. H. Janzen, 1975, reference 16, chapter 10; D. A. Levin, 1976, reference 6, above; R. E. Schultes and T. Swain, 1976, The Plant Kingdom:A Virgin Field for New Biodynamic Constituents, in N. J. Finer, editor, *The Recent Chemistry of Natural Products*: 133–71, Philip Morris Inc., New York.

15. T. Swain, editor, 1972, *Plants in the Development of Modern Medicine*, Harvard University Press, Cambridge, Mass.

16. T. Eisner, 1983, Cornell University, Ithaca, N. Y., personal communication.

17. M. L. Oldfield, 1981, reference 1, chapter 10.

18. N. R. Farnsworth, 1977, The Current Importance of Plants as a Source of Drugs, in D. S. Seigler, editor, *Crop Resources*: 61–74, Academic Press, New York; N. R. Farnsworth and R. W. Morris, 1976, Higher Plants—The Sleeping Giant of Drug Development, *American Journal of Pharmacy* 148(2):46–52; M. L. Oldfield, 1981, reference 1, chapter 10.

19. L. V. Asolkar and R. Chadha, 1979, *Diosgenin and Other Steroid Drug Precursors*, Publication and Information Directorate, CSIR, New Delhi, India; J. F. Morton, 1977, *Major Medicinal Plants*, Charles C. Thomas, Springfield, Illinois.

20. N. R. Farnsworth, 1969, Drugs from the Sea, *Tile and Till* 55.

21. W. H. Lewis and M. P. F. Elvin-Lewis, 1977, reference 9, above.

22. N. Applezweig, 1980, Steroid Drugs from Botanical Sources: Future Prospects, in E. Campos Lopez, editor, *Renewable Resources, A Systematic Approach*: 369–78, Academic Press, New York.

23. L. V. Asolkar and R. Chadha, 1979, reference 19, above; and D. D. Soejarto, A. S. Bingel, M. Slaytor and N. R. Farnsworth, 1978, Fertility-Regulating Agents from Plants, *Bulletin of World Health Organization* 56:343–52.

24. P. Crabbe, 1979, Mexican Plants and Human Fertility, *Courier* (UNESCO monthly) May 1979:33–34; N. R. Farnsworth, 1978, Indigenous Plants for Fertility Regulation, in *Proceedings of IUPAC Eleventh International Symposium on Chemistry of Natural Products* 4(2):475–89, International Union for Pure and Applied Chemistry, Brussels, Belgium; N. R. Farnsworth and J. M. Farley, 1980, *Traditional Medicine Programmes of the World Health Organization*, Department of Pharmacognosy and Pharmacology, College of Pharmacy, University of Illinois at the Medical Center, Chicago, Illinois.

25. World Health Organization, 1981, *Report of Task Force on Anti-Fertility Plants, Especially from Tropical Forests*, World Health Organization, Geneva, Switzerland.

26. D. Lee, 1980, *The Sinking Ark*, Heinemann Educational Books (Asia) Ltd., Kuala Lumpur, Malaysia.

27. M. L. Oldfield, 1981, reference 1, chapter 10.

28. J. F. Morton, 1977, reference 19, above.

29. M. L. Oldfield, 1981, reference 1, chapter 10.

30. N. R. Farnsworth and G. A. Cordell, 1976, A Review of Some Biologically Active Compounds Isolated from Plants as Reported in the 1974–1975 Literature, *Lloydia* 39(6):420–55.

31. R. E. Schultes and A. Hofmann, 1979, *Plants of the Gods*, McGraw-Hill Book Co., New York.

32. R. E. Schultes, 1980, reference 1, above.

33. Interagency Primate Steering Committee, 1978, *National Primate Plan*, U.S. Department of Health, Education and Welfare, Washington D. C.; National Academy of Sciences, 1975, *Nonhuman Primates: Usage and Availability for Biomedical Programs*, National Research Council, Washington D. C.; M. L. Oldfield, 1981, reference 1, chapter 10; M. R. N. Prasad and T. C. Anand Kumar, 1977, *Use of Non-Human Primates in Biomedical Research*, Indian National Science Adacemy, New Delhi, India.

34. M. L. Oldfield, 1981, reference 1, chapter 10.

35. E. S. Ayensu, 1978, *Medicinal Plants of West Africa,* Reference Publications Inc., Algonac, Michigan, and 1981, *Medicinal Plants of the West Indies,* Reference Publications Inc., Algonac, Michigan; Government of Thailand and UNESCO, 1981, *Proceedings of Fourth Asian Symposium on Medicinal Plants and Spices,* UNESCO, Paris, France; Government of the Philippines, 1980, *Medicinal Plants of Philippines Forests,* Department of Agriculture and Natural Resources, Manila, Philippines.
36. M. Jacobs, 1982, The Study of Minor Forest Products, *Flora Malesiana Bulletin* 35:3768–82.

12. Forest Industrial Complexes

1. R. A. Buchannan, F. H. Otey, and G. E. Hamerstrand, 1980, Multi-Use Botanochemical Crops: An Economic Analysis and Feasibility Study, *I and E. C. Product Research and Development* (publication of American Chemical Society):489–96; J. A. Duke, 1981, *The Gene Revolution,* Office of Technology Assessment, U. S. Congress, Washington D. C.; I. S. Goldstein, 1981, Chemicals from Biomass: Present Status, *Forest Products Journal* 31(10):63–68; O. R. Gottlieb and W. B. Mors, 1980, Potential Utilization of Brazilian Wood Extractives, *Journal of Agricultural and Food Chemistry* 28:196–215; E. S. Lipinsky, 1981, Chemicals from Biomass: Petrochemical Substitution Options, *Science* 212:1465–71; M. L. Oldfield, 1981, reference 1, chapter 10; B. O. Palsson *et al.*, 1981, Biomass as a Source of Chemical Feedstocks: An Economic Evaluation, *Science* 213:513–17; L. H. Princen, 1979, New Crop Development for Industrial Oils, *Journal of the American Oil Chemists' Society* 56(9):845–48; E. H. Pryde, L. H. Princen, and K. D. Mukherjee, editors, 1981, *New Sources of Fats and Oils,* American Oil Chemists' Society, Champaign, Illinois; R. E. Schultes, 1980, The Amazonia as a Source of New Economic Plants, *Economic Botany* 33(3):259–66; S. C. Wag and J. B. Huffman, 1981, Botanochemicals: Supplements to Petrochemicals, *Economic Botany* 35(4):369–82. For a fine overall survey of industrial materials and other natural resources available from tropical forests, see J. A. Lasschuit and F. A. C. M. van Eerd, 1983, *Minor Forest Products* and *Non-Timber Products of the Forest,* University of Wageningen, Netherlands.
2. J. Dransfield, 1981, The Biology of Asiatic Rattans in Relation to the Rattan Trade and Conservation, in H. Synge, editor, *The Biological Aspects of Rare Plant Conservation:* 179–86, John Wiley and Sons Ltd., Chichester, U. K.; International Development Centre, 1980, *Rattan: Report of a Workshop, Singapore 4–6 June, 1979,* International Development Research Centre, Ottawa, Canada.
3. M. J. Balick, 1979, Amazonian Oil Palms of Promise: A Survey, *Economic Botany* 33(1):11–28; M. J. Plotkin and R. E. Schultes, 1982, *Conservation and Ethnobotany in Tropical South America,* Report to World Wildlife Fund-US, Harvard Botanical Museum, Cambridge, Mass.
4. J. R. Hunter, 1981, Tendu *(Diospyros melanoxylon)* Leaves, Bidi Cigarettes, and Resource Management, *Economic Botany* 35(4):450–59; M. Jacobs, 1982, The Study of Minor Forest Products, *Flora Malesiana Bulletin* 35: 3768–82.
5. J. A. Duke, 1981, reference 1, above; L. H. Princen, 1979, reference 1, above; E. H. Pryde *et al.*, reference 1, above.
6. A. Tcheknavorian-Asenbauer and R. O. B. Wijesekera, 1982, *Medicinal and Aromatic Plants for Industrial Development,* United Nations Industrial Development Organization, Vienna, Austria.
7. M. J. Balick, 1979, reference 3, above; M. J. Plotkin and R. E. Schultes, 1982, reference 3, above; and see R. E. Schultes, 1980, reference 1, above.
8. E. P. Imle, 1978, Hevea Rubber: Past and Future, *Economic Botany* 32:264–77; A. B. Man and D. Blandford, 1980, *The Outlook for Natural Rubber in the 1980s,* Department of Agricultural Economics, Cornell University, Ithaca, New York; J. K. Templeton, 1978, *Natural Rubber: Organizations and Research in Producing Countries,* International Agricultural Development Service, New York.

9. R. E. Schultes, 1980, reference 1, above.
10. G. T. Prance, 1977, The Phytogeographic Subdivisions of Amazonia and Their Influence on the Selection of Biological Reserves, in G. T. Prance and T. S. Elias, editors, *Extinction is Forever:* 195–213, New York Botanical Garden, New York.
11. M. Jacobs, 1982, reference 2, above.
12. A. Tcheknavorian-Asenbauer and R. O. B. Wijesekera, 1982, reference 6, above.
13. M. Jacobs, 1982, reference 2, above.
14. T. Gupta and A. Guleria, 1982, *Non-Wood Forest Products in India,* Indian Institute of Management, Ahmedabad, and Oxford and IBH Publishing Company, New Delhi, India.
15. R. A. Buchanan and F. H. Otey, 1979, Multi-Use Oil- and Hydrocarbon-Producing Crops in Adaptive Systems for Food, Material and Energy Production, *Biosources Digest* 6:176–202.

13. Green Gasoline and Other Energy Forms

1. D. O. Hall, G. W. Barnard, and P. A. Moss, 1982, *Biomass for Energy in the Developing Countries,* Pergamon Press, Oxford, U. K.; Office of Technology Assessment, 1980, *Energy from Biological Processes,* Westview Press, Boulder, Colorado; P. B. L. Srivastava et al., 1982, *Tropical Forests:Source of Energy Through Optimization and Diversification,* Penerbit Universiti Pertanian Malaysia, Serdang, Selangor, Malaysia.
2. D. E. Earl, 1975, *Forest Energy and Economic Development,* Clarendon Press, Oxford, U. K.; R. Iman, 1977, *Silvicultural Biomass Farms,* 6 vol., MITRE Corporation, McLean, Virginia; J. W. Tatom, 1981, Pyrolysis Experience in Developing Countries, in P. F. Bente, editor, *Proceedings of Bio-Energy 1980:* 180–83, Bio-Energy Council, Washington D. C.
3. National Academy of Sciences, 1977, *Leucaena:Promising Forage and Tree Crop for the Tropics,* National Research Council, Washington D. C.
4. K. K. Chin and T. N. Gohj, 1978, *Bioconversion of Solar Energy and Methane Production from Water Hyacinth,* Institute of Gas Technology, Chicago, Illinois; Dynatech Research and Development Corporation, 1978, *Cost Analysis of Aquatic Biomass Systems,* U.S. Department of Energy, Washington D. C.; B. C. Wolverton, and R. C. McDonald, 1979, Water Hyacinth Productivity and Harvesting Studies, *Economic Botany* 33(1):1–10.
5. R. A. Buchanan, et al., 1978, Hydrocarbon- and Rubber-Producing Crops, *Economic Botany* 32:131–53; H. M. Benedict et al., *A Review of Current Research on Hydrocarbon Production by Plants,* Solar Energy Research Institute, Golden, Colorado; M. Calvin, 1980, Hydrocarbons from Plants: Analytical Methods and Observations, *Naturwissenschaften* 67:525–33; J. D. Johnson and T. W. Hinman, 1980, Oils and Rubber from Arid Land Plants, *Science* 208:460–64; J. L. Jones et al., 1979, *Extraction of Hydrocarbon Liquids from Euphorbia Type Plants,* Stanford Research Institute International, Menlo Park, California.
6. M. Calvin, 1980, reference 5, above.
7. C. A. Arroyo, 1981, Fuel for Home Use, *Canopy International* 7(7):15.
8. M. J. Plotkin and R. E. Schultes, 1982, reference 3, chapter 12.

14. Environmental Services

1. Population Center Foundation, 1978, *Images from the Future: The Philippines In the Year 2000,* Population Center Foundation, Manila, Philippines; D. M. Roth, 1983, Philippine Forests and Forestry, 1565–1920, in Richard P. Tucker and J. F. Richards,

eds., *Global Deforestation and the Nineteenth-Century World Economy:* 30–49, Duke University Press, Durham, North Carolina.

2. L. K. Ba, 1977, *Bio-Economics of Trees in Native Malayan Forest,* Department of Botany, University of Malaya, Kuala Lumpur, Malaysia; J. G. Daniel and A. Kulasingam, 1974, Problems Arising From Large-Scale Forest Clearing for Agricultural Use, *Malaysian Forester* 37(3):152–60.

3. H. Dosso, *et al.,* 1981, The Tai Project: Land Use Problems in a Tropical Forest, *Ambio* 10(2–3):120–25.

4. C. J. Barrow, 1980, Development in Peninsular Malaysia: Environmental Problems and Conservation Measures, *Third World Planning Review* 2(1):7–25; C. Toebes and G. K. Seng, 1975, *Some Hydrological Effects of Land Use Changes in Peninsular Malaysia,* Ministry of Agriculture and Rural Development, Kuala Lumpur, Malaysia.

5. M. Falkenmark and G. Lindh, 1976, *Water for a Starving World,* Westview Press, Boulder, Colorado; Economic and Social Commission for Asia and the Pacific, 1980, *Review of Water Resources Development in the Region,* Committee on Natural Resources of Economic and Social Commission for Asia and the Pacific, Bangkok, Thailand; United Nations, 1978, *Water Development and Management: Proceedings of the United Nations Water Conference,* 1977, Pergamon Press, Elmsford, New York; and J. Williams and L.S. Hamilton, 1982, *Watershed Forest Influences in the Tropics and Subtropics: A Selected Annotated Bibliography,* East-West Environment and Policy Institute, East-West Center, Honolulu, Hawaii.

6. R. K. Gupta, 1980, *Bibliography of the Himalayas,* Indian Documentation Service, Gurgaon, India; K. M. Gupta and D. Bandhu, editors, 1979, *Man and Forest: A New Dimension in The Himalaya,* Day and Tomorrow's Printers and Publishers, New Delhi, India; R. Revelle and V. Lakshminarayana, 1975, The Ganges Water Machine, *Science* 188:611–16; J. S. Lall, editor, 1981, *The Himalaya:Aspects of Change,* Oxford University Press, New York; M. M. Pant, 1977, Forestry Sector—Its Contribution to Gross National Product, *Indian Forester* 103(11):739–69; H. C. Rieger, editor, 1977, *Himalayan Mountain Ecosystems,* Max Mueller Bhavan, New Delhi, India; T. Singh, editor, 1980, *Studies in Himalayan Ecology and Development Strategies,* The English Bookstore, New Delhi, India.

7. Centre for Science and Environment, 1982, *State of India's Environment—1982: A Citizen's Report,* Centre for Science and Environment, New Delhi, India.

8. A. T. Abbas, 1976, *Problems of Water Resource Development in Bangladesh,* United Nations Water Conference Secretariat, United Nations, New York.

9. R. G. Barady, 1982, *Draft Environmental Profile of Burma,* National Parks Service and Man and the Biosphere Program, Washington D. C.

10. T. Hatch and C. P. Lim, 1978, *Shifting Cultivation in Sarawak,* Soils Division of Research Branch, Department of Agriculture, Sarawak.

11. R. W. Kates, 1980, Climate and Society: Lessons from Recent Events, *Weather* 35(a):17–25.

12. N. C. Brady, 1975, Rice Responds to Science, in A. W. A. Brown, editor, *Crop Productivity Research Imperatives:* 62–96, Michigan State University, East Lansing, Michigan; W. R. Gasser, 1981, *Survey of Irrigation in Eight Asian Nations,* Economics and Statistics Service, U. S. Department of Agriculture, Washington D. C.

13. A. H. Gentry and J. Lopez-Parodi, 1980, Deforestation and Increased Flooding of the Upper Amazon, *Science* 210:1354–56.

14. K. Hewitt, 1981, *Seismic Risk and Mountain Environments: An Interpretation of the Geography of Earthquake Disasters,* Department of Geography, Wilfrid Laurier University, Waterloo, Canada.

15. R. S. Ayers and D. W. Westcott, 1976, *Water Quality for Agriculture,* Food and Agriculture Organization, Rome, Italy; U. Colombo, D. G. Johnson, and T. Shishido, 1978, *Reducing Malnutrition in Developing Countries: Increasing Rice Production in South and Southeast Asia,* the Trilateral Commission, New York; M. Falkenmark and G. Lindh, 1976, reference 5, above; W. R. Gasser, 1981, reference 12, above; United Nations, 1978, reference 5, above.

16. U. Colombo *et al.,* 1978, reference 15, above.

17. W. R. Gasser, 1981, reference 12, above.
18. M. Falkenmark and G. Lindh, 1976, reference 5, above; and United Nations, 1978, reference 5, above.
19. J. C. Cool, 1980, *Stability and Survival—The Himalayan Challenge*, Ford Foundation, New York.
20. L. Daryadi, 1981, *Forestry and Agriculture: A Key Connection in Java*, Department of Forestry, Jakarta, Indonesia; O. Soematwoto, 1979, *Ecological and Environmental Impacts of Energy Use in Asian Developing Countries, with Particular Reference to Indonesia*, Institute of Ecology, Bandung, Indonesia; A. Sumitro, editor, 1979, *Population and Environment Planning for Communities Practicing Shifting Cultivation*, Gadjah Mada University, Yogyakarta, Indonesia.
21. A. T. Abbas, 1976, reference 8, above; and H. C. Rieger, 1977, reference 6, above.
22. U. S. Agency for International Development, 1978, *Panama Project Paper: Watershed Management*, U. S. Agency for International Development, Washington D. C.; D. M. Windsor and A. S. Rand, 1980, *Decreasing Rainfall at Inland Stations in Central Panama: A Challenge to Assumptions of Climatic Stability*, Smithsonian Tropical Research Institute, Balboa, Panama.
23. U. S. Agency for International Development, 1979, *Environmental and Natural Resources Management in Developing Countries*, and 1980, *Proceedings of U.S. A.I.D. Asia Bureau Conference on Energy, Forestry and Environment*, U. S. Agency for International Development, Washington D.C.; United Nations, 1981, *Survey of Energy Resources*, Secretariat Paper for United Nations Energy Conference, United Nations, New York.
24. West Pakistan Water and Power Development Authority, 1977, *Mangla Watershed Management Study*, West Pakistan Water and Power Development Authority, Islamabad, Pakistan.
25. Indus Basin Research Assessment Group, 1978, *Research Issues Affecting Indus Basin Agricultural Development Policy: Report*, Ministry of Agriculture, Islamabad, Pakistan.
26. W. M. Fleming, 1979, *Environmental and Economic Impacts of Watershed Conservation on a Major Reservoir Project in Ecuador*, Environmental Improvement Division, Santa Fe, New Mexico.
27. United Nations, 1981, reference 23, above.
28. *World Bank, 1980, Energy in Developing Countries*, World Bank, Washington D.C.
29. A. T. White and S. M. Wells, 1982, Coral Reefs in the Philippines, *Oryx* 16(5):445–51.
30. F. B. Davey and M. Graham, editors, 1982, *Bivalve Culture in Asia and the Pacific*, International Development Research Centre, Ottawa, Canada.
31. World Health Organization, 1978, *Public Health and Water in the Developing World*, World Health Organization, Geneva, Switzerland; World Bank, 1982, *Water Supply and Waste Disposal*, World Bank, Washington D. C.
32. G. M. Howe, 1977, *A World Geography of Human Diseases*, Academic Press, New York.
33. S. R. Aiken and M. R. Moss, 1975, Man's Impact on the Tropical Rainforest of Peninsular Malaysia, *Biological Conservation* 8(3): 213–29; C. J. Barrow, 1980, Development in Peninsular Malaysia: Environmental Problems and Conservation Measures, *Third World Planning Review* 2(1):7–25; S. T. Mok, 1978, *Impact of Forest Resources Development in the Environment of Malaysia*, Forestry Department of Peninsular Malaysia, Kuala Lumpur, Malaysia; S. M. Nor and H. T. Tang, 1980, *Forest Conservation in Malaysia: A Reappraisal*, Forest Research Institute, Kepong, Malaysia; C. Toebes and G. K. Seng, 1975, reference 4, above.
34. P. Phisphumvidhi, 1981, *Water Resources for Water Supply in Bangkok, Thailand*, Metropolitan Water Works Authority, Bangkok, Thailand.
35. P. B. Durst, 1981, Problems Facing Reforestation in the Philippines, *Journal of Forestry* October 1981: 686–88, 699.
36. R. Krishna, 1981, Tropical Cyclones in Fiji November 1969 to April 1980, *Fiji Meteorological Survey Publication* no. 2, Fiji Meteorological Service, Nandi, Fiji; Economic and Social Commission for Asia and the Pacific, 1980, reference 5, above;

World Weather Watch Department of the World Meteorological Organization, 1980, Cyclones and Other Freak Weather Phenomena in the Humid Tropics, *Water Resources Journal* (June 1980), publication of Economic and Social Commission for Asia and the Pacific.

37. Economic and Social Commission for Asia and the Pacific, 1980, reference 5, above.
38. World Weather Watch Department of the World Meteorological Organization, 1980, reference 36, above.

15. Climatic Linkages

1. O. Fraenzle, 1979, The Water Balance of the Tropical Rain Forest of Amazonia and the Effects of Human Impact, *Applied Science and Development* 13:88–117; H. Lettau, K. Lettau and L. C. B. Molion, 1979, Amazonia's Hydrologic Cycle and the Role of Atmospheric Recycling in Assessing Deforestation Effects, *Monthly Weather Review* 107:227–38; E. Salati, A. D. Olio, E. Matsui, and J. R. Gat, 1979, Recycling of Water in the Amazon Basin: An Isotopic Study, *Water Resources Research* 15(5):1250–58.
2. R. E. Dickinson, 1981, Effects of Tropical Deforestation on Climate, *Studies in Third World Societies* 14:411–42; A. Henderson-Sellers, 1981, The Effects of Land Clearance and Agricultural Practices on Climate, *Studies in Third World Societies* 14:443–86; R. T. Pinker, O. E. Thompson, and R. F. Eck, 1980, The Albedo of a Tropical Evergreen Forest, *Quarterly Journal of Royal Meteorological Society* 106–551–558; G. L. Potter *et al.*, 1975, Possible Climatic Impact of Tropical Deforestation, *Nature* 258:697–98, and 1981, Albedo Change by Man: Test of Climatic Effects, *Nature* 291:47–50; and C. Sagan, O. B. Toon and J. B. Pollack, 1979, Anthropogenic Albedo Changes and the Earth's Climate, *Science* 206:1363–68.
3. R. A. Houghton *et al.*, 1983, Changes in the Carbon Content of Terrestrial Biota and Soils Between 1860 and 1980: A Net Release of CO_2 to the Atmosphere, *Ecological Monographs;* G. M. Woodwell *et al.*, 1983, Global Deforestation: Contribution to Atmospheric Carbon Dioxide Science 222:1081–86.
4. S. Brown and A. E. Lugo, 1982, The Storage and Production of Organic Matter in Tropical Forests and Their Role in the Global Carbon Cycle, *Biotropica* 14:151–187.
5. W. Bach, J. Pankrath, and S. H. Schneider, editors, 1981, *Food-Climate Interactions*, D. Riedel Publishing Company, Dordrecht, Netherlands; S. S. Batie and R. G. Healy, editors, 1980, *The Future of American Agriculture as a Strategic Resource*, The Conservation Foundation, Washington D. C.; R. A. Bryson and T. J. Murray, 1977, *Climates of Hunger*, University of Wisconsin Press, Madison, Wisconsin; J. Gribbin, 1981, The Politics of Carbon Dioxide, *New Scientist* April 9, 1981:82–84; J. E. Newman, 1980, Climate Change Impacts on the Growing Season of the North American Corn Belt, *Biometeorology* 7:128–42; L. M. Thompson, 1979, *Climate Change and World Grain Production*, Chicago Council on Foreign Relations, Chicago, Illinois.
6. N. J. Rosenberg, 1981, the Increasing Carbon Dioxide Concentration in the Atmosphere and the Implication for Agricultural Productivity, *Climatic Change* 3:265–80; S. H. Wittwer, 1980, Carbon Dioxide and Climatic Change: An Agricultural Perspective, *Journal of Soil and Water Conservation* 35:116–20.
7. W. W. Kellogg and R. Schware, 1981, *Climate Change and Society: Consequences of Increasing Atmospheric Carbon Dioxide*, Westview Press, Boulder, Colorado; S. H. Schneider and R. S. Chen, 1980, Carbon Dioxide Warming and Coastline Floodline: Physical Factors in Climatic Impact *Annual Review of Energy* 5:107–14.
8. C. F. Cooper, 1983, Carbon Storage in Managed Forests, *Canadian Journal of Forestry Research.* 13:155–66; F. J. Dyson, 1976, Can We Control the Carbon Dioxide in the Atmosphere? *Energy* 2:287–91; P. J. Stewart, 1978, Forestry for Carbon Dioxide Fixation, *Commonwealth Forestry Review* 57(4):263–66.
9. S. H. Schneider and R. S. Chen, 1980, reference 7, above.

16. Role of Tropical Forest Countries

1. C. Chandrasekharan, 1980, *Forests and Forestry in Thailand*, Regional Office for Asia and the Pacific, Food and Agriculture Organization, Bangkok, Thailand.
2. Population Center Foundation of the Philippines, 1978, *Images From the Future: The Philippines in the Year 2000*, Population Center Foundation, Manila, Philippines; Population Center Foundation of the Philippines, 1980, *Population/Environment Planning for Asian Forest Communities Practicing Shifting Cultivation: The Philippine Component*, Population Center Foundation of the Philippines, published in conjunction with FAO and UN FPA, Manila, Philippines.
3. K. M. Gupta and D. Bandhu, editors, 1979, *Man and Forest: A New Dimension in the Himalaya*, Day and Tomorrow's Printers and Publishers, New Delhi, India; M. M. Pant, 1980, The Impact of Social Forestry on the National Economy of India, *The International Tree Crops Journal* 1:69–92.
4. S. M. Nor and H. T. Tang, 1980, *Forest Conservation in Malaysia—A Reappraisal*, Forest Research Institute, Kepong, Kuala Lumpur, Malaysia.
5. K. N. Brooks, H. M. Gregersen, E. R. Berglund, and M. Tayaa, 1982, Economic Evaluation of Watershed Projects—An Overview Methodology and Application, *Water Resources Bulletin* 18(2):245–50; R. A. Carpenter, editor, 1982, *Assessing Tropical Forest Lands: Their Suitability For Sustainable Uses*, Tycooly Press, Dublin, Republic of Ireland; H. M. Gregerson and A. H. Contreras, 1979, *Economic Analysis of Forestry Projects*, Food and Agriculture Organization, Rome, Italy.
6. J. I. Zerbe, J. L. Whitmore, H. E. Wahlgren, J. F. Laundrie, and K. A. Christopherson, 1980, *Forestry Activities and Deforestation Problems in Developing Countries*, Forest Products Laboratory, Madison, Wisconsin. For a lengthy assessment of what we need to do, with much pioneering analysis, see Office of Technology Assessment, 1983, *Technologies to Sustain Tropical Forest Resources*, Office of Technology Assessment, U. S. Congress, Washington D. C.
7. C. L. J. Van der Meer, 1981, *Rural Development in Northern Thailand*, State University of Groningen, Netherlands.
8. T. Hatch and C. P. Lim, 1978, *Shifting Cultivation in Sarawak: Report Based Upon a Workshop on Shifting Cultivation*, Kuching, 7–8 December, 1978, Soils Division of Research Branch, Department of Agriculture, Sarawak.
9. O. H. Frankel and M. E. Soule, editors, 1982, *Conservation and Evolution*, Cambridge University Press, New York; M. E. Soule and B. A. Wilcox, editors, 1980, *Conservation Biology*, Sinauer Associates Inc., Sunderland, Mass.; T. C. Whitmore, 1983, *Tropical Rain Forests of the Far East*, second edition, revised, Clarendon Press, Oxford, U. K.
10. Lord Medway and D. R. Wells, 1971, Diversity and Density of Birds and Mammals at Kuala Lumpar, Pahang, *Malayan Nature Journal* 24:238–47.
11. G. B. Wetterberg *et al.*, 1976, *An Analysis of Nature Conservation Priorities in the Amazon*, Brazilian Institute for Forestry Development, Brasilia, Brazil.

17. Support From Outsiders

1. M. Goulding, 1979, *The Fishes and the Forest*, University of California Press, Berkeley, California.
2. World Bank and Food and Agriculture Organization, 1981, *Forestry Research Needs in Developing Countries—Time for a Reappraisal?* World Bank, Washington D. C., and Food and Agriculture Organization, Rome, Italy.
3. P. H. Raven, editor, 1980, *Research Priorities in Tropical Biology*, National Research Council, Washington D.C.
4. World Bank, 1982, *World Development Report*, World Bank, Washington D.C.
5. Curiously enough, a parallel strategy was being worked out, at virtually the same time that I was completing this chapter, by John S. Spears, Forestry Adviser to the World Bank, who, in a Discussion Paper entitled *Saving the Tropical Forest Eco-System*,

proposes an International Tropical Forest Conservation and Development Fund. Spears's proposal contains more or less the same list of programs and projects envisaged in this chapter, and his estimate of the cost to the "outside world" is also $1 billion per year for ten years. At the same time, Spears's figure of $1 billion per year reflects the "opportunity costs" of locking away one-tenth of the tropical forest biome in the form of parks and reserves in order to protect a majority of the forests' species. Spears calculates that, without parks and reserves, the area in question (1 million square kilometers) could be exploited for commercial timber over a period of 50 years, at a rate of 10 cu. m. per hectare, and could generate net foreign-exchange earnings of $50 per cu. m.—hence, $1 billion per year.

18. The Global Heritage

1. R. P. Tucker and J. F. Richards, editors, 1983, *Global Deforestation and the Nineteenth-Century World Economy*, Duke University Press, Durham, North Carolina.
2. A. H. Harcourt and D. Fossey, 1981, The Virunga Gorillas: Decline of an Island Population, *African Journal of Ecology* 19:83–97
3. Much of this analysis is based on the exceptionally illuminating writings of J. C. Westoby, former director of the Forest Resources Division under FAO; see, for example, Forest Industries for Socioeconomic Development, *Commonwealth Forestry Review* 58: 107–17. I readily acknowledge my debt to the visionary wisdom and basic common sense of Jack Westoby. I can think of no better way to conclude this list of hundreds of references than by citing this insightful observer of the tropical forest scene.

Epilogue to the 1992 Edition

1. N. Myers, 1991, The World's Forests and Human Populations: The Environmental Interconnections, in K. Davis and M.S. Bernstam, eds., *Resources, Environment and Population: Present Knowledge, Future Options:* 237–51. The Population Council, New York; W.J. Peters and L.F. Neunschwander, 1988, *Slash-and-Burn Farming in Third World Forests,* University of Idaho Press, Moscow, Idaho; D.A. Schumann and W.L. Partridge, editors, 1989, *The Human Ecology of Tropical Land Settlement in Latin America,* Westview Press, Boulder, Colorado; and W.C. Thiesenhusen, editor, 1989, *Searching for Agrarian Reform in Latin America,* Unwin Hyman, Boston, Massachusetts.
2. N. Myers, 1992, Tropical Forests: The Policy Challenge, in *The Environmentalist* (in press).
3. D.W. Pearce, 1990, *Ecological Restoration for Forest Conservation: The Perspective of Environmental Economics,* Department of Economics, University College, London University, U.K.
4. M.T. Katzman and W.G. Cale, 1990, Tropical Forest Preservation Using Economic Incentives, *BioScience* 40: 827–32.
5. N. Myers, 1989, The Greenhouse Effect: A Tropical Forestry Response, *Biomass* 18: 73–78; see also G. Marland, 1988, *The Prospect of Solving the CO_2 Problem Through Global Reforestation,* Carbon Dioxide Research Division, U.S. Department of Energy, Washington D.C.; and S. Postel and L. Heise, 1988, *Reforesting the Earth,* Worldwatch Institute, Washington, D.C.

Index*

* Prologue and Epilogue are not included.

DATE DUE

NOV 1 3 1997		
MR 12 '99		
GAYLORD		PRINTED IN U.S.A.